Regularization of Inverse Problems

Mathematics and Its Applications

Managing Editor:

M. HAZEWINKEL

Centre for Mathematics and Computer Science, Amsterdam, The Netherlands

Volume 375

Regularization of Inverse Problems

by

Heinz W. Engl
*Johannes Kepler University,
Linz, Austria*

Martin Hanke
*University of Karlsruhe,
Karlsruhe, Germany*

and

Andreas Neubauer
*Johannes Kepler University,
Linz, Austria*

KLUWER ACADEMIC PUBLISHERS
DORDRECHT / BOSTON / LONDON

A C.I.P. Catalogue record for this book is available from the Library of Congress.

ISBN-13: 978-0-7923-6140-4   e-ISBN-13: 978-94-009-1740-8
DOI: 10.1007/978-94-009-1740-8

Published by Kluwer Academic Publishers,
P.O. Box 17, 3300 AA Dordrecht, The Netherlands.

Sold and distributed in North, Central and South America
by Kluwer Academic Publishers,
101 Philip Drive, Norwell, MA 02061, U.S.A.

In all other countries, sold and distributed
by Kluwer Academic Publishers,
P.O. Box 322, 3300 AH Dordrecht, The Netherlands.

All Rights Reserved
© 2000 Kluwer Academic Publishers

Softcover reprint of the hardcover 1st edition 2000

No part of the material protected by this copyright notice may be reproduced or
utilized in any form or by any means, electronic or mechanical,
including photocopying, recording or by any information storage and
retrieval system, without written permission from the copyright owner.

Andreas Neubauer dedicates his efforts on this book to his daughter *Lisa Maria* (May 10, 1992 – October 20, 1994).

Contents

Preface   1

1. Introduction: Examples of Inverse Problems   3
    1.1. Differentiation as an Inverse Problem 4
    1.2. Radon Inversion (X-Ray Tomography) 7
    1.3. Examples of Inverse Problems in Physics 10
    1.4. Inverse Problems in Signal and Image Processing 12
    1.5. Inverse Problems in Heat Conduction 18
    1.6. Parameter Identification 23
    1.7. Inverse Scattering 25

2. Ill-Posed Linear Operator Equations   31
    2.1. The Moore-Penrose Generalized Inverse 32
    2.2. Compact Linear Operators: Singular Value Expansion 36
    2.3. Spectral Theory and Functional Calculus 42

3. Regularization Operators   49
    3.1. Definition and Basic Results 49
    3.2. Order Optimality 55
    3.3. Regularization by Projection 63

4. Continuous Regularization Methods   71
    4.1. A-priori Parameter Choice Rules 71
    4.2. Saturation and Converse Results 80
    4.3. The Discrepancy Principle 83
    4.4. Improved A-posteriori Rules 89
    4.5. Heuristic Parameter Choice Rules 100
    4.6. Mollifier Methods 112

5. Tikhonov Regularization   117
    5.1. The Classical Theory 117
    5.2. Regularization with Projection 126
    5.3. Maximum Entropy Regularization 134
    5.4. Convex Constraints 140

6. Iterative Regularization Methods   154
    6.1. Landweber Iteration 154
    6.2. Accelerated Landweber Methods 160
    6.3. The ν-Methods 166

7. The Conjugate Gradient Method — 177
- 7.1. Basic Properties — 177
- 7.2. Stability and Convergence — 181
- 7.3. The Discrepancy Principle — 186
- 7.4. The Number of Iterations — 191

8. Regularization With Differential Operators — 197
- 8.1. Weighted Generalized Inverses — 197
- 8.2. Regularization with Seminorms — 202
- 8.3. Examples — 207
- 8.4. Hilbert Scales — 210
- 8.5. Regularization in Hilbert Scales — 215

9. Numerical Realization — 221
- 9.1. Derivation of the Discrete Problem — 221
- 9.2. Reduction to Standard Form — 224
- 9.3. Implementation of Tikhonov Regularization — 228
- 9.4. Updating the Regularization Parameter — 233
- 9.5. Implementation of Iterative Methods — 237

10. Tikhonov Regularization of Nonlinear Problems — 241
- 10.1. Introduction — 241
- 10.2. Convergence Analysis — 243
- 10.3. A-posteriori Parameter Choice Rules — 249
- 10.4. Regularization in Hilbert Scales — 253
- 10.5. Applications — 256
- 10.6. Convergence of Maximum Entropy Regularization — 262

11. Iterative Methods for Nonlinear Problems — 277
- 11.1. The Nonlinear Landweber Iteration — 277
- 11.2. Newton Type Methods — 285

A. Appendix — 289
- A.1. Weighted Polynomial Minimization Problems — 289
- A.2. Orthogonal Polynomials — 291
- A.3. Christoffel Functions — 295

Bibliography — 299

Index — 319

Preface

In the last two decades, the field of inverse problems has certainly been one of the fastest growing areas in applied mathematics. This growth has largely been driven by the needs of applications both in other sciences and in industry. In Chapter 1, we will give a short overview over some classes of inverse problems of practical interest. Like everything in this book, this overview is far from being complete and quite subjective.

As will be shown, inverse problems typically lead to mathematical models that are not well-posed in the sense of Hadamard, i.e., to *ill-posed problems*. This means especially that their solution is unstable under data perturbations. Numerical methods that can cope with this problem are the so-called *regularization methods*. This book is devoted to the mathematical theory of regularization methods. For linear problems, this theory can be considered to be relatively complete and will be described in Chapters 2 – 8. For nonlinear problems, the theory is so far developed to a much lesser extent. We give an account of some of the currently available results, as far as they might be of lasting value, in Chapters 10 and 11.

Although the main emphasis of the book is on a functional analytic treatment in the context of operator equations, we include, for linear problems, also some information on numerical aspects in Chapter 9.

Since all of the authors of this book have also made original contributions to the field, our view and emphasis might be slightly subjective. Therefore, we mention the following other books on the same subject and general references which the reader might consult for aspects not treated here: [20, 22, 103, 107, 131, 177, 183, 194, 195, 271, 273, 278]. Many more references will be given in the text.

We thank Doris Nikolaus for LaTeXing parts of the manuscript, and Dr. Wilhelm Grever and Dr. Gerhard Landl for their careful reading of and their helpful remarks on a preliminary draft of this book.

Each of us cooperated with several coauthors, who definitely influenced our understanding of the field. Special thanks are due to them.

Finally, we thank Michiel Hazewinkel for his invitation to write this book and the staff of Kluwer for their patience.

We gratefully acknowledge financial support from the Christian Doppler Society (Wien) and the Austrian Fonds zur Förderung der wissenschaftlichen Forschung (projects S32/03, P7869, and P10866).

Heinz W. Engl, Martin Hanke and Andreas Neubauer

1. Introduction: Examples of Inverse Problems

In this introductory section, we give, after a general discussion of the term "inverse problems", examples of various classes of inverse problems arising in various application fields. As we will see, *linear inverse problems* frequently lead to *integral equations of the first kind*, which is why such equations play an important role in the study of inverse problems, as we will see throughout a large part of this book. On the other hand, many basic *inverse problems* are inherently nonlinear even if the corresponding *direct problem* is linear. This fact is our motivation for devoting a significant part of this book to the mathematics of nonlinear inverse problems.

When using the term *inverse problem*, one immediately is tempted to ask "inverse to what?". Following J.B. Keller [150], one calls *two* problems *inverse to each other* if the formulation of one problem involves the other one. For mostly historic reasons, one might call one of these problems (usually the simpler one or the one which was studied earlier) the *direct problem*, the other one the *inverse problem*. However, if there is a *real-world* problem behind the mathematical problem studied, there is, in most cases, a quite natural distinction between the direct and the inverse problem. E.g., if one wants to predict the future behaviour of a physical system from knowledge of its present state and the physical laws (including concrete values of all relevant physical parameters), one will call this the *direct problem*. Possible *inverse problems* are the determination of the present state of the system from future observations (i.e., the calculation of the evolution of the system backwards in time) or the identification of physical parameters from observations of the evolution of the system (*parameter identification*).

There are, from the applications point of view, two different motivations for studying such inverse problems: first, one wants to *know* past states or parameters of a physical system. Second, one wants to find out how to influence a system via its present state or via parameters in order to *steer* it to a desired state in the future.

Thus, one might say that *inverse problems are concerned with determining causes for a desired or an observed effect*.

As we will see, such inverse problems most often do not fulfill Hadamard's postulates of *well-posedness* (see Chapter 2). They might not have a solution in the strict sense, solutions might not be unique and/or might not depend continuously on the data. Mathematical problems having these undesirable properties are called *ill-posed problems* (*improperly posed problems*) and pose (mostly because of the discontinuous dependence of solutions on the data) severe numerical difficulties. While the study of concrete inverse problems frequently involves the question how to enforce uniqueness by additional information or assumptions, not much can be said about this in a general context. The aspect of lack of stability and its restoration by appropriate methods (*regularization methods*), however, can be treated in sufficient generality. The theory of regularization methods is well-developed for linear inverse problems and at least emerging for nonlinear problems and forms the core of this

4   1. Introduction: Examples of Inverse Problems

book.

There is a vast literature on inverse and ill-posed problems. In addition to the books already quoted in the Preface, we mention, as general references,

- the following monographs: [13, 32, 45, 95, 136, 225, 230, 282, 288],

- the following conference proceedings: [14, 36, 43, 72, 78],[81] (which emphasizes inverse problems arising in industry), [83, 138, 223, 241, 242, 265, 289],

- the journals *Inverse Problems* (Institute of Physics Publ.), *Inverse Problems in Engineering* (Gordon & Breach), and *Journal of Inverse and Ill-Posed Problems* (VSP).

Many more references will be given in the appropriate sections.

1.1. Differentiation as an Inverse Problem

Two mathematical problems inverse to each other are differentiation and integration. A-priori, it is not clear which of these problems should be the direct problem and which one the inverse problem. However, as we will now see, differentiation has (as opposed to integration) the properties of an ill-posed problem. Moreover, since, as we will see later, many inverse problems involve at some step differentiation of the data, differentiation might be viewed as the "inverse problem", although in most calculus courses, it is treated first.

Let $f \in C^1[0,1]$ be any function, $\delta \in (0,1)$, $n \in \mathbb{N}$ ($n \geq 2$) be arbitrary, and define

$$f_n^\delta(x) := f(x) + \delta \sin \frac{nx}{\delta}, \quad x \in [0,1]. \tag{1.1}$$

Then

$$(f_n^\delta)'(x) = f'(x) + n \cos \frac{nx}{\delta}, \quad x \in [0,1]. \tag{1.2}$$

Now, in the uniform norm,

$$\|f - f_n^\delta\|_\infty = \delta,$$

but

$$\|f' - (f_n^\delta)'\|_\infty = n.$$

Hence, if we consider f and f_n^δ as the exact and perturbed data, respectively, then for an arbitrarily small data error δ, the error in the result, namely the derivative, can be arbitrarily large, namely n. Hence, the derivative does not depend continuously on the data with respect to the uniform norm. Of course, we could enforce continuous dependence by measuring the data error in the C^1-norm. However, this would be a sort of cheating, since then, we would call a data error small if the error in the function values and in the values of the derivative, which is exactly what we want to compute, would be small.

1.1. Differentiation as an Inverse Problem

For later reference, note that f' solves the simple integral equation of the first kind

$$(Kx)(s) := \int_0^s x(t)\,dt = f(s) - f(0)\,, \qquad (1.3)$$

which is solvable in $C[0,1]$ only if $f \in C^1[0,1]$. The corresponding direct problem would be to compute f from x, i.e., integration, which is a stable process on $C[0,1]$. Note that integration is a smoothing process, i.e., highly oscillatory errors in x (e.g., of the form $n\cos(nx/\delta)$ as they appeared in (1.2)) are damped out (to $\delta\sin(nx/\delta)$) and have a very small effect on the data for the inverse problem. This smoothing is responsible for the fact that errors of small amplitude, but high frequency, create large oscillations in the solution of the inverse problem. These considerations are not restricted to this concrete problem: whenever a direct problem has smoothing properties one has to expect the appearance of oscillations coming from small data perturbations (of high frequency) in the solution of the inverse problem. This effect is the more pronounced the stronger smoothing the direct problem is.

Why (or, under what circumstances) can we differentiate a function in spite of these problems? We have to be able to exclude the presence of data errors of arbitrarily high frequency, e.g., of the form as in (1.1); this can be done, e.g., if we know a bound for f''. In the example above, such a bound would give a bound for n in terms of δ, thus coupling the amplitude and the frequency of the possible data errors. A functional analytic argument is the following:

If we consider the operator K as defined in (1.3) on $C[0,1]$, then it is a continuous linear injective operator, whose inverse (defined on $C^1[0,1]$, considered as a subspace of $C[0,1]$) is unbounded. However, if we restrict K to the set $\{x \in C^1[0,1] \mid \|x\|_\infty + \|x'\|_\infty \leq \gamma\}$, which is compact in $C[0,1]$ due to the Arzela-Ascoli Theorem, then the inverse of this restricted operator is continuous on its range, as the inverse of a continuous bijective (not necessarily linear) map defined on a compact set is again continuous. Thus, we can "restore stability" by assuming an a-priori bound for f' and f''.

The stability problems addressed must appear somehow when computing the derivative via difference quotients: let f be the function we want to differentiate, f^δ its noisy version with

$$\|f - f^\delta\|_\infty \leq \delta\,.$$

We want to use the central difference quotient with step size h. If $f \in C^2[0,1]$, Taylor expansion yields

$$\frac{f(x+h) - f(x-h)}{2h} = f'(x) + O(h)\,,$$

while for $f \in C^3[0,1]$,

$$\frac{f(x+h) - f(x-h)}{2h} = f'(x) + O(h^2)\,.$$

Thus, the accuracy of the central difference quotient depends on the smoothness of

1. Introduction: Examples of Inverse Problems

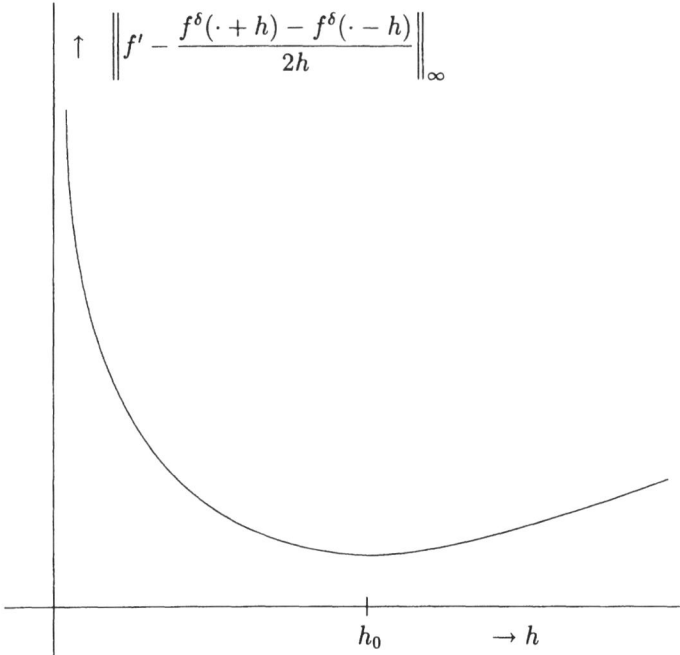

Figure 1.1: Total error depending on h

the exact data. Instead of f', we are actually computing
$$\frac{f^\delta(x+h) - f^\delta(x-h)}{2h} \sim \frac{f(x+h) - f(x-h)}{2h} + \frac{\delta}{h}.$$
Thus, the total error behaves like
$$O(h^\nu) + \frac{\delta}{h}, \tag{1.4}$$
where $\nu = 1$ or 2 if $f \in C^2[0,1]$ or $f \in C^3[0,1]$, respectively. For a fixed error level δ, it looks as in Figure 1.1.

If h becomes too small, the total error increases due to the error term δ/h, the propagated data error. Of course, if h is too large, then the approximation error becomes too large. There is an *optimal* discretization parameter h_0, which can, however, not be computed explicitly, since it depends on unavailable information about the exact data, e.g., their smoothness. However, one can at least estimate the asymptotic behaviour of h_0 if h is chosen as a power of δ, i.e.,
$$h \sim \delta^\mu,$$
then one can minimize (1.4) by taking $\mu = 1/2$ or $\mu = 1/3$, which results in a behaviour of the total error as $O(\sqrt{\delta})$ or $O(\delta^{\frac{2}{3}})$ for $f \in C^2[0,1]$ or $f \in C^3[0,1]$,

respectively. Thus, even in the best possible case ($\nu > 2$ is obviously not possible in (1.4)) and for an optimal choice of h, we obtain only a convergence rate of $O(\delta^{\frac{2}{3}})$, where δ denotes the data error, i.e., there is an intrinsic loss of information. This rate cannot be improved unless f is a quadratic polynomial [106]. If μ is not chosen optimally, i.e., if the discretization parameter h and the noise level are not linked appropriately, this loss of information becomes more severe.

Early references to the treatment of numerical differentiation by regularization are [11, 12].

In this example, we saw some effects that are typical for ill-posed problems:

- amplification of *high frequency* errors

- restoration of stability by using *a-priori information*

- two error terms of different nature, one for the approximation error, the other one for the propagation of the data error, adding up to a total error as in Figure 1.1

- the appearance of an optimal discretization parameter, whose choice depends on a-priori information

- loss of information even under optimal circumstances.

1.2. Radon Inversion (X-Ray Tomography)

An inverse problem that has been widely studied because of its importance in, e.g., medical applications arises in *Computerized Tomography* (CT). We consider the two-dimensional situation:

Let $D \subseteq \mathbb{R}^2$ be a compact domain with a spatially varying density f. In medical applications, D symbolizes a cross-section of the human body; in nondestructive testing, D is a cross-section of the material to be tested. The aim is to recover the density f from X-ray measurements in the plane where D lies. These X-rays travel along lines, which are parameterized by their normal vector $w \in \mathbb{R}^2$ ($\|w\| = 1$) and their distance $s > 0$ from the origin (see Figure 1.2).

If one assumes that the decay $-\Delta I$ of an X-ray beam along a distance Δt is proportional to the intensity I, the density f, and to Δt, one obtains

$$\Delta I(sw + tw^\perp) = -I(sw + tw^\perp)f(sw + tw^\perp)\Delta t, \tag{1.5}$$

where $w^\perp$ is a unit vector orthogonal to w. By letting Δt tend to 0 in (1.5), one obtains

$$\frac{d}{dt}I(sw + tw^\perp) = -I(sw + tw^\perp)f(sw + tw^\perp). \tag{1.6}$$

We denote by $I_L(s, w)$ and $I_0(s, w)$ the intensity of the X-ray beam measured at the detector and at the emitter, respectively, where detector and emitter are connected

8   1. Introduction: Examples of Inverse Problems

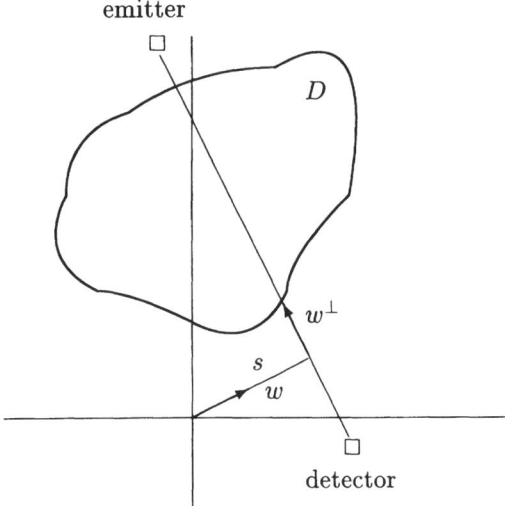

Figure 1.2: Computerized Tomography

by the line parameterized by s and w and are located outside of D (so that we can as well assume that they are located "at infinity"). Then, (1.6) has the solution

$$\log I_L(s,w) - \log I_0(s,w) = -\int_{\mathbb{R}} f(sw + tw^\perp)\, dt,$$

so that the density f is related to the measured quantities I_L and I_0 via

$$(Rf)(s,w) := \int_{\mathbb{R}} f(sw + tw^\perp)\, dt = -\log \frac{I_L(s,w)}{I_0(s,w)}, \quad w \in \mathbb{R}^2, \|w\| = 1, s > 0; \quad (1.7)$$

thus, the inverse problem of determining the density distribution f from X-ray measurements amounts to solving this integral equation of the first kind.

The integral operator R is called the *Radon transform* after the Austrian mathematician Johann Radon, who considered the problem of reconstructing a function of two variables from its line integrals already in 1917 [231].

There are many papers and some books on the Radon transform and its inversion; we refer to [207] and [183, Chapter 6].

A special case that is of interest in material testing is that D is a circle of radius ρ and f is axisymmetric with respect to its center (say, the origin). Then it suffices to use X-rays or, if D symbolizes a material absorbing light in way proportional to an absorption coefficient f, light rays parallel to the x-axis, i.e., to consider (1.7) only for $w_0 = (0, \pm 1)$; additional directions do not provide additional information. Thus, we assume that

$$f(s,w) = F(s), \quad 0 < s \leq \rho, \|w\| = 1,$$

1.2. Radon Inversion (X-Ray Tomography)

with a suitable function F, which is now to be reconstructed. Let

$$g(s) := -\log \frac{I_L(s, w_0)}{I_0(s, w_0)}$$

denote the measurements in this situation. Now, for $0 < s \leq \rho$,

$$(Rf)(s, w_0) = 2 \int_s^\rho \frac{rF(r)}{\sqrt{r^2 - s^2}} dr,$$

so that we obtain from (1.7) an *Abel integral equation* of the first kind

$$\int_s^\rho \frac{rF(r)}{\sqrt{r^2 - s^2}} dr = \frac{g(s)}{2}, \quad 0 < s \leq \rho. \tag{1.8}$$

Note that the kernel in (1.8) has a singularity. (1.8) can be explicitly inverted; if $g(\rho) = 0$,

$$F(r) = -\frac{1}{\pi} \int_r^\rho \frac{g'(s)}{\sqrt{s^2 - r^2}} ds.$$

This inversion formula involves g', i.e., the data have to be differentiated. According to Section 1.1, this is an indication for the ill-posedness of the problem to recover F from g. However, after differentiating g, the result is put into an integral operator and thus smoothed again. But note that the kernel of this integral operator is singular, so that this smoothing does not annihilate the instability introduced by differentiation completely, but only partly. In the following way this can be made more precise:

Via the substitution $t = r^2$ and forming the adjoint equation, (1.8) can be related to the classical Abel integral equation

$$(Ax)(s) = \frac{1}{\sqrt{\pi}} \int_0^s \frac{x(t)}{\sqrt{s-t}} dt.$$

It turns out that

$$(A^2 x)(s) = \int_0^s x(t) \, dt,$$

so that differentiation is the inverse of A^2. Hence, inverting A can be thought of as differentiation of half order.

Note that also in the two-dimensional case, an explicit inversion formula for the Radon transform (already due to J. Radon [231]) is available, which also involves differentiation of the data.

Abel equations appear in many applications (cf. [97]). E.g., (1.8) also appears in plasma physics in the problem of determining a circularly symmetric radiation intensity of a gas enclosed in a cylinder from measurements of the emitted radiation from outside of the cylinder (see [161]). This is a typical *remote sensing* problem.

1.3. Examples of Inverse Problems in Physics

The journal *Inverse Problems* regularily contains many papers formulating and solving inverse problems arising in physics. We mention only some quite arbitrarily chosen examples which lead to integral equations of the first kind:

A widely studied problem in physical chemistry is *time resolved fluorescence*: there, a substance to be analyzed is illuminated by a short light-pulse (produced, e.g., by a laser) and absorbs energy (photons) as some molecules switch over into an excited state. When the molecules fall back into their original state, the photons are emitted again. Denoting by $l(t)$ the (time dependent) intensity of the light-pulse and by $f(t)$ the probability density of a certain molecule excited at time $t_0 = 0$ to emit a photon at time t (called the *fluorescence response*), the observed fluorescence intensity at time t is

$$h(t) = \int_0^t l(s) f(t-s) \, ds \, . \tag{1.9}$$

It is known that for simple (nonnegative) molecular systems containing only one type of (optically active) elements the fluorescence response has the form $f(t) = e^{-\frac{t}{\tau}}$, where τ is the *fluorescence lifetime*. In more complicated systems, however, the lifetimes are distributed, i.e.,

$$f(t) = \int_0^{T_0} a(\tau) e^{-\frac{t}{\tau}} \, d\tau \, , \tag{1.10}$$

where the function a (the *lifetime distribution*) contains the information relevant for the physicist. Here, T_0 is an upper bound for the lifetimes that are known a-priori.

In order to get the desired information, i.e., the function a, from the measurements h one has to solve the integral equations of the first kind (1.9) and (1.10) simultaneously, i.e., one has to solve the single integral equation of the first kind

$$\int_0^t \int_0^{T_0} l(t-s) a(\tau) e^{-\frac{s}{\tau}} \, d\tau ds = h(t) \, . \tag{1.11}$$

For further information on the physical background and on applications of this problem see [175] and the papers quoted there, as well as [143, 266, 279].

The paper [175] studies numerical algorithms for solving (1.11). From the numerical experiments reported there, one can clearly see the instability of "naive" algorithms and the need for using regularization methods.

Note that (1.10) is closely related to the Laplace transform; the reconstruction of a function x from its Laplace transform Lx can also be seen as an integral equation of the first kind:

$$\int_0^\infty e^{-st} x(t) \, dt = (Lx)(s) \, .$$

Laplace inversion appears in many applications and is, if performed numerically in a straightforward way, highly unstable. The integral in (1.9), on the other hand, is a convolution integral, so that solving (1.9) can be seen as *deconvolution*, another problem that appears frequently in applications, e.g., in the measurement of physical quantities by real instruments, i.e., by instruments with finite resolution:

1.3. Examples of Inverse Problems in Physics

Assume, e.g., that $x(s)$ is the energy content of a light beam (consisting of a whole range of frequencies) at frequency s. Only an ideal measuring instrument (with infinite resolution) would allow to measure x directly. In reality, the measuring instrument, if calibrated to a frequency s, measures the energy at this frequency s, but also (to a lesser extent) the energy at some neighbouring frequencies $s - \tau$. Thus, instead of $x(s)$, the instrument measures

$$\int_{-\infty}^{+\infty} k(s,\tau) x(s-\tau) \, d\tau \, ,$$

where $k(s,.)$ has a peak at $\tau = 0$ and decays away from $\tau = 0$; the faster this decay, the better the instrument. Ideally, $k(s,\tau)$ would be a δ-distribution centered at $\tau = 0$. Thus, we have to reconstruct x from (noisy) measurements of y in the first-kind integral equation

$$\int_{-\infty}^{+\infty} k(s,\tau) x(s-\tau) \, d\tau = y(s) \, .$$

Another area where integral equations of the first kind frequently appear is rheology (see [132] for a survey). We mention the problem of determining the molecular weight distribution of a polymer in solution from measurable rheological quantities. A classical equation in this context is Fujita's equation

$$\int_0^{m_{max}} \frac{\lambda m e^{-\lambda m \xi}}{1 - e^{-\lambda m}} f(m) \, dm = u(\xi)$$

for determining the molecular weight distribution $f(m)$ from the equilibrium concentration of the solution in a centrifuge. Here, u is the quotient of the equilibrium and initial concentrations at location ξ, and λ contains physical parameters and the angular velocity of the rotation of the centrifuge; see [91] for details.

The problems of determining molecular weight distributions of flexible polymers from linear viscoelastic material functions or from light scattering data (*photon correlation spectroscopy*) also lead to integral equations of the first kind and are discussed in [190].

Recently, inverse problems were also investigated in connection with high temperature superconductivity [49].

Another context where integral equations of the first kind naturally appear are inverse potential problems, i.e., problems which are inverse to direct problems whose solutions can be calculated via surface or volume potentials. Such problems appear, e.g., in inverse scattering as will be discussed in Section 1.7, but also in other applications, like in geophysics:

Let $\Omega \subseteq \mathbb{R}^3$ symbolize a part of the earth, ρ be the density. The gravitational potential at $x \notin \Omega$ is then given by

$$\phi(x) := \frac{G}{4\pi} \int_\Omega \frac{\rho(y)}{\|x-y\|} \, dy \, , \tag{1.12}$$

where G is the gravitational constant. In geophysics, one is interested in finding ρ (or, at least, obtaining some information about ρ) from gravity measurements; ϕ itself cannot be measured, but derivatives of differences of ϕ can. Via (1.12), these measurements lead to integral equations of the first kind. In geophysics, one usually assumes that ρ assumes several constant, but different values. It is of special interest, e.g., in prospecting to know where ρ has jumps. See [192] and [78] for inverse problems in geophysics and geodesy.

Potential theory is also used for representing the solution of Maxwell's equations; to give a simple example, the z-component of the electric field generated by a cylindrical electrode (around the z-axis) of length $2L$ and radius R is determined by

$$E(z) = \frac{R}{2\varepsilon} \int_{-L}^{L} \frac{q(a)}{\sqrt{(R^2 + (z-a)^2)^3}} \, da, \qquad (1.13)$$

where ε is the dielectric constant and q is the charge distribution. The *synthesis problem* of distributing charges in such a way that a prescribed electric field is generated is therefore described by (1.13), considered as an integral equation of the first kind for q; see [3]. The monograph [238] is solely devoted to inverse problems for Maxwell's equations.

For a problem from nondestructive testing which can be interpreted as reconstructing a spatially varying magnetic permeability from magnetic field measurements and whose linearization also leads to an integral equation of the first kind see [73].

In recent years, inverse problems in transport theory have attracted considerable interest: consider, e.g., a flow through a scattering medium; from the transmitted or reflected flow, one wants to determine the structure of the medium. If interaction between the flow and the medium is neglected, then one arrives at problems like the inversion of the Radon transform. If, however, collisions are taken into account, then the direct problem has to be modelled in the framework of kinetic theory. For corresponding inverse problems see, e.g., [15, 16, 57, 126].

1.4. Inverse Problems in Signal and Image Processing

The story of the Hubble Space Telescope (see [4] for more information) found considerable interest even in the newspapers. Launched in 1990 as a joint project of the North American and European space agencies NASA and ESA, this optical observatory was designed for a breakthrough in astronomy by providing space images with enormous spatial resolution. Unfortunately, soon after the launch, engineers discovered a manufacturing error in the main mirror of the telescope, leading to severe spherical artefacts in the images sent back to Earth. Therefore, before the telescope could eventually be fixed in late 1993, the astronomers improved the blurred images by numerical reconstruction, i.e., by solving an inverse problem.

1.4. Inverse Problems in Signal and Image Processing

The mathematical model of blurring is

$$Tx(s,t) = \int_{\mathbb{R}^2} k(s,t;s',t')x(s',t')\,ds'dt' = y(s,t), \qquad (1.14)$$

where x and y are the true and the blurred 2D images, respectively, and k is the blurring function, cf. [170]. Thus, deblurring amounts to solving the integral equation (1.14). To be able to cope with the enormous amount of data (the main camera of Hubble has a detector with 800×800 pixels, so that the discretized quantities x and y consist of almost a million unknowns), the kernel function k is usually assumed to be spatially invariant, i.e.,

$$k(s,t;s',t') = h(s-s',t-t'), \qquad s,t,s',t' \in \mathbb{R}. \qquad (1.15)$$

The function h is called *point spread function*: it represents the image of a point source (a delta function), and the final image y is obtained via superposition, by spreading with h the original pixel intensities in x over the region of the image. For example, if an image is taken by a camera out of focus, each grain (s,t) of film is exposed by all light sources (s',t') from a circular region $|(s,t)-(s',t')| \leq r$, i.e.,

$$h(s,t) = \begin{cases} 1, & s^2+t^2 \leq r^2, \\ 0, & \text{elsewhere}. \end{cases}$$

In many image restoration applications, a crucial problem is the determination of a suitable point spread function. Concerning the Hubble space telescope, it turned out that, on the one hand, the blurring function was to some extent spatially varying, and on the other hand, the point spread function had a comparatively large support. We remark that in astronomical images the modelling of data errors in the imaging process is also quite complicated: the output of the individual electronic sensors is subject to a signal dependent amount of additive noise.

In mathematical terms, the solution of a convolution equation over $\mathbb{R}^2$ like (1.14), (1.15) can be stated explicitly by means of Fourier transformations. Denoting the (2D) Fourier transforms of x, y, and h, by $\hat{x}$, $\hat{y}$, and $\hat{h}$, respectively, the Convolution Theorem states that

$$\hat{y} = \hat{h}\hat{x}, \qquad (1.16)$$

to be valid for all frequencies $(\xi,\eta) \in \mathbb{R}^2$. Here, we have adopted the physical normalization of the Fourier transform, i.e.,

$$Ff(\xi,\eta) := \hat{f}(\xi,\eta) = \int_{\mathbb{R}^2} e^{-\xi s - \eta t} f(s,t)\,dsdt. \qquad (1.17)$$

Therefore, formally, all one has to do is to compute Fourier transforms, and solve (1.16) for $\hat{x}$ by dividing through $\hat{h}$.

The numerical difficulties stem from the fact that h is usually a function with compact support, or at least absolutely integrable over $\mathbb{R}^2$. By the Riemann-Lebesgue Lemma this implies that $\hat{h}$ is continuous, and vanishing asymptotically as (ξ,η)

goes to infinity. This means that the division $\hat{y}/\hat{h}$ is unstable, and high frequency data error components in y are magnified without bound when solving (1.16) for $\hat{x}$. As a consequence, straightforward inversion methods will lead to highly oscillating artefacts in x.

An immediate option for curing this instability problem is to remove high frequencies from y, and to compute a so-called *bandlimited* reconstruction of x by resolving (1.16) only for the remaining frequencies $\mathcal{F} := \{(\xi,\eta) \mid |(\xi,\eta)| \leq \tau\}$, say. This is reminiscent of the method of truncated spectral expansion, which we consider in Example 4.8. If $\hat{h}(\xi,\eta) \neq 0$ for all $(\xi,\eta) \in \mathcal{F}$, then this leads to stable approximations with respect to $\mathcal{L}^2$ in the sense that

$$\|x\|_{\mathcal{L}^2} \leq \left\|\frac{1}{\hat{h}}\right\|_{\mathcal{F}} \|y\|_{\mathcal{L}^2} \, ;$$

here, $\|\cdot\|_{\mathcal{F}}$ denotes the maximum norm over $\mathcal{F}$. Moreover, we can readily determine the bound

$$\left\|\frac{1}{\hat{h}}\right\|_{\mathcal{F}} \|y^\delta - y\|_{\mathcal{L}^2} \tag{1.18}$$

for the propagation of the error $y^\delta - y$ in the data. In other words, the smaller the data error is, the more frequencies of x can be resolved within a certain accuracy. In order to find a suitable truncation parameter τ, one can plot the Fourier spectrum of y, and determine the point where the signal in y dissolves in the noise level. More sophisticated parameter selection methods will be discussed in the framework of Chapter 4.

Recall that we have seen in Section 1.1 that the quality of the reconstructions may benefit from a-priori known smoothness properties of the exact solution x. Here we observe the same phenomenon. By the Paley-Wiener Theorems, the Fourier transform of $\hat{x}$ decays the more rapidly, the smoother x is, and hence, the larger is the portion of energy of x contained in the ball $\mathcal{F}$ of maintained frequencies. Consequently, with the same truncation level, smoother functions can be approximated more accurately. This can be manifested by rigorous estimates: note that the $\mathcal{L}^2$-norm of $(\xi^2+\eta^2)^{s/2}\hat{x}(\xi,\eta)$ can be bounded in terms of the Sobolev norm $\mathcal{H}^s$ of x, and hence, the $\mathcal{L}^2$-approximation error for x introduced by the truncation of frequencies beyond $\mathcal{F}$ can be bounded by

$$\tau^{-s}\|x\|_{\mathcal{H}^s} \, .$$

Together with the bound (1.18) for the propagated data error this leads to an a-priori computable bound, which can be optimized by an appropriate choice of τ. Note that the propagated data error generally increases with τ, whereas the approximation error decreases with τ, similar to Section 1.1

In astronomical imaging, however, the true image x has hardly any global smoothness properties which do affect the asymptotic behaviour of the Fourier transform. Instead, astronomers use other a-priori informations such as nonnegativity of the pixel values to be reconstructed. Such knowledge may affect the choice of regularization technique to be employed: for instance, nonnegativity may be incorporated

1.4. Inverse Problems in Signal and Image Processing

as a convex constraint, or by maximizing an entropy-like functional, which leads to nonnegative reconstructions per se. We will deal with such alternatives in Sections 5.3 and 5.4.

Besides these two possibilities to incorporate prior knowledge one can use the fact that images typically have some kind of local smoothness rather than global smoothness. For instance, astronomical images may have large areas with very small intensity, almost identically zero. Modern technologies built on wavelet decompositions provide tools for regularized image restoration, which may benefit from local smoothness properties. We shall not pursue this any further in this book, but rather refer to [56, 188] for recent work on regularization through wavelet approximations.

Another fundamental problem in image restoration, and in signal processing in general, is that of extrapolating a band- or time-limited signal. For example, any real measuring device will only be able to resolve a signal within a certain range $\mathcal{F}$ of frequencies. On the other hand, signals (or what we consider them to be) are typically presumed to have compact support in time, i.e., they start at some given time and end after some time interval of length $2T$, say. We conclude, again from the Paley-Wiener Theorems, that the Fourier transform $\hat{f}$ of such a signal f is an entire function of ξ. Hence, $\hat{f}$ cannot vanish on any open subset of the real line, and we are faced with the problem of extrapolating the values of $\hat{f}$ beyond the frequency interval $\mathcal{F}$ in which it is given. This problem of extrapolating an analytic function, however, is extremely ill-posed and therefore numerically unstable. Because of its prevalent role in information theory, we shall exemplarily describe one reconstruction algorithm which is frequently encountered in this context: the *Gerchberg-Papoulis algorithm* (cf. [105] for a description of this method in a mathematical framework).

Assume that the signal f has compact support in the time interval $[-T,T]$, and its Fourier transform $\hat{f} = Ff$ is known in the set $\mathcal{F} = [-\Omega, \Omega]$ for some $T, \Omega > 0$. Denoting by P_τ the orthogonal projector in $\mathcal{L}^2$ which sets a function to zero outside the interval $[-\tau, \tau]$, we can introduce the bandlimiting operator B_Ω by

$$B_\Omega = F^{-1} P_\Omega F.$$

Assume further that we are given the data $\tilde{g} = P_\Omega \hat{f}$ in the frequency domain. Then, our problem can be formulated as an operator equation

$$B_\Omega f = F^{-1} \tilde{g}.$$

Since f is living in the time interval $[-T,T]$, we have $P_T f = f$, and hence, the previous equation can be reformulated as $B_\Omega P_T f = F^{-1}\tilde{g}$, and a further multiplication with P_T from the left yields

$$Kf := P_T B_\Omega P_T f = P_T F^{-1} \tilde{g} =: h, \qquad (1.19)$$

with a selfadjoint and positive (semi)definite operator K in $\mathcal{L}^2$. Being the product of three orthogonal projections, K is bounded with $\|K\| \leq 1$.

16   1. Introduction: Examples of Inverse Problems

From the convolution theorem (1.16) it follows that the bandlimiting operator B_Ω can be written as

$$B_\Omega f(t) = \int_\mathbb{R} \check{\chi}(t-s) f(s)\, ds,$$

where χ is the characteristic function of the interval $[-\Omega, \Omega]$, and $\check{\chi} = F^{-1}\chi$ is its inverse Fourier transform, related to the so-called *sinc function*

$$\operatorname{sinc} t = \frac{\sin t}{t}$$

via

$$\check{\chi}(t) = \frac{\Omega}{\pi} \operatorname{sinc} \Omega t.$$

In other words, considered as an equation in $\mathcal{L}^2[-T, T]$, we can rewrite (1.19) as an integral equation of the first kind,

$$Kf(t) = \frac{\Omega}{\pi} \int_{-T}^{T} \operatorname{sinc}(\Omega(t-s)) f(s)\, ds = h(t), \qquad -T \leq t \leq T. \tag{1.20}$$

Note that the sinc function is an entire function, and hence K a compact operator in $\mathcal{L}^2[-T, T]$ (cf. Section 2.2).

The Gerchberg-Papoulis algorithm proceeds in an iterative manner as follows: in a first step the inverse Fourier transform of the given data $\tilde{g}$ is computed; then, since f has continuous support and $F^{-1}\tilde{g}$ has not (for the same reason for which $\hat{f}$ cannot be compactly supported), $F^{-1}\tilde{g}$ is restricted to $[-T, T]$, which yields h, and this restriction $f_0 = h$ is taken as a first guess of f. Afterwards, the Fourier transform of f_0 is computed, so that its values can be compared with those of $\hat{f}$ in $[-\Omega, \Omega]$, i.e., with $\tilde{g}$. If

$$\tilde{g} - P_\Omega F f_0 \tag{1.21}$$

is zero, then we are done, i.e., $B_\Omega(f - f_0)$ vanishes, and hence $f_0 = f$ by the uniqueness theorem for analytic functions, and by the invertibility of the Fourier transformation. If (1.21) is not zero, then we proceed with some "iterative refinement", i.e., we compute for $k = 1, 2, \ldots$,

$$f_k = f_{k-1} + P_T F^{-1}(\tilde{g} - P_\Omega F f_{k-1}) = f_{k-1} + (h - K f_{k-1}). \tag{1.22}$$

We observe immediately that the solution f is a fixed point of this iteration, and for the errors $e_k = f_k - f$ we obtain that

$$e_k = e_{k-1} - P_T B_\Omega e_{k-1} = (I - K) e_{k-1}, \qquad k = 1, 2, \ldots.$$

Since K is selfadjoint and positive definite, with $\|K\| \leq 1$, the operator $I - K$ is nonexpansive, and we can conclude that $\{f_k\}$ remains bounded as $k \to \infty$. That f_k actually converges to f in $\mathcal{L}^2$ follows from the results to be presented in Chapter 6. For practical means the Gerchberg-Papoulis algorithm is not very attractive, since its convergence is slow. However, using the conjugate gradient method as described in Chapter 7 the convergence can be significantly accelerated.

1.4. Inverse Problems in Signal and Image Processing

The theory to be developed in the subsequent parts of the book strongly depends on the spectral properties of the given operator. We mention that the spectrum of the particular operator $K = P_T B_\Omega P_T$ from (1.19) is very well understood due to the works of Landau, Pollak, Slepian, Widom, and others; we refer to the survey article [255] by Slepian. The eigenfunctions of K are the so-called *prolate spheroidal wave functions* that play an important role in potential theory. There are infinitely many eigenvalues λ_n of K, all of them strictly between zero and one, but converging to zero for $n \to \infty$, with an exponential decay. Besides this cluster of eigenvalues at the origin, there is another (finite) cluster near $\lambda = 1$, and a comparatively small amount of eigenvalues spread throughout the open interval $(\varepsilon, 1-\varepsilon)$ for any positive parameter ε.

To be more precise, we quote from [171] that asymptotically (i.e., for large ΩT) for any $0 < \alpha < 1$ the number of eigenvalues larger than α is

$$\frac{2}{\pi}\Omega T + \frac{1}{\pi^2}\log\frac{1-\alpha}{\alpha}\log(\Omega T) + o(\log(\Omega T)) \quad \text{as} \quad \Omega T \to \infty. \tag{1.23}$$

Given no further knowledge about the function h, this number may be considered as a measure for the information content in the problem (1.19): only the information from the eigenspaces corresponding to eigenvalues greater than some small $\varepsilon > 0$ (depending on the level of noise in the data, cf. Chapter 4) can be expected to be reconstructed in a stable way. According to (1.23) this is a subspace of dimension slightly bigger than $2\Omega T/\pi$. In a way, this is a generalization of the well-known Shannon-Whittaker Sampling Theorem, which states that for a bandlimited function with bandwidth Ω we cannot resolve details (in time) smaller than the Nyquist rate

$$\Delta t = \frac{\pi}{\Omega}. \tag{1.24}$$

This restriction of possible spatial resolution plays a key role in a number of inverse problems, e.g., in scattering theory [86], and X-Ray tomography [207].

We conclude this section with an extension of (1.20) originating from confocal scanning light microscopy, and leading to an integral equation (1.14) with a point-spread function, which is not spatially invariant. For simplicity, we present this problem in 1D.

As described in more detail in [105], the integral equation (1.20) is a standard model in Fourier optics describing the result of an imaging system (in 1D) with aperture width Ω. In confocal scanning light microscopy (cf. [24]) the situation is further complicated by the fact that the object is not uniformly illuminated; rather, incident light is focused onto the object plane by a second (confocal) lens, the illumination lens. Therefore, if h is the point-spread function of this second lens centered at the origin, the illuminated object becomes $h(t)f(t)$. Assuming next that the collector lens and the illumination lens have the same point-spread functions $h(t) = \text{sinc}\,\Omega t$, the image formed by the microscope is

$$Kf(t) = \int_{\mathbb{R}} h(t-s)h(s)f(s)\,ds. \tag{1.25}$$

18   1. Introduction: Examples of Inverse Problems

It is easily seen that

$$\int_{\mathbb{R}^2}\bigl(h(t-s)h(s)\bigr)^2\,dtds = \int_{\mathbb{R}} h^2(s)\int_{\mathbb{R}} h^2(t-s)\,dtds = \|h\|_{\mathcal{L}^2}^4,$$

which means that K is an operator of Hilbert-Schmidt type, and hence, compact. From the detailed spectral analysis of K in [98] it can be seen that the numerical reconstruction of the solution f of (1.25) will lead to quantitatively similar stability problems as the problem of numerical differentiation considered in Section 1.1.

To enhance the reconstruction process, in a scanning microscope the object is translated by certain values τ, say. This leads to additional data

$$g(t,\tau) = \int_{\mathbb{R}} h(t-s)h(s)f(s+\tau)\,ds.$$

Note that for fixed τ the function $g(\cdot,\tau)$ is bandlimited with bandwidth Ω. Hence, from the Shannon Sampling Theorem one might expect that the Nyquist rate Δt of (1.24) determines the resolution in f that can be reconstructed from these data. However, as argued in [24], one can actually increase the ultimate resolution in f to *half* the Nyquist rate, by taking *all* values $g(t,\tau)$ into account, where τ runs through all multiples of $\Delta t/2$, i.e., if the specimen is shifted by this amount for each new scanning position. In the literature this effect is known as *superresolution*.

Finally, given exact data $g(t,\tau)$ for all $(t,\tau) \in \mathbb{R}^2$, we have the reconstruction formula (cf. [52])

$$f(s) = \int_{\mathbb{R}} g(t, s - \frac{t}{2})\,dt, \qquad (1.26)$$

valid for bandlimited functions f with bandwidth 2Ω. The orthogonal complement of the set of bandlimited functions with bandwidth 2Ω belongs to the nullspace of K, and hence is "invisible" from the data. Note that the ill-posedness of (1.25) is now reflected by the fact that the integral (1.26) need not exist, and even need not be defined for $g \in \mathcal{L}^2(\mathbb{R}^2)$. See [52] for the construction of regularized approximations of the reconstruction formula (1.26).

1.5. Inverse Problems in Heat Conduction

The classical (well-posed) problem in heat conduction is the *direct problem* of computing the temperature evolution in a body with known thermal parameters, given the initial temperature and the temperature or the heat flux on the whole boundary. Several important types of *inverse problems* arise:

- the problem of determining the initial temperature from later measurements, mathematically spoken, the *backwards heat equation*. This problem will be treated at the end of this section and in Example 2.9.

- the problem of determining thermal parameters of the material from temperature measurements; see Section 1.6.

1.5. Inverse Problems in Heat Conduction

- the problem of determining the temperature on an inaccessible part of the boundary from heat flux and temperature measurements on other parts of the boundary. We discuss a model problem for this *sideways heat equation*:

We pose the problem of determining

$$f(t) := u(1,t) \quad \text{for } t \in \mathbb{R} \tag{1.27}$$

from measurements of

$$g(t) := u(0,t) \quad \text{for } t \in \mathbb{R} \tag{1.28}$$

and the information that the left-hand boundary of $\Omega := [0,1]$ is insulated, i.e., that

$$\frac{\partial u}{\partial x}(0,t) = 0 \quad \text{for } t \in \mathbb{R},$$

where u is assumed to fulfill the linear heat equation

$$\frac{\partial u}{\partial x^2} = \frac{\partial u}{\partial t} \quad \text{in } [0,1] \times \mathbb{R}. \tag{1.29}$$

Assuming that for all $x \in [0,1]$, $u(x,\cdot) \in \mathcal{L}^2(\mathbb{R})$, we can treat this problem by taking Fourier transforms: for $v = v(x,t)$, we denote by $\hat{v}$ its Fourier transform with respect to t, i.e.,

$$\hat{v}(x,w) := \frac{1}{\sqrt{2\pi}} \int_{-\infty}^{+\infty} e^{-iwt} v(x,t) \, dt, \quad w \in \mathbb{R}.$$

Since $\widehat{\frac{\partial v}{\partial t}} = iw\hat{v}$, (1.29) becomes

$$\frac{\partial^2 \hat{u}}{\partial x^2} = iw\hat{u} \quad \text{in } [0,1] \times \mathbb{R},$$

which implies (together with (1.27) and (1.28)) that

$$\hat{u}(x,w) = \frac{\cosh(x\sqrt{iw})\hat{f}(w)}{\cosh(\sqrt{iw})}, \quad x \in [0,1], w \in \mathbb{R},$$

and hence

$$\hat{f}(w) = \cosh(\sqrt{iw})\hat{g}(w), \quad w \in \mathbb{R}. \tag{1.30}$$

A formal solution for the problem of determining f from g can thus be obtained using the inverse Fourier transform

$$f(t) = \frac{1}{\sqrt{2\pi}} \int_{-\infty}^{+\infty} e^{iwt} \cosh(\sqrt{iw})\hat{g}(w) \, dw, \quad t \in \mathbb{R}. \tag{1.31}$$

However, (1.31) makes sense only if the integral exists in some reasonable sense, which is the case if the right-hand side of (1.30) is an $\mathcal{L}^2$-function. But, since $|\cosh\sqrt{iw}| = (\sinh^2\sqrt{w/2} + \cos^2\sqrt{w/2})^{\frac{1}{2}}$ goes to infinity exponentially as $|w| \to \infty$,

1. Introduction: Examples of Inverse Problems

this can only be the case if $\hat{g}$ decays very rapidly as $|w| \to \infty$, i.e., if g is very smooth. Even then, arbitrarily small errors in the data g can lead to arbitrarily large errors in the result f: if g^δ is such that $\|g - g^\delta\|_{\mathcal{L}^2(\mathbb{R})} = \delta$, then also $\|\hat{g} - \hat{g}^\delta\|_{\mathcal{L}^2(\mathbb{R})} = \delta$. The result corresponding to the perturbed data g^δ is then determined by

$$\hat{f}^\delta(w) = \cosh\sqrt{iw}\,\hat{g}^\delta(w)\,, \qquad w \in \mathbb{R}\,.$$

Now, let $w_0 \in \mathbb{R}$ be arbitrary, and define g^δ via its Fourier transform by

$$\hat{g}^\delta(w) = \begin{cases} \hat{g}(w)\,, & w \notin [w_0, w_0+1]\,, \\ \hat{g}(w) + \delta\,, & w \in [w_0, w_0+1]\,. \end{cases}$$

Then, $\|g - g^\delta\|_{\mathcal{L}^2(\mathbb{R})} = \|\hat{g} - \hat{g}^\delta\|_{\mathcal{L}^2(\mathbb{R})} = \delta$. For the corresponding error in the result, we have then that

$$\begin{aligned}\|f - f^\delta\|^2_{\mathcal{L}^2(\mathbb{R})} &= \|\hat{f} - \hat{f}^\delta\|^2_{\mathcal{L}^2(\mathbb{R})} = \delta^2 \int_{w_0}^{w_0+1} |\cosh\sqrt{iw}|^2\, dw \\ &\geq \delta^2 \int_{w_0}^{w_0+1} \sinh^2\sqrt{\frac{w}{2}}\, dw \geq \delta^2 \sinh^2\sqrt{\frac{w_0}{2}}\,,\end{aligned}$$

so that for $w_0 > 0$,

$$\|f - f^\delta\|_{\mathcal{L}^2(\mathbb{R})} \geq \frac{\delta}{2} \exp\sqrt{\frac{w_0}{2}} \qquad (1.32)$$

holds. Thus, the data error δ is amplified essentially with the factor $\exp\sqrt{w_0/2}$ if it is concentrated around frequencies of w_0. High frequency components of the error are magnified very much, which we have also seen in the different context of Section 1.1.

Thus, if we cannot exclude high frequency errors in our data, arbitrarily small data errors can lead to arbitrarily large errors in the result.

As in Section 1.1, a-priori information about the unknown solution f can stabilize the problem. One can show (see [80]) that, if f, $f_\delta \in C^1(\mathbb{R}) \cap \mathcal{L}^2(\mathbb{R})$ and

$$\|f'\|_{\mathcal{L}^2(\mathbb{R})} \leq E \quad \text{and} \quad \|(f^\delta)'\|_{\mathcal{L}^2(\mathbb{R})} \leq E\,, \qquad (1.33)$$

then

$$|f(t) - f^\delta(t)| = O\left(\frac{1}{\log\frac{1}{\delta}}\right)$$

holds uniformly in t. Thus, the a-priori assumption (1.33) gives *logarithmic stability*, which is still not much, since it means (asymptotically) that in order to reduce the error in the result to 50 %, one has to square the error in the data. Under stronger a-priori assumptions, Hölder stability can also be obtained (see [80]).

We have seen here that the inherent reason for ill-posedness lies in the (strong) smoothing properties of the operator describing the *direct* problem, i.e., of the operator mapping f onto g, which results in amplification of high frequency error. Again, stability can be restored by an a-priori assumption.

Problems of this type appear in many practical problems, e.g., in connection with cooling processes in continuous casting or hot rolling of steel. There, however, the thermal parameters (heat conductivity, specific heat capacity) are temperature-dependent, so that the governing partial differential equation is a nonlinear parabolic equation. See, e.g., [28, 77, 99, 174] for inverse problems in steel processing and [23, 83, 198] for inverse problems in heat conduction and in other diffusion processes.

We close this section by considering the *backwards heat equation* on the whole space; our aim is to show its ill-posedness, to relate it to an integral equation of the first kind, and to outline a classical method for stabilizing certain inverse problems by using a-priori information, the *method of logarithmic convexity*, for this simple situation:

Let $\Omega \subseteq \mathbb{R}^3$ be the region that a body, whose boundary is kept at constant temperature 0, occupies. Putting all physical parameters to 1, the temperature $u = u(x,t)$ obeys the linear heat equation

$$\Delta u = \frac{\partial u}{\partial t} \quad \text{in } \Omega \times [0,T] \tag{1.34}$$

with Dirichlet boundary condition

$$u = 0 \quad \text{on } \partial\Omega \times [0,T]. \tag{1.35}$$

We want to determine the initial temperature distribution $u(x,0)$ $(x \in \Omega)$ from measurements of the final temperature

$$f(x) := u(x,T) \quad \text{for } x \in \Omega. \tag{1.36}$$

It can be shown (cf. [42]) that no solution of this inverse problem exists unless f is analytic. Again, the operator representing the "direct problem" (mapping $u(\cdot,0)$ onto f) has strong smoothing properties. Even if we restrict our interest to those f for which a solution exists, this (unique, as one can show) solution does not depend continuously on the data, as the following construction shows: let $\lambda_1, \lambda_2, \lambda_3, \ldots$ be the eigenvalues for the Dirichlet problem for Ω, $\phi_1, \phi_2, \phi_3, \ldots \in \mathcal{L}^2(\Omega)$ the corresponding normalized eigenvalues, i.e., for all $k \in \mathbb{N}$, $\|\phi_k\|_{\mathcal{L}^2(\Omega)} = 1$ and

$$\Delta\phi_k + \lambda_k \phi_k = 0 \quad \text{in } \Omega$$
$$\phi_k = 0 \quad \text{on } \partial\Omega.$$

Let

$$u_k(x,t) := \frac{1}{\lambda_k} \phi_k(x) \exp(\lambda_k(T-t)), \quad x \in \Omega, t \in [0,T].$$

Then

$$(\Delta u_k)(x,t) = \frac{1}{\lambda_k}(\Delta\phi_k)(x)\exp(\lambda_k(T-t))$$
$$= -\phi_k(x)\exp(\lambda_k(T-t)) = \frac{\partial}{\partial t}u_k(x,t),$$

i.e., u_k fulfills (1.34) and, by construction, also (1.35). With $f_k := \phi_k/\lambda_k$, u_k solves (1.34) – (1.36). Since $\lambda_k \to +\infty$, $\lim_{k\to\infty} \|f_k\|_{\mathcal{L}^2(\Omega)} = 0$, but $\lim_{k\to\infty} \|u_k(\cdot,0)\|_{\mathcal{L}^2(\Omega)} = \lim_{k\to\infty} \exp(\lambda_k T)/\lambda_k = +\infty$. Thus, if we consider f_k as perturbations of $f = 0$ with $\mathcal{L}^2$-error $1/\lambda_k$, the corresponding error in the solution of the inverse problem is amplified exponentially, namely by the factor $\exp(\lambda_k T)$. Especially, the solution of the inverse problem does not depend continuously on the data.

The exponential amplification of the error appears already for arbitrarily short time $T > 0$, but becomes worse as T increases.

This inverse problem can be represented by an integral equation of the first kind. E.g., for the one-dimensional unbounded case (i.e., $\Omega = \mathbb{R}$), the final temperature $u(\cdot, T)$ is related to the initial temperature $u(\cdot, 0)$ via

$$\frac{1}{2\sqrt{\pi T}} \int_{-\infty}^{+\infty} u(s,0) \exp\left(-\frac{(x-s)^2}{4T}\right) ds = u(x,T), \qquad (1.37)$$

which is a *convolution equation* with a very smooth kernel. For the case of a bounded domain see Example 2.9.

Again, a-priori information can be used to stabilize (or *regularize*) the problem. Since the method of proof for showing this, the *method of logarithmic convexity*, has many applications (see, e.g., [156, 225]) to inverse problems in partial differential equations, we give the proof for one specific a-priori assumption, namely that for the initial temperature $u(\cdot, 0)$ an $\mathcal{L}^2$-bound

$$\|u(\cdot,0)\|_{\mathcal{L}^2(\Omega)} \leq M \qquad (1.38)$$

is known: for $t \in [0,T]$, let

$$F(t) := \int_\Omega u^2(x,t)\, dx. \qquad (1.39)$$

If $F(0) = 0$, then $u(\cdot, 0) = 0$ and hence, because of (1.35), also $u(\cdot, t) = 0$ for all $t \in [0,T]$. Now, assume that $F(0) \neq 0$. Then also $F(t) \neq 0$ for all $t \in [0,T]$ by uniqueness for the backwards heat equation. Since $F(t) = \langle u(\cdot,t), u(\cdot,t) \rangle$, we have

$$F'(t) = 2\left\langle \frac{\partial u}{\partial t}(\cdot, t), u(\cdot, t) \right\rangle$$

and

$$F''(t) = 2\left\langle \frac{\partial^2 u}{\partial t^2}(\cdot, t), u(\cdot, t) \right\rangle + 2\left\langle \frac{\partial u}{\partial t}(\cdot, t), \frac{\partial u}{\partial t}(\cdot, t) \right\rangle$$

where $\langle \cdot, \cdot \rangle$ denotes the $\mathcal{L}^2(\Omega)$-inner product. We compute the first term in this expression for F'' using (1.34), (1.35) (which also implies that $\frac{\partial u}{\partial t} = 0$ on $\partial\Omega$) and Green's identity

$$\int_\Omega (u\Delta v - v\Delta u)\, dV = \int_{\partial\Omega} \left(u\frac{\partial u}{\partial n} - v\frac{\partial u}{\partial n}\right) dS$$

(valid for C^2-functions u, v):

$$\left\langle \frac{\partial^2 u}{\partial t^2}(\cdot, t), u(\cdot, t) \right\rangle = \left\langle \Delta\left(\frac{\partial u}{\partial t}\right)(\cdot, t), u(\cdot, t) \right\rangle$$

$$= \int_\Omega u(x,t) \Delta\left(\frac{\partial u}{\partial t}\right)(x,t)\, dx = \int_\Omega (\Delta u)(x,t) \frac{\partial u}{\partial t}(x,t)\, dx$$

$$= \int_\Omega \left(\frac{\partial u}{\partial t}\right)^2 (x,t)\, dx = \left\langle \frac{\partial u}{\partial t}(\cdot, t), \frac{\partial u}{\partial t}(\cdot, t) \right\rangle.$$

Hence, it follows with the Cauchy-Schwarz-inequality that for all $t \in [0, T]$,

$$F(t)F''(t) - (F'(t))^2 = 4\|u(\cdot,t)\|^2_{\mathcal{L}^2(\Omega)} \left\|\frac{\partial u}{\partial t}(\cdot,t)\right\|^2_{\mathcal{L}^2(\Omega)} - 4\left\langle \frac{\partial u}{\partial t}(\cdot,t), u(\cdot,t) \right\rangle^2 \geq 0,$$

so that also

$$(\log F)''(t) = \frac{F''(t)F(t) - (F'(t))^2}{F(t)^2} \geq 0. \qquad (1.40)$$

Since (1.40) implies that $\log F$ is a convex function, this method of proof bears the name of *logarithmic convexity*.

The convexity of $\log F$ now implies for $t \in [0,T]$ that

$$(\log F)(t) \leq \frac{T-t}{T}(\log F)(0) + \frac{t}{T}(\log F)(T)$$

and hence

$$F(t) \leq F(0)^{\frac{T-t}{T}} F(T)^{\frac{t}{T}}. \qquad (1.41)$$

Now, (1.36), (1.38), (1.39) and (1.41) immediately imply that for all $t \in [0,T]$,

$$\|u(\cdot,t)\|_{\mathcal{L}^2(\Omega)} \leq M^{1-\frac{t}{T}} \|f\|^{\frac{t}{T}}_{\mathcal{L}^2(\Omega)}.$$

This means that for any $t \in (0,T]$, the temperature $u(\cdot, t)$ depends in a Hölder continuous way on the final temperature $f = u(\cdot, T)$. The Hölder exponent t/T becomes smaller as t decreases to 0, for $t = 0$, the stability result vanishes. Note that Hölder continuity even with arbitrarily small Hölder exponent is still (asymptotically) better than the logarithmic stability we obtained for the sideways heat equation considered above. In practical situations, the assumption of having an a-priori bound for the temperature is of course reasonable.

1.6. Parameter Identification

In physical or technical applications, one often has the situation that the physical laws governing the process are known, but quantitative information about physical parameters is not available. E.g., if one considers heat conduction in a material

24   1. Introduction: Examples of Inverse Problems

occupying a three-dimensional domain Ω whose temperature is kept at 0 at the boundary, the temperature distribution u after sufficiently long time can be modelled by

$$\begin{aligned} -\operatorname{div}(a(x,y,z)\nabla u) &= f(x,y,z), \quad (x,y,z) \in \Omega, \\ u &= 0 \qquad\qquad \text{on } \partial\Omega, \end{aligned} \quad (1.42)$$

where f denotes internal heat sources and a is the (spatially varying) heat conductivity.

The question is then how to determine a from internal measurements of the temperature u or from measurements of, e.g., the heat flux $a\dfrac{\partial u}{\partial n}$ at the boundary $\partial\Omega$.

The first question to be addressed in such a context is, whether the data contain the information one is looking for, i.e., if in (1.42) a is uniquely determined by u. In general, the answer is negative, since obviously, in subregions where u is constant, a can be arbitrary.

To get more insight, we consider the one-dimensional version of (1.42) of determining the coefficient $a = a(x)$ in

$$-(a(x)u_x)_x = f(x), \qquad x \in [0,1],$$

(with appropriate boundary conditions) from measurements of u. If u_x vanishes nowhere, a can be computed explicitly as

$$a(x) = \frac{1}{u_x(x)}\left[a(0)u_x(0) - \int_0^x f(s)\,ds\right] \quad (1.43)$$

and is therefore uniquely determined. Thus, in order to compute a, one has to differentiate the data u, which is an ill-posed problem as explained in Section 1.1.

In addition, there is another effect of instability coming from the division by u_x in (1.43): in regions where u_x is small, errors, e.g., in f are amplified, which is not surprising since where u_x vanishes, a cannot be determined at all, so that some instability has to be expected where u_x is small. This is a nonlinear effect, while the ill-posedness involved with differentiation of the data comes from the fact that also the linearized problem is ill-posed. We emphasize that, in general, parameter identification is a nonlinear inverse problem even if the underlying equation is (for a known parameter) a linear equation. This is, in addition to the problems considered in Sections 1.5 and 1.7, a strong motivation for studying nonlinear inverse and ill-posed problems in a more general and abstract framework as we will do in Chapters 10 and 11.

In the multi-dimensional case, (1.42) can also be written as a first-order hyperbolic equation for a, namely

$$-\langle \nabla a, \nabla u \rangle - a\Delta u = f. \quad (1.44)$$

From this, conditions for uniqueness of the parameter a can be derived. Again, note the differentiation of the data for setting up the equation (1.44).

Recent monographs on parameter identification especially in partial differential equations are [21] and [135].

In recent years, the problem of determining material parameters from *boundary measurements* has attracted considerable interest. For example, it is of interest to determine a in (1.42) from the *Dirichlet-to-Neumann map*, which is the operator mapping Dirichlet boundary data, which make the direct problem (i.e., the problem to compute u with the parameter a given) well-posed, into the resulting Neumann data. In applications, a might be an electrical conductivity, the Dirichlet-to-Neumann map transforms voltage to current. This is the problem of *impedance tomography*, which is important both in medical applications and in the nondestructive testing of materials; from the vast literature on this subject we quote [5, 7, 59, 79, 89, 90, 133, 134, 137, 158, 159, 167, 189, 200, 243, 257, 258, 259, 262, 263, 287], thereby, referring to various different aspects of this field, both theoretical and computational.

1.7. Inverse Scattering

A practically important class of inverse problems are *inverse scattering problems*, where information about an unknown object (e.g., a body, an inhomogeneity in a material, a potential) is to be recovered from measurements of waves or fields scattered by this object. We first give an example that can also be seen as a parameter identification problem: we consider the problem of identifying a spatially varying *acoustic profile* (the reciprocal of the sound speed) described by a function $n = n(x)$ which equals 1 outside some compact set. To this end, a time harmonic wave U^i given by

$$U^i(x,t) = e^{ikt}u^i(x) \tag{1.45}$$

is sent in. Here, u^i is the *velocity potential*, whose gradient with respect to x represents the speed of the wave motion (see below). Since this *incident wave* would be observed if the inhomogeneity in n were not there, i.e., if $n \equiv 1$ everywhere, U^i obeys the wave equation

$$\frac{\partial^2 U^i}{\partial t^2} = \Delta U^i,$$

which (by (1.45)) yields the *reduced wave equation* or *Helmholtz equation*

$$\Delta u^i + k^2 u^i = 0 \tag{1.46}$$

for the spatial component of the incident wave. Due to the inhomogeneity, the truly observed wave $U(x,t) = e^{ikt}u(x)$ obeys

$$\frac{\partial^2 U}{\partial t^2} = \frac{1}{n^2}\Delta U$$

and hence

$$\Delta u + k^2 n^2 u = 0. \tag{1.47}$$

1. Introduction: Examples of Inverse Problems

The *scattered wave* is defined by

$$u^s := u - u^i. \tag{1.48}$$

With

$$f := 1 - n^2,$$

where this unknown function f has compact support, one obtains from (1.46), (1.47) and (1.48) that

$$\Delta u^s + k^2 u^s = k^2 f(u^i + u^s). \tag{1.49}$$

The *inverse scattering problem* consists now in computing the unknown function f in (1.49) from u^i and u^s. Note that usually, u^i and u^s can only be measured far away from the support of f, so that a direct computation of f from (1.49) (which would involve differentiation of the data u^s) is not possible.

Note that the problem of determining f in (1.49) from u^i and u^s is nonlinear, even though the governing equations are linear. However, in many applications it is reasonable to assume that the scattering wave is much smaller than the incident wave and to replace (1.49) by the *Born-(Rytov-)approximation*

$$\Delta u^s + k^2 u^s = k^2 f u^i; \tag{1.50}$$

the problem of determining f from (1.50) is now linear and leads to an integral equation of the first kind for f as follows: since the fundamental solution of the Helmholtz equation is given by

$$G(x,y) := -\frac{1}{4\pi} \frac{e^{ik|x-y|}}{|x-y|}, \quad x \neq y \in \mathbb{R}^3,$$

u^s as given by (1.50) can be represented as

$$u^s(x) = -\frac{k^2}{4\pi} \int_{\text{supp } f} \frac{e^{ik|x-y|}}{|x-y|} u^i(y) f(y) \, dy \tag{1.51}$$

for x outside of the support of f; (1.51) is an integral equation of the first kind (with a rather smooth kernel) for f.

We now turn to an inverse scattering problem where an obstacle (for simplicity in $\mathbb{R}^2$), which is (as opposed to the situation just described) not penetrated by the wave, is to be identified from measurements of an acoustic wave scattered by the obstacle.

The perturbations of a homogeneous isotropic stationary medium with sound speed c by an acoustic wave can be determined from the *velocity potential* $U = U(x,t)$ ($x \in \mathbb{R}^2, t \in \mathbb{R}$), which determines the velocity field via

$$v = \nabla_x U;$$

the pressure perturbations in the medium are determined by the (linearized) momentum equation, which reduces to

$$p = -\rho_0 \frac{\partial U}{\partial t},$$

where ρ_0 is the density of the unperturbed medium. The (linearized) equation for mass conservation then yields the linear wave equation

$$\frac{\partial^2 U}{\partial t^2} = c^2 \Delta U.$$

The space-dependent part u of a time-harmonic wave (in the usual complex notation, where it is understood that, physically, the real part is to be taken)

$$U(x,t) := u(x)e^{-iwt} \tag{1.52}$$

with frequency $w > 0$ fulfills again the *Helmholtz equation*

$$\Delta u + k^2 u = 0, \tag{1.53}$$

with the *wave number* $k = w/c$.

If $D \subseteq \mathbb{R}^2$ is an open, bounded, simply connected set considered as an obstacle for the wave, then the behaviour of the wave is described by (1.53) in $\mathbb{R}^2 \setminus D$ and boundary conditions on ∂D (together with a condition at infinity, see (1.55)), i.e., by an *exterior boundary value problem*. The boundary condition on ∂D depends on the physical properties of the obstacle D. Because of (1.51) and (1.52), $p = iw\rho_0 u(x)e^{-iwt}$, so that prescribing u at ∂D corresponds to prescribing the pressure. In general, for locally reacting surfaces, the (outward) normal component of the velocity field, i.e., $\frac{\partial u}{\partial n}$, is proportional to the pressure, the proportionality factor being $-\rho_0 cz$, where z is called the *acoustic impedance*. This leads to *Robin (or impedance) boundary conditions* of the type

$$\frac{\partial u}{\partial n}(x) + \lambda(x)u(x) = 0, \qquad x \in \partial D, \tag{1.54}$$

with $\lambda := iw/cz$. For $\lambda = 0$, this reduces to a Neumann boundary condition, namely that the normal velocity of the wave vanishes at ∂D; this motivates to call such an obstacle *sound-hard*. In contrast, for a *sound-soft* obstacle, the pressure vanishes at ∂D, which leads to a homogeneous Dirichlet boundary condition for u (or to the limiting case of (1.54) as $\lambda \to +\infty$).

Again, the total field u is usually decomposed into

$$u := u^i + u^s,$$

where u^i is the *incident field*, i.e., the field that would be there, if the obstacle D were not present; u^i is an entire solution of (1.53), i.e., a solution in all of $\mathbb{R}^2$. The *scattered field* u^s also fulfills (1.53) (in $\mathbb{R}^2 \setminus \overline{D}$) and represents the perturbation generated by D; u^s cannot be measured directly, but only via subtracting u^i from the measurement u computationally.

In the absence of dissipation, both *incoming* and *outgoing* waves u^s are possible solutions. The (physically relevant) outgoing solutions satisfy the *Sommerfeld radiation condition*

$$\frac{\partial u^s}{\partial r} - iku^s = o(|x|^{-\frac{1}{2}}) \quad \text{as} \quad r := |x| \to \infty, \tag{1.55}$$

28   1. Introduction: Examples of Inverse Problems

uniformly for all directions $x/|x|$.

The direct scattering problem is now to compute u^s when D is given, i.e., to solve the exterior boundary value problem

$$\Delta u + k^2 u = 0 \quad \text{in } \mathbb{R}^2 \setminus \overline{D} \tag{1.56}$$

together with (1.55) and

$$\frac{\partial u}{\partial n} = 0 \quad \text{on } \partial D \tag{1.57}$$

(sound-hard obstacle) or

$$u = 0 \quad \text{on } \partial D \tag{1.58}$$

(sound-soft obstacle).

These exterior boundary value problems can be reduced to integral equations over ∂D as will now be outlined (see, e.g., [44]). The fundamental solution of (1.56) (fulfilling (1.55)) in $\mathbb{R}^2$ is given by

$$\gamma(x, y) := \frac{i}{4} H_0^{(1)}(k|x-y|), \quad x \neq y.$$

where $H_0^{(1)}$ is the Hankel function of the first kind of order 0. For a continuous density $\phi \in C(\partial D)$, one defines the *single layer potential* as

$$(S\phi)(x) := \int_{\partial D} \phi(y) \gamma(x, y)\, ds(y), \quad x \in \mathbb{R}^2, \tag{1.59}$$

and the *double layer potential* as

$$(D\phi)(x) := \int_{\partial D} \phi(y) \frac{\partial}{\partial n(y)} \gamma(x, y)\, ds(y), \quad x \in \mathbb{R}^2. \tag{1.60}$$

For $x \in \partial D$, these integrals are singular, but if $\partial D \in C^2$, then the integral operators S and D are weakly singular, which is due to the fact that $\gamma(x, y)$ has a logarithmic singularity at $x \neq y$. Both $S\phi$ and $D\phi$ solve (1.53) in $\mathbb{R}^2 \setminus \partial D$. The behaviour of (1.59) and (1.60) as x approaches or crosses ∂D is crucial:

$S\phi$ is continuous in $\mathbb{R}^2$, especially across ∂D, but the normal derivative of $S\phi$ jumps across ∂D:

$$\left(\frac{\partial}{\partial n} S\phi\right)_\pm (x) = \int_{\partial D} \phi(y) \frac{\partial}{\partial n(x)} \gamma(x, y)\, ds(y) \mp \frac{1}{2}\phi(x), \quad x \in \partial D, \tag{1.61}$$

holds, where the subscripts $+$ and $-$ mean limits as x approaches ∂D from the exterior or interior, respectively.

For the double layer potential, the normal derivative is continuous (at least in normal direction) across ∂D, while for the values the jump relation

$$(D\phi)_\pm(x) = \int_{\partial D} \phi(y) \gamma(x, y)\, ds(y) \pm \frac{1}{2} \phi(x), \quad x \in \partial D, \tag{1.62}$$

1.7. Inverse Scattering

holds. These jump relations can be used to represent boundary conditions at ∂D. For example, if we look at the exterior Neumann problem for (1.53), i.e., at (1.53) in the exterior domain of ∂D with the boundary condition

$$\frac{\partial u}{\partial n} = g \text{ at } \partial D$$

with a given function g, then in the *integral equation method*, one represents the solution u as a single layer potential

$$u = S\phi$$

with an unknown density ϕ. Now, (1.55) and (1.61) (with the subscript $+$) imply that ϕ has to fulfill the integral equation

$$\phi(x) - 2\int_{\partial D} \phi(y)\frac{\partial}{\partial n(x)}\gamma(x,y)\,ds(x) = -2g(x), \qquad x \in \partial D.$$

For a Dirichlet problem, a representation of the solution as a double layer potential also leads to an integral equation of the second kind.

We now turn to inverse scattering problems. For a comprehensive review both about theory and numerical methods see [45], cf. also [232].

One class of numerical methods tries to solve an inverse problem iteratively by some kind of optimization technique; the big disadvantage is that in each iteration, one has to solve the direct problem (repeatedly) in order to compute the value of the objective functional (and, e.g., its gradient). *Direct methods* for solving inverse problems avoid such an iteration; to find such methods, one has to derive an equation which directly links the solution of the inverse problem with the available data. Following [165], we describe here the basic steps of one such method without going into any proof details. For another direct method for inverse acoustic scattering see [46, 47, 48]; numerical comparisons of these two methods are presented in [153].

Usually, the data in an inverse scattering problem is the *far field pattern* u_∞ in the representation

$$u^s(x) = \frac{e^{ik|x|}}{\sqrt{|x|}}\left[u_\infty\left(\frac{x}{|x|}\right) + O\left(\frac{1}{|x|}\right)\right] \quad \text{as } |x| \to \infty. \qquad (1.63)$$

for the scattered field.

The far field pattern is defined on the unit circle and describes the direction-dependent component of the leading term in an expansion of the scattered field far away from the obstacle (where the other terms are negligible). The inverse problem considered is the determination of location of the sound-soft scattering obstacle $D \subseteq \mathbb{R}^2$ from measurements of the far-field pattern (for one incoming wave).

The first step is to assume that a closed curve Γ wholly contained in D is known; this is some a-priori information about the location and size of D. There is also a technical condition on Γ to make the integral equation method work, namely that

1. Introduction: Examples of Inverse Problems

the homogeneous Dirichlet problem for (1.53) in the interior domain of Γ has only the trivial solution. This condition can be fulfilled by proper choice of Γ.

Now, one represents the scattered field as a single-layer potential with a density $\varphi \in \mathcal{L}^2(\Gamma)$ over Γ:

$$u^s(x) := (S_\Gamma \varphi)(x) := \int_\Gamma \varphi(y) \gamma(x,y) \, ds(y) \,.$$

It turns out that the corresponding far-field pattern is given by

$$(F\varphi)(x) = \frac{e^{\frac{i\pi}{4}}}{\sqrt{8\pi k}} \int_\Gamma \varphi(y) e^{-ik\langle x,y \rangle} \, ds(y) \,, \qquad \|x\| = 1 \,,$$

which can be derived from (1.63) using asymptotic expressions for the Hankel function $H_0^{(1)}$.

The *far-field operator* F is injective; its range is a dense, but proper subset of $\mathcal{L}^2(\{x \in \mathbb{R}^2 \mid \|x\| = 1\})$. The integral operator F has an analytic kernel and is hence heavily smoothing. The first step of the method now consists of inverting F, i.e., of solving

$$\frac{e^{\frac{i\pi}{4}}}{\sqrt{8\pi k}} \int_\Gamma \varphi(y) e^{-ik\langle x,y \rangle} \, ds(y) = u_\infty(x) \,, \qquad \|x\| = 1 \,. \tag{1.64}$$

Note that this is an integral equation of the first kind again. If φ denotes a solution, then, since by (1.58) the total field vanishes on ∂D, the basic idea for the second step is to find ∂D as zero-set of $u^i + u^s = u^i + S_\Gamma \varphi$. However, since (1.64) is highly unstable, a regularization method has to be used for approximately solving this equation. The final method can then be described as combining this regularization method and a defect minimization for finding a parameterized curve were $u^i + S_\Gamma \varphi$ vanishes into one step. We close this outline here, since our main purpose was to show that, also in inverse scattering, integral equations of the first kind play a role.

2. Ill-Posed Linear Operator Equations

In this chapter, we discuss well-posedness and ill-posedness of linear operator equations and provide some basic tools for dealing with such equations from the theory of generalized inverses and from spectral theory.

As we have seen in Chapter 1, inverse problems frequently lead to mathematical problems which do not fulfill Hadamard's definition of *well-posedness*, i.e., for which one of the following properties does not hold:

For all admissible data, a solution exists.	(2.1)
For all admissible data, the solution is unique.	(2.2)
The solution depends continuously on the data.	(2.3)

Of course, this is not a precise mathematical definition; to make it precise in a concrete situation, one has to specify the notion of a solution, which data are considered admissible, and which topology is used for measuring continuity. These specifications have to be such that they are appropriate for the concrete problem. One can always make a problem (technically) well-posed by making these specifications in an artificial way (cf., e.g., Section 1.1, where continuous dependence can be enforced by measuring the data error in the C^1-norm, which, however, is not appropriate for the problem of differentiation).

If a problem is ill-posed, one is usually not too much concerned with the violation of (2.1), although of course also existence of a solution (for exact data) is an important requirement. (2.1) can usually be enforced by relaxing the notion of a solution at least for exact data, while for perturbed data, the problem has to be "regularized" and hence changed anyway.

Violation of (2.2) is considered to be much more serious. If a problem has several solutions, one either has to decide which one is of interest (e.g., the one with smallest norm, which is appropriate for some, but not all, applications) or one has to check the model for completeness and, if possible, feed in additional information. It might happen that in a practical problem, the available data (even if measured exactly at infinitely many points) simply do not determine the quantity which is sought for (cf., e.g., the situation of Section 1.6, where the parameter a is not uniquely determined by the solution u where $\nabla u = 0$). Then one has to "invent" additional measurements. The question of uniqueness is relevant in inverse problems where one looks for a cause for an *observed* effect; if one just wants to find a cause for a *desired* effect, then one is usually content or even happy with having a variety of possible solutions, since then, one can try to pick one which fulfills some additional criteria.

Even if (2.2) is fulfilled when the data are measured "everywhere" (i.e., when they are a function, at uncountably many points, which is of course practically impossible), non-uniqueness is usually introduced by the need for discretization.

Violation of (2.3) creates serious numerical problems: if one wants to approximate a problem whose solution does not depend continuously on the data by a "traditional" numerical method as one would use for a well-posed problem, then one has to expect that the numerical method becomes unstable. A (partial) remedy for this is the use of "regularization methods", although one has to keep in mind that no mathematical trick can make an inherently unstable problem stable. All that a regularization method can do is to recover partial information about the solution as stably as possible. The "art" of applying regularization methods will always be to find the right compromise between accuracy and stability.

We consider linear operator equations of the form

$$Tx = y, \qquad (2.4)$$

where T is a bounded linear operator between Hilbert spaces $\mathcal{X}$ and $\mathcal{Y}$. We will call y *attainable* if

$$y \in \mathcal{R}(T)$$

holds. Then, (2.1) is equivalent to every $y \in \mathcal{Y}$ being attainable. (2.2) holds if and only if $\mathcal{N}(T) = \{0\}$, and if (2.1) and (2.2) hold, so that T^{-1} exists, (2.3) is equivalent to the continuity (or boundedness) of T^{-1}.

It would be too restrictive to assume that y is attainable and that $\mathcal{N}(T) = \{0\}$: even if y is not attainable, one might still be interested in some generalized solution of (2.4), i.e., some element which solves (2.4) in some approximate sense. Also, if $\mathcal{N}(T) \neq \{0\}$, so that solutions of (2.4) are not unique, one might be interested in a specific solution satisfying additional requirements. This generalized notion of solution will be provided via the concept of a *generalized inverse* of T; its continuity will then be relevant for (2.3).

2.1. The Moore-Penrose Generalized Inverse

Here, we provide the basic theory of best-approximate solutions of linear operator equations and of the (Moore-Penrose) generalized inverse of linear operators between Hilbert spaces. For a comprehensive treatment, see [101, 202].

Definition 2.1. *Let* $T : \mathcal{X} \to \mathcal{Y}$ *be a bounded linear operator.*

(i) $x \in \mathcal{X}$ *is called* least-squares solution *of* $Tx = y$ *if*

$$\|Tx - y\| = \inf\{\|Tz - y\| \mid z \in \mathcal{X}\}. \qquad (2.5)$$

(ii) $x \in \mathcal{X}$ *is called* best-approximate solution *of* $Tx = y$ *if* x *is a least-squares solution of* $Tx = y$ *and*

$$\|x\| = \inf\{\|z\| \mid z \text{ is least-squares solution of } Tx = y\} \qquad (2.6)$$

holds.

2.1. The Moore-Penrose Generalized Inverse

The best-approximate solution (which, as we will see, is unique) thus is defined as the least-squares solution of minimal norm. In many applications, (2.6) is not appropriate for uniqueness, but has to be replaced by the minimization of $\|Lx\|$, where L is usually a differential operator. We will consider this case in Section 8.1

As we will see, the notion of a best-approximate solution is closely related to the *Moore-Penrose (generalized) inverse* of T, which will turn out to be the *solution operator* mapping y onto the best-approximate solution of $Tx = y$.

We define the Moore-Penrose inverse in an operator-theoretic way by restricting the domain and range of T in such a way that the resulting restricted operator is invertible; its inverse will then be extended to its maximal domain:

Definition 2.2. *The* Moore-Penrose (generalized) inverse $T^\dagger$ *of* $T \in \mathcal{L}(\mathcal{X}, \mathcal{Y})$ *is defined as the unique linear extension of* $\tilde{T}^{-1}$ *to*

$$\mathcal{D}(T^\dagger) := \mathcal{R}(T) \dotplus \mathcal{R}(T)^\perp \tag{2.7}$$

with

$$\mathcal{N}(T^\dagger) = \mathcal{R}(T)^\perp, \tag{2.8}$$

where

$$\tilde{T} := T|_{\mathcal{N}(T)^\perp} : \mathcal{N}(T)^\perp \to \mathcal{R}(T). \tag{2.9}$$

$T^\dagger$ is well-defined: since $\mathcal{N}(\tilde{T}) = \{0\}$ and $\mathcal{R}(\tilde{T}) = \mathcal{R}(T)$, $\tilde{T}^{-1}$ exists. Due to (2.8) and the requirement that $T^\dagger$ is linear, for any $y \in \mathcal{D}(T^\dagger)$ with the unique representation $y = y_1 + y_2$, $y_1 \in \mathcal{R}(T)$, $y_2 \in \mathcal{R}(T)^\perp$, $T^\dagger y$ has to be $\tilde{T}^{-1} y_1$.

Proposition 2.3. *Let now (and below) P and Q be the orthogonal projectors onto $\mathcal{N}(T)$ and $\overline{\mathcal{R}(T)}$, respectively. Then $\mathcal{R}(T^\dagger) = \mathcal{N}(T)^\perp$, and the four "Moore-Penrose equations" hold:*

$$TT^\dagger T = T, \tag{2.10}$$
$$T^\dagger T T^\dagger = T^\dagger, \tag{2.11}$$
$$T^\dagger T = I - P, \tag{2.12}$$
$$TT^\dagger = Q|_{\mathcal{D}(T^\dagger)}. \tag{2.13}$$

Proof. Because of the definition of $T^\dagger$, for all $y \in \mathcal{D}(T^\dagger)$,

$$T^\dagger y = \tilde{T}^{-1} Q y = T^\dagger Q y, \tag{2.14}$$

so that $T^\dagger y \in \mathcal{R}(\tilde{T}^{-1}) = \mathcal{N}(T)^\perp$. For all $x \in \mathcal{N}(T)^\perp$, $T^\dagger T x = \tilde{T}^{-1} \tilde{T} x = x$. This proves that $\mathcal{R}(T^\dagger) = \mathcal{N}(T)^\perp$. Now, for $y \in \mathcal{D}(T^\dagger)$, (2.14) implies that $TT^\dagger y = TT^\dagger Q y = T\tilde{T}^{-1} Q y = \tilde{T}\tilde{T}^{-1} Q y = Q y$, since $\tilde{T}^{-1} Q y \in \mathcal{N}(T)^\perp$. Hence, (2.13) holds. By definition of $T^\dagger$, we have for all $x \in \mathcal{X}$:

$$T^\dagger T x = \tilde{T}^{-1} T [P x + (I - P) x] = \tilde{T}^{-1} T P x + \tilde{T}^{-1} \tilde{T} (I - P) x = (I - P) x,$$

34   2. Ill-Posed Linear Operator Equations

which implies (2.12). Now, (2.12) implies $TT^\dagger T = T(I - P) = T - TP = T$, hence (2.10); (2.14) and (2.13) imply (2.11). ∎

As can be seen from the proof, (2.12) and (2.13) imply (2.10) and (2.11). On the other hand, the Moore-Penrose equations uniquely characterize $T^\dagger$; for this, it would even be sufficient to assume that $TT^\dagger$ and $T^\dagger T$ are orthogonal projectors instead of requiring (2.12) and (2.13). Any operator fulfilling (2.10) or (2.11) is called an *inner inverse* or an *outer inverse* of T, respectively.

Proposition 2.4. *The Moore-Penrose generalized inverse $T^\dagger$ has a closed graph $gr(T^\dagger)$. Furthermore, $T^\dagger$ is bounded (i.e., continuous) if and only if $\mathcal{R}(T)$ is closed.*

Proof. First we show that

$$\{(y_1, \tilde{T}^{-1}y_1) \mid y_1 \in \mathcal{R}(T)\} = \{(Tx, x) \mid x \in \mathcal{X}\} \cap (\mathcal{Y} \times \mathcal{N}(T)^\perp). \tag{2.15}$$

Let $y_1 \in \mathcal{R}(T)$, $x := \tilde{T}^{-1}y_1$; by the definition of $\tilde{T}$, $x \in \mathcal{N}(T)^\perp$, and due to (2.13), $Tx = TT^\dagger y_1 = y_1$. Hence, $(y_1, \tilde{T}^{-1}y_1) = (Tx, x) \in \mathcal{Y} \times \mathcal{N}(T)^\perp$.

If $x \in \mathcal{N}(T)^\perp$, $y_1 := Tx$ (hence, $y_1 \in \mathcal{R}(T)$), then $\tilde{T}^{-1}y_1 = T^\dagger Tx = x$, so that $(y_1, \tilde{T}^{-1}y_1) = (Tx, x)$. Thus, (2.15) holds.

By definition of $T^\dagger$, we have for the graph of $T^\dagger$:

$$\begin{aligned} gr(T^\dagger) &= \{(y, T^\dagger y) \mid y \in \mathcal{D}(T^\dagger)\} \\ &= \{(y_1 + y_2, \tilde{T}^{-1}y_1) \mid y_1 \in \mathcal{R}(T), y_2 \in \mathcal{R}(T)^\perp\} \\ &= \{(y_1, \tilde{T}^{-1}y_1) \mid y_1 \in \mathcal{R}(T)\} + (\mathcal{R}(T)^\perp \times \{0\}), \end{aligned}$$

which implies with (2.15) that

$$gr(T^\dagger) = [\{(Tx, x) \mid x \in \mathcal{X}\} \cap (\mathcal{Y} \times \mathcal{N}(T)^\perp)] + [\mathcal{R}(T)^\perp \times \{0\}]. \tag{2.16}$$

The spaces on the right-hand side of (2.16) are closed and orthogonal to each other in $\mathcal{Y} \times \mathcal{X}$, so that also their sum $gr(T^\dagger)$ is closed.

To prove the second assertion of this proposition, assume that $\mathcal{R}(T)$ is closed, so that $\mathcal{D}(T^\dagger) = \mathcal{Y}$. Because of the Closed Graph Theorem, $T^\dagger$ is bounded. Conversely, let $T^\dagger$ be bounded. Then $T^\dagger$ has a unique continuous extension $\overline{T^\dagger}$ to $\mathcal{Y}$. From (2.13) and the continuity of T we conclude that $T\overline{T^\dagger} = Q$. Hence, for $y \in \overline{\mathcal{R}(T)}$, $y = Qy = T\overline{T^\dagger}y \in \mathcal{R}(T)$. Thus, $\overline{\mathcal{R}(T)} \subseteq \mathcal{R}(T)$, so that $\mathcal{R}(T)$ is closed. ∎

We will now provide the connection between least-squares solutions and the Moore-Penrose inverse:

Theorem 2.5. *Let $y \in \mathcal{D}(T^\dagger)$. Then, $Tx = y$ has a unique best-approximate solution, which is given by*

$$x^\dagger := T^\dagger y \,.$$

The set of all least-squares solutions is $x^\dagger + \mathcal{N}(T)$.

2.1. The Moore-Penrose Generalized Inverse

Proof. Let
$$\mathcal{S} := \{z \in \mathcal{X} \mid Tz = Qy\}.$$
Since $y \in \mathcal{D}(T^\dagger) = \mathcal{R}(T) \dotplus \mathcal{R}(T)^\perp$, $Qy \in \mathcal{R}(T)$, so that $\mathcal{S} \neq \emptyset$. As the orthogonal projector Q is also a metric projector, we have for all $z \in \mathcal{S}$ and $x \in \mathcal{X}$:
$$\|Tz - y\| = \|Qy - y\| \le \|Tx - y\|.$$
Hence, all elements of $\mathcal{S}$ are least-squares solutions of $Tx = y$. Conversely, let z be a least-squares solution of $Tx = y$. Then,
$$\|Qy - y\| \le \|Tz - y\| = \inf\{\|u - y\| \mid u \in \mathcal{R}(T)\} = \|Qy - y\|,$$
so that Tz is the closest element to y in $\mathcal{R}(T)$, i.e., $Tz = Qy$. Hence, we have shown that
$$\mathcal{S} = \{x \in \mathcal{X} \mid x \text{ is least-squares solution of } Tx = y\} \neq \emptyset.$$
Now let $\bar{z}$ be the element of minimal norm in the closed linear manifold $\mathcal{S} = T^{-1}(\{Qy\})$. Since then $\mathcal{S} = \bar{z} + \mathcal{N}(T)$, it suffices to show that
$$\bar{z} = T^\dagger y. \tag{2.17}$$
As the element of minimal norm in $\bar{z} + \mathcal{N}(T)$, $\bar{z}$ is orthogonal to $\mathcal{N}(T)$, i.e., $\bar{z} \in \mathcal{N}(T)^\perp$. This implies that $\bar{z} = (I - P)\bar{z} = T^\dagger T\bar{z} = T^\dagger Qy = T^\dagger TT^\dagger y = T^\dagger y$, i.e., (2.17). ∎

The best-approximate solution can be characterized by the *Gaußian normal equation*:

Theorem 2.6. *Let $y \in \mathcal{D}(T^\dagger)$. Then $x \in \mathcal{X}$ is a least-squares solution of $Tx = y$ if and only if the normal equation*
$$T^*Tx = T^*y \tag{2.18}$$
holds.

Proof. x is least-squares solution of $Tx = y$ if and only if Tx is the closest element in $\mathcal{R}(T)$ to y, which is equivalent to $Tx - y \in \mathcal{R}(T)^\perp = \mathcal{N}(T^*)$, i.e., to $T^*(Tx - y) = 0$ and thus to (2.18). ∎

It follows from Theorem 2.6 that $T^\dagger y$ is the solution of $T^*Tx = T^*y$ of minimal norm, i.e.,
$$T^\dagger = (T^*T)^\dagger T^*.$$
One can show that $\mathcal{D}(T^\dagger)$ as defined in (2.7) is the natural domain of definition for $T^\dagger$ in the sense that, if $y \notin \mathcal{D}(T^\dagger)$, no least-squares solution of $Tx = y$ exists. Thus, as opposed to the finite-dimensional case, the introduction of the concept of a best-approximate solution, although it enforces uniqueness, does not always lead

to a solvable problem and, as we have just seen, also is no remedy for the lack of continuous dependence in general.

For least-squares solutions which minimize $\|Lx\|$ for some suitable linear operator L and the associated concept of generalized inverse, the *weighted generalized inverse*, see Section 8.1.

Operator equations with a *compact* linear operator will be of special importance. Before discussing properties of their Moore-Penrose inverse, we need the concept of the *singular value expansion* of such operators.

2.2. Compact Linear Operators: Singular Value Expansion

Compact linear operators are of special interest, since integral operators are compact under suitable assumptions. E.g., if $\Omega \subseteq \mathbb{R}^n$ is compact and Jordan measurable and the kernel k is either in $\mathcal{L}^2(\Omega \times \Omega)$ or weakly singular, i.e., k is continuous on $\{(s,t) \in \Omega \times \Omega \mid s \neq t\}$ and for all $s \neq t \in \Omega$,

$$|k(s,t)| \leq \frac{M}{|s-t|^{n-\varepsilon}}$$

with $M > 0, \varepsilon > 0$, then

$$\begin{aligned} K : \mathcal{L}^2(\Omega) &\to \mathcal{L}^2(\Omega) \\ x &\mapsto (Kx)(s) := \int_\Omega k(s,t)x(t)\,dt \end{aligned} \quad (2.19)$$

is compact.

If we assume a bounded linear operator to be compact, we will use the symbol K instead of T for this operator as a reminder of this assumption.

For a selfadjoint compact linear operator, the notion of an *eigensystem* plays an important role: an eigensystem $(\lambda_n; v_n)$ consists of all non-zero eigenvalues λ_n and a corresponding complete orthogonal set of eigenvectors v_n. With such an eigensystem, the operator K can be diagonalized as follows:

$$Kx = \sum_{n=1}^\infty \lambda_n \langle x, v_n \rangle v_n$$

holds for all $x \in \mathcal{X}$. This diagonalization can be used to (formally) solve the equation

$$Kx = \lambda x + y.$$

If K is not selfadjoint, no eigenvalues (and hence no eigensystem) need to exist. However, by exploiting the connection between (2.4) and the normal equation (2.18), one can construct a substitute for an eigensystem, a *singular system*, which can be used in the same way for the non-selfadjoint case:

For any compact linear operator $K : \mathcal{X} \to \mathcal{Y}$, a *singular system* $(\sigma_n; v_n, u_n)$ is defined as follows:

2.2. Compact Linear Operators: Singular Value Expansion

If $K^* : \mathcal{Y} \to \mathcal{X}$ denotes the adjoint of K (defined via the requirement that for all $x \in \mathcal{X}$ and $y \in \mathcal{Y}$, $\langle Kx, y \rangle = \langle x, K^*y \rangle$ holds), then the $\{\sigma_n^2\}_{n \in \mathbb{N}}$ are the non-zero eigenvalues of the selfadjoint operator K^*K (and also of KK^*), written down in decreasing order with multiplicity, $\sigma_n > 0$, the $\{v_n\}_{n \in \mathbb{N}}$ are a corresponding complete orthonormal system of eigenvectors of K^*K (which spans $\overline{\mathcal{R}(K^*)} = \overline{\mathcal{R}(K^*K)}$), and the $\{u_n\}_{n \in \mathbb{N}}$ are defined via vectors

$$u_n := \frac{Kv_n}{\|Kv_n\|}.$$

The $\{u_n\}_{n \in \mathbb{N}}$ are a complete orthonormal system of eigenvectors of KK^* and span $\overline{\mathcal{R}(K)} = \overline{\mathcal{R}(KK^*)}$, and the following formulas hold:

$$Kv_n = \sigma_n u_n, \tag{2.20}$$

$$K^* u_n = \sigma_n v_n, \tag{2.21}$$

$$Kx = \sum_{n=1}^{\infty} \sigma_n \langle x, v_n \rangle u_n, \quad x \in \mathcal{X}, \tag{2.22}$$

$$K^* y = \sum_{n=1}^{\infty} \sigma_n \langle y, u_n \rangle v_n, \quad y \in \mathcal{Y}, \tag{2.23}$$

where these infinite series converge in the Hilbert space norms of $\mathcal{X}$ and $\mathcal{Y}$, respectively; (2.22) and (2.23) are called *singular value expansion* and are the infinite-dimensional analogues of the well known *singular value decomposition* of a matrix.

If (and only if) K has a finite-dimensional range, K has only finitely many singular values, so that all infinite series involving singular values degenerate to finite sums. If K is an integral operator of the form

$$\begin{aligned} K : \mathcal{L}^2(\Omega) &\to \mathcal{L}^2(\Omega) \\ x &\mapsto (Kx)(s) := \int_\Omega k(s,t) x(t) \, dt, \end{aligned}$$

where $\Omega \subseteq \mathbb{R}^n$ is compact and Jordan-measurable with positive measure, and k is an $\mathcal{L}^2$-kernel, this happens if and only if the kernel k is *degenerate*, i.e., has the form

$$k(s,t) = \sum_{i=1}^{n} \varphi_i(s) \psi_i(t), \quad s, t \in \Omega,$$

with $n \in \mathbb{N}$ and $\varphi_i, \psi_i \in \mathcal{L}^2(\Omega)$.

If there are infinitely many singular values, they accumulate (only) at 0, i.e.,

$$\lim_{n \to \infty} \sigma_n = 0.$$

The range $\mathcal{R}(K)$ is closed if and only if it is finite-dimensional, so that in the (generic) case of infinitely many singular values, $\mathcal{R}(K)$ is non-closed. This can be seen as follows:

38   2. Ill-Posed Linear Operator Equations

If $\mathcal{R}(K)$ is closed, then it is complete, so that (by Banach's Open Mapping Theorem)
$$K|_{\mathcal{N}(K)^\perp} : \mathcal{N}(K)^\perp \to \mathcal{R}(K)$$
is continuously invertible. Then, however
$$K\left(K|_{\mathcal{N}(K)^\perp}\right)^{-1} = I_{\mathcal{R}(K)}$$
is compact and, hence, $\dim \mathcal{R}(K) < \infty$. Together with Proposition 2.4, this implies

Proposition 2.7. *Let $K : \mathcal{X} \to \mathcal{Y}$ be compact, $\dim \mathcal{R}(K) = \infty$. Then $K^\dagger$ is a densely defined unbounded linear operator with closed graph.*

Hence, for a compact linear operator with non-closed range (e.g., for an integral operator with a non-degenerate $\mathcal{L}^2$-kernel), the best-approximate solution of (2.4) does not depend continuously on the right-hand side; the equation is ill-posed.

Using a singular system, one can find a series representation of the Moore-Penrose inverse of a compact linear operator. Note that, if the singular system contains only finitely many elements (i.e., if $\mathcal{R}(K)$ is finite-dimensional), then all infinite series containing singular functions have to be interpreted as finite sums.

Theorem 2.8. *Let $(\sigma_n; v_n, u_n)$ be a singular system for the compact linear operator K, $y \in \mathcal{Y}$. Then we have:*

(i) $y \in \mathcal{D}(K^\dagger) \iff \sum_{n=1}^{\infty} \dfrac{|\langle y, u_n \rangle|^2}{\sigma_n^2} < \infty$.

(ii) For $y \in \mathcal{D}(K^\dagger)$,
$$K^\dagger y = \sum_{n=1}^{\infty} \frac{\langle y, u_n \rangle}{\sigma_n} v_n. \qquad (2.24)$$

Proof. Let $y \in \mathcal{D}(K^\dagger)$, i.e., $Qy \in \mathcal{R}(K)$. The orthogonal projector Q onto $\overline{\mathcal{R}(K)}$ can be written as
$$Q = \sum_{n=1}^{\infty} \langle \cdot, u_n \rangle u_n,$$
since the $\{u_n\}$ span $\overline{\mathcal{R}(K)}$. Since $Qy \in \mathcal{R}(K)$, there exists an $x \in \mathcal{X}$ with $Kx = Qy$; without loss of generality, we can assume that $x \in \mathcal{N}(K)^\perp$. Since the $\{v_n\}$ span $\overline{\mathcal{R}(K^*)} = \mathcal{N}(K)^\perp$, $x = \sum_{n=1}^{\infty} \langle x, v_n \rangle v_n$, so that we have

$$\sum_{n=1}^{\infty} \langle y, u_n \rangle u_n = Kx = \sum_{n=1}^{\infty} \langle x, v_n \rangle K v_n = \sum_{n=1}^{\infty} \sigma_n \langle x, v_n \rangle u_n.$$

Thus, for all $n \in \mathbb{N}$,
$$\langle y, u_n \rangle = \sigma_n \langle x, v_n \rangle \qquad (2.25)$$

must hold. Since, as a sequence of Fourier coefficients, $(\langle x, v_n \rangle) \in l^2$, so that, by (2.25), also $(\langle y, u_n \rangle / \sigma_n) \in l^2$, the condition in (i) follows. Conversely, assume that this condition holds; by the Riesz-Fischer Theorem from functional analysis,

$$x := \sum_{n=1}^{\infty} \frac{\langle y, u_n \rangle}{\sigma_n} v_n \in \mathcal{X}.$$

We have

$$Kx = \sum_{n=1}^{\infty} \frac{\langle y, u_n \rangle}{\sigma_n} Kv_n = \sum_{n=1}^{\infty} \langle y, u_n \rangle u_n = Qy,$$

especially $Qy \in \mathcal{R}(K)$ and hence $y \in \mathcal{D}(K^\dagger)$. Since the $\{v_n\}$ span $\mathcal{N}(K)^\perp$, $x \in \mathcal{N}(K)^\perp$. Now, $\{z \in \mathcal{X} \mid Kz = Qy\} = K^\dagger y + \mathcal{N}(K)$. Since x lies both in this set and in $\mathcal{N}(K)^\perp$, x is the element of minimal norm in this set, i.e, $x = K^\dagger Qy = K^\dagger y$. Thus, (2.24) holds. ∎

The condition in Theorem 2.8 (i) for the existence of a best-approximate solution is called the *Picard criterion*. It says that a best-approximate solution of $Kx = y$ exists only if the (generalized) Fourier coefficients $(\langle y, u_n \rangle)$ with respect to the singular functions u_n decay fast enough relative to the singular values σ_n. This condition is reminiscent of the situation discussed in Section 1.5, where a solution existed only if the Fourier transform of the data decayed fast enough. The operator in that inverse heat conduction problem was not compact and had a continuous spectrum. We will now give an example for another inverse heat conduction problem, namely the backwards heat equation on a bounded real interval, which leads to a first-kind integral equation with a compact integral operator, and illustrate Theorem 2.8 for that problem. Before doing so, we make some remarks concerning stability: (2.24) shows how errors in y affect the result $K^\dagger y$: error components (with respect to the basis $\{u_n\}$) which correspond to large singular values are harmless. However, error components which correspond to small singular values σ_n are amplified by the (then large) factors $1/\sigma_n$, so that those are dangerous. If $\dim \mathcal{R}(K) < \infty$, e.g., if K is an integral operator with a degenerate kernel, then there are only finitely many singular values, so that these amplification factors are at least bounded, although they might still be unacceptably large. If, however, $\dim \mathcal{R}(K) = \infty$, then $\lim_{n \to \infty} \sigma_n = 0$ holds, so that data errors of a fixed size can be amplified arbitrarily much, namely by the factors $1/\sigma_n$, which increase without bound. E.g., if $y_{\delta,n} := y + \delta u_n$, then $\|y_{\delta,n} - y\| = \delta$, but, due to (2.24),

$$K^\dagger y - K^\dagger y_{\delta,n} = \frac{\langle \delta u_n, u_n \rangle}{\sigma_n} v_n$$

and hence

$$\|K^\dagger y - K^\dagger y_{\delta,n}\| = \frac{\delta}{\sigma_n} \to \infty \quad \text{as} \quad n \to \infty.$$

A special case of this effect appeared already in Section 1.1 for the simple integral equation (1.3); for a non-compact operator, an analogous effect showed up in Section 1.5 (cf. (1.32)).

40   2. Ill-Posed Linear Operator Equations

Of course, the instability in (2.24) (and the solvability condition of Picard's criterion) becomes the more severe the faster the singular values decay. This makes it possible to quantify the *degree of ill-posedness* of $Kx = y$: usually, a problem is called *mildly (modestly) ill-posed* if $\sigma_n = O(n^{-\alpha})$ for some $\alpha \in \mathbb{R}^+$, and *severely ill-posed* otherwise, e.g., if $\sigma_n = O(e^{-n})$ holds. In the latter case, the amplification factor of a data error in the n-th Fourier component with respect to the singular functions increases at least as e^n, which makes it impossible to permit more than a few degrees of freedom in the data, i.e., one has to make sure that the data contain only information in the first few Fourier components and to filter out all higher components.

For numerical differentiation, i.e., for the integral equation (1.3), $\sigma_n := O(n^{-1})$, while for the Abel equation (1.8), which modeled the axisymmetric case of computerized tomography, $\sigma_n = O(n^{-\frac{1}{2}})$; roughly, that rate also holds in general for computerized tomography in two dimensions, so that these problems are mildly ill-posed. However, "incomplete data problems" in computerized tomography, where X-ray measurements are available only for some directions, are severely ill-posed (cf. [183, 207], where the singular value expansion for the Radon transform can be found).

Now we turn to the example announced above:

Example 2.9. We consider the one-dimensional version of the *backwards heat equation* (cf. Section 1.5), i.e., the following problem: we consider the heat equation

$$\frac{\partial u}{\partial t}(x,t) = \frac{\partial^2 u}{\partial x^2}(x,t), \quad x \in [0, \pi], \ t \geq 0, \qquad (2.26)$$

with homogeneous Dirichlet boundary conditions

$$u(0,t) = u(\pi,t) = 0, \quad t \geq 0, \qquad (2.27)$$

and assume that the final temperature

$$f(x) := u(x, 1), \quad x \in [0, \pi], \qquad (2.28)$$

is given with $f(0) = f(\pi) = 0$; we want to determine the initial temperature

$$v_0(x) := u(x, 0), \quad x \in [0, \pi]. \qquad (2.29)$$

The functions $\varphi_n(x) := \sqrt{\frac{2}{\pi}} \sin(nx)$ are a complete orthonormal system in $\mathcal{L}^2[0, \pi]$ and eigenfunctions of $\frac{d^2}{dx^2}$ on $[0, \pi]$ with homogeneous Dirichlet boundary conditions. Thus, $v_0 \in \mathcal{L}^2[0, \pi]$ can be expanded as

$$v_0(x) = \sum_{n=1}^{\infty} c_n \varphi_n(x), \quad x \in [0, \pi], \quad \text{with} \quad c_n = \sqrt{\frac{2}{\pi}} \int_0^\pi v_0(\tau) \sin(n\tau) \, d\tau. \qquad (2.30)$$

2.2. Compact Linear Operators: Singular Value Expansion

Now we make the following Ansatz by separation of variables for the solution of the direct problem (2.26), (2.27), (2.29):

$$u(x,t) := \sum_{n=1}^{\infty} a_n(t)\varphi_n(x), \quad x \in [0,\pi],\ t \geq 0. \tag{2.31}$$

The following arguments are formal, but can be rigorously justified: by (2.31) and (2.26),

$$\sum_{n=1}^{\infty} a'_n(t)\varphi_n(x) = \sum_{n=1}^{\infty} a_n(t)\varphi''_n(x)$$

and hence (since $\varphi''_n = -n^2\varphi_n$)

$$\sum_{n=1}^{\infty} a'_n(t)\varphi_n(x) = -\sum_{n=1}^{\infty} n^2 a_n(t)\varphi_n(x). \tag{2.32}$$

Since $\{\varphi_n\}$ is an orthonormal system, it follows from (2.32) that the a_n have to solve the initial value problems

$$a'_n(t) = -n^2 a_n(t), \quad t \geq 0,$$
$$a_n(0) = c_n,$$

where the initial conditions follow from (2.29) and (2.30). Hence, for $n \in \mathbb{N}$,

$$a_n(t) = c_n e^{-n^2 t}, \quad t \geq 0.$$

Thus, $u(x,t) = \sum_{n=1}^{\infty} c_n e^{-n^2 t}\varphi_n(x)$, so that by (2.28),

$$f(x) = \sum_{n=1}^{\infty} c_n e^{-n^2}\varphi_n(x)$$
$$= \frac{2}{\pi}\sum_{n=1}^{\infty}\int_0^{\pi} v_0(\tau)\sin(n\tau)\,d\tau\, e^{-n^2}\sin(nx).$$

With

$$k(x,\tau) := \frac{2}{\pi}\sum_{n=1}^{\infty} e^{-n^2}\sin(n\tau)\sin(nx)$$

we thus have

$$\int_0^{\pi} k(x,\tau) v_0(\tau)\,d\tau = f(x). \tag{2.33}$$

Thus, the inverse problem is equivalent to solving the integral equation of the first kind (2.33). Note that a singular system for the integral operator in (2.33) is given by

$$\left(e^{-n^2};\ \sqrt{\tfrac{2}{\pi}}\sin(nx),\ \sqrt{\tfrac{2}{\pi}}\sin(nx) \right).$$

Since the singular functions are complete in $\mathcal{L}^2[0,\pi]$, we have that $\mathcal{N}(K) = \mathcal{N}(K^*) = \{0\}$ and that $\mathcal{D}(K^\dagger) = \mathcal{R}(K)$ is dense in $\mathcal{L}^2[0,\pi]$. Thus, it follows from

42   2. Ill-Posed Linear Operator Equations

Theorem 2.8 that (2.33) and hence our inverse problem is (uniquely) solvable if and only if

$$\sum_{n=1}^{\infty} e^{2n^2} |f_n|^2 < \infty \qquad (2.34)$$

holds, where the

$$f_n := \sqrt{\tfrac{2}{\pi}} \int_0^\pi f(\tau) \sin(n\tau)\, d\tau$$

are the (classical) Fourier coefficients of f. In this case, the solution is given by

$$v_0(x) = \sqrt{\tfrac{2}{\pi}} \sum_{n=1}^{\infty} e^{n^2} f_n \sin(nx). \qquad (2.35)$$

(2.34) and (2.35) show that this inverse problem is extremely ill-posed: a solution exists only for such f for which the Fourier coefficients $\{f_n\}$ decay extremely rapidly (much faster than $\{e^{-n^2}\}$), i.e., for very smooth f. A small error in the n-th Fourier coefficient is amplified by the factor e^{n^2}! Thus, already an error of, say, 10^{-8} in the fifth Fourier coefficient of the data leads to an error of about 10^3 in the initial temperature. Thus, one can consider at most about three degrees of freedom in the data and has to filter out everything else. This is the necessary "compromise between accuracy and stability" mentioned above in this situation. It is not a valid question to ask for five Fourier coefficients of the solution if one has (as always in practice) even slight noise in the data, this is not a "well-posed question" as defined by Sabatier [240]. Filtering out all information (and all noise) contained in the fourth and higher Fourier coefficients is some form of regularization, which makes the remaining problem of determining the first three Fourier coefficients of the initial temperature stable. No mathematical trick can recover information about the higher Fourier coefficients which is completely dominated by the extremely high frequency amplified data noise.

2.3. Spectral Theory and Functional Calculus

For the construction and analysis of regularization methods, we will need the notion of a *function of a selfadjoint operator*. The definition and properties of such operator functions are done in the framework of *functional calculus*, which we motivate for the case of a compact operator first:

Let $(\sigma_n; v_n, u_n)$ be a singular system for a compact linear operator K. Since $(\sigma_n^2; v_n)$ is an eigensystem for the selfadjoint compact operator K^*K,

$$K^*Kx = \sum_{n=1}^{\infty} \sigma_n^2 \langle x, v_n \rangle v_n \qquad (2.36)$$

holds, which will be written below as

$$K^*Kx = \int \lambda\, dE_\lambda x. \qquad (2.37)$$

2.3. Spectral Theory and Functional Calculus

What does (2.37) mean? First, for $\lambda \in \mathbb{R}$ and $x \in \mathcal{X}$, we define

$$E_\lambda x := \sum_{\substack{n=1 \\ \sigma_n^2 < \lambda}}^{\infty} \langle x, v_n \rangle v_n \quad (+Px), \tag{2.38}$$

where P is the orthogonal projector onto $\mathcal{N}(K^*K)$ and is meant to appear in (2.38) only for $\lambda > 0$. For all λ, E_λ is an orthogonal projector and projects onto

$$\mathcal{X}_\lambda := \operatorname{span}\{v_n \mid n \in \mathbb{N}, \sigma_n^2 < \lambda\} \quad (+\mathcal{N}(K^*K), \text{ if } \lambda > 0).$$

For $\lambda \leq 0$, $E_\lambda = 0$. Since the $\{v_n\}$ span $\overline{\mathcal{R}(K^*K)}$,

$$\mathcal{X}_\lambda = \overline{\mathcal{R}(K^*K)} + \mathcal{N}(K^*K) = \mathcal{X}$$

for $\lambda > \sigma_1^2$, so that $E_\lambda = I$ for $\lambda > \sigma_1^2$. For all $\lambda \leq \mu$ and $x \in \mathcal{X}$,

$$\begin{aligned}\langle E_\lambda x, x \rangle &= \sum_{\substack{n=1 \\ \sigma_n^2 < \lambda}}^{\infty} |\langle x, v_n \rangle|^2 \, (+\|Px\|^2) \\ &\leq \sum_{\substack{n=1 \\ \sigma_n^2 < \mu}}^{\infty} |\langle x, v_n \rangle|^2 \, (+\|Px\|^2) = \langle E_\mu x, x \rangle,\end{aligned}$$

which is a kind of monotonicity of this so-called *spectral family* E_λ. Note that E_λ is piecewise constant and has jumps at $\lambda = \sigma_n^2$ (and at $\lambda = 0$ if and only if $\mathcal{N}(K) = \mathcal{N}(K^*K) \neq \{0\}$) of "height"

$$\sum_{\substack{n=1 \\ \sigma_n^2 = \lambda}}^{\infty} \langle \cdot, v_n \rangle v_n,$$

i.e., the orthogonal projectors onto the span of all eigenvectors corresponding to the (possibly multiple) eigenvalue σ_n^2 is added. Now, recall that the integral with respect to a piecewise constant weight function is defined as the sum over all function values of the integrand where the weight function has jumps, multiplied with the heights of these jumps. This fits into the general measure-theoretic concept of an integral, all important results of integration theory like the Lebesgue Dominated Convergence Theorem and integration by parts hold (cf., e.g., [129]). This motivates the notations

$$\int_{-\infty}^{+\infty} f(\lambda) \, dE_\lambda x := \sum_{n=1}^{\infty} f(\sigma_n^2) \langle x, v_n \rangle v_n, \tag{2.39}$$

$$\int_{-\infty}^{+\infty} f(\lambda) \, d\langle E_\lambda x, y \rangle := \sum_{n=1}^{\infty} f(\sigma_n^2) \langle x, v_n \rangle \langle y, v_n \rangle, \tag{2.40}$$

$$\int_{-\infty}^{+\infty} f(\lambda) \, d\|E_\lambda x\|^2 := \sum_{n=1}^{\infty} f(\sigma_n^2) |\langle x, v_n \rangle|^2 \tag{2.41}$$

for a (piecewise) continuous function f and $x, y \in \mathcal{X}$. Note that the limits of integration could also be 0 and $\sigma_1^2 + \varepsilon = \|K\|^2 + \varepsilon$ for any $\varepsilon > 0$, f has to be

44   2. Ill-Posed Linear Operator Equations

defined and (piecewise) continuous only there. If we want to emphasize this, we use the notation $\int_0^{\|K\|^2+}$. If we integrate over the whole relevant domain, we sometimes omit the limits of integration altogether.

For $f = id$, (2.39) reduces to

$$\int \lambda \, dE_\lambda x = \sum_{n=1}^\infty \sigma_n^2 \langle x, v_n \rangle v_n,$$

so that by (2.36)

$$\int \lambda \, dE_\lambda = K^*K$$

holds, which can be written in a more complicated way as

$$\int id(\lambda) \, dE_\lambda = id(K^*K),$$

where id denotes the identity map. This, in turn, motivates the definition

$$f(K^*K) := \int f(\lambda) \, dE_\lambda := \sum_{n=1}^\infty f(\sigma_n^2) \langle \cdot, v_n \rangle v_n \qquad (2.42)$$

of a (piecewise) continuous function f of a selfadjoint compact operator K^*K. $f(K^*K)$ is again a bounded selfadjoint linear operator on $\mathcal{X}$. This definition is compatible with other possible definitions in the sense that, e.g., for a polynomial

$$p(\lambda) = \sum_{k=0}^n a_k \lambda^k, \quad p(K^*K) = \sum_{k=0}^n a_k (K^*K)^k,$$

and that, if a sequence $\{p_n\}$ of polynomials converges uniformly to f, then for all $x \in \mathcal{X}$, $\{p_n(K^*K)x\}$ converges to $f(K^*K)x$. From this argument, one sees that for all (piecewise) continuous functions f,

$$f(K^*K)K^* = K^* f(KK^*) \qquad (2.43)$$

holds, where $f(KK^*)$ is defined analogously to $f(K^*K)$ using the spectral family F_λ defined by

$$F_\lambda y := \sum_{\substack{n=1 \\ \sigma_n^2 < \lambda}}^\infty \langle y, u_n \rangle u_n \, (+I - Q), \qquad (2.44)$$

where $I - Q$ is the orthogonal projector onto $\mathcal{N}(KK^*) = \mathcal{R}(KK^*)^\perp$.

Finally, if $f(K^*K)$ is defined via (2.42), then for $x, y \in \mathcal{X}$,

$$\langle f(K^*K)x, y \rangle = \int f(\lambda) \, d\langle E_\lambda x, y \rangle \qquad (2.45)$$

and

$$\|f(K^*K)x\|^2 = \int f^2(\lambda) \, d\|E_\lambda x\|^2 \qquad (2.46)$$

2.3. Spectral Theory and Functional Calculus

hold. From these formulas, it follows that

$$\|f(K^*K)\| \leq \sup|f|, \qquad (2.47)$$

and that

$$\|f(K^*K)K^*\| \leq \sup\{\sqrt{\lambda}|f(\lambda)|\}, \qquad (2.48)$$

where both suprema are taken over the smallest interval containing the spectrum of K^*K, i.e., over $[0, \|K\|^2]$. All these results can be easily deduced from the respective singular value expansions, so that for compact operators, the use of the "integrals with respect to a spectral family" is merely a notation.

Now, we present the basic results of functional calculus also for non-compact (possibly unbounded) linear selfadjoint operators, A, in a Hilbert space, $\mathcal{X}$. For details and proofs, we refer to [50, 127].

Definition 2.10. *A family $\{E_\lambda\}_{\lambda \in \mathbb{R}}$ of orthogonal projectors in $\mathcal{X}$ is called a spectral family or a resolution of the identity if it satisfies the following conditions:*

(i) $\quad E_\lambda E_\mu = E_{\min\{\lambda,\mu\}}, \quad \lambda, \mu \in \mathbb{R},$

(ii) $\quad E_{-\infty} = 0, E_{+\infty} = I,$ *where* $E_{\pm\infty} x = \lim_{\lambda \to \pm\infty} E_\lambda x$ *for all* $x \in \mathcal{X},$

(iii) $\quad E_{\lambda-0} = E_\lambda,$ *where* $E_{\lambda-0} x = \lim_{\varepsilon \to 0^+} E_{\lambda-\varepsilon} x$ *for all* $x \in \mathcal{X}.$

As in (2.39), it is possible to define an integral with respect to a spectral family.

Proposition 2.11. *Let $f : \mathbb{R} \to \mathbb{R}$ be a continuous function. Then the limit of the Riemann sum*

$$\sum_{i=1}^{n} f(\xi_i)(E_{\lambda_i} - E_{\lambda_{i-1}})x,$$

exists in $\mathcal{X}$ for $\max_{1 \leq i \leq n} |\lambda_i - \lambda_{i-1}| \to 0$, where $-\infty < a = \lambda_0 < \ldots < \lambda_n = b < \infty$, $\xi_i \in (\lambda_{i-1}, \lambda_i]$, and is denoted by

$$\int_a^b f(\lambda)\,dE_\lambda x.$$

Definition 2.12. *For any given $x \in \mathcal{X}$ and any continuous function f on $\mathbb{R}$, the integral $\int_{-\infty}^{+\infty} f(\lambda)\,dE_\lambda x$ is defined as the limit in $\mathcal{X}$ if it exists, of $\int_a^b f(\lambda)\,dE_\lambda x$ when $a \to -\infty$ and $b \to +\infty$.*

Since condition (i) in Definition 2.10 is equivalent to

$$\langle E_\lambda x, x \rangle \leq \langle E_\mu x, x \rangle \quad \text{for all } x \in \mathcal{X} \text{ and } \lambda \leq \mu,$$

the function $\lambda \mapsto \langle E_\lambda x, x \rangle = \|E_\lambda x\|^2$ is monotonically increasing and due to condition (iii) in Definition 2.10 also continuous from the left. Hence, it defines a measure on $\mathbb{R}$, denoted by $d\|E_\lambda x\|^2$. The following connection holds:

Proposition 2.13. *For $x \in \mathcal{X}$ and $f : \mathbb{R} \to \mathbb{R}$ a continuous function, the following conditions are equivalent:*

$$\int_{-\infty}^{+\infty} f(\lambda)\, dE_\lambda x \quad \text{exists},$$

$$\int_{-\infty}^{+\infty} f^2(\lambda)\, d\|E_\lambda x\|^2 < \infty.$$

If A is a selfadjoint operator in $\mathcal{X}$, a special spectral family exists, which can be used to represent A and functions of A as integrals:

Proposition 2.14. *Let A be a selfadjoint operator in $\mathcal{X}$. Then there exists a unique spectral family $\{E_\lambda\}_{\lambda \in \mathbb{R}}$ such that*

$$\mathcal{D}(A) = \{x \in \mathcal{X} \mid \int_{-\infty}^{+\infty} \lambda^2\, d\|E_\lambda x\|^2 < \infty\}$$

and

$$Ax = \int_{-\infty}^{+\infty} \lambda\, dE_\lambda x, \qquad x \in \mathcal{D}(A).$$

We use the symbolic notation

$$A = \int_{-\infty}^{+\infty} \lambda\, dE_\lambda.$$

The spectral family in Proposition 2.14 is called *spectral decomposition of A* or *the spectral family of A*.

It is now possible to define functions of a selfadjoint operator (cf. (2.42)).

Definition 2.15. *Let A be a selfadjoint operator in $\mathcal{X}$ with spectral family $\{E_\lambda\}_{\lambda \in \mathbb{R}}$. Moreover, let $\mathcal{M}_0$ denote the set of all functions measurable with respect to the measure $d\|E_\lambda x\|^2$ for all $x \in \mathcal{X}$. Then $f(A)$ is the operator defined by the formula*

$$f(A)x = \int_{-\infty}^{+\infty} f(\lambda)\, dE_\lambda x, \quad x \in \mathcal{D}(f(A)),$$

where

$$\mathcal{D}(f(A)) = \{x \in \mathcal{X} \mid \int_{-\infty}^{+\infty} f^2(\lambda)\, d\|E_\lambda x\|^2 < \infty\}.$$

Note that in particular $\mathcal{M}_0$ contains all piecewise continuous functions.

Proposition 2.16. *Let A be a selfadjoint operator in $\mathcal{X}$ with spectral family $\{E_\lambda\}_{\lambda \in \mathbb{R}}$ and let $f, g \in \mathcal{M}_0$.*

2.3. Spectral Theory and Functional Calculus

(i) If $x \in \mathcal{D}(f(A))$ and $y \in \mathcal{D}(g(A))$, then

$$\langle f(A)x, g(A)y \rangle = \int_{-\infty}^{+\infty} f(\lambda)g(\lambda)\, d\langle E_\lambda x, y\rangle.$$

(ii) If $x \in \mathcal{D}(f(A))$, then $f(A)x \in \mathcal{D}(g(A))$ if and only if $x \in \mathcal{D}((gf)(A))$; furthermore,

$$g(A)f(A)x = (gf)(A)x.$$

(iii) If $\mathcal{D}(f(A))$ is dense in $\mathcal{X}$, then $f(A)$ is selfadjoint.

(iv) $f(A)$ commutes with E_λ for all $\lambda \in \mathbb{R}$.

There is a strong connection between the spectral family of A and its spectrum $\sigma(A)$:

Proposition 2.17. *Let A be a selfadjoint operator in $\mathcal{X}$ with spectral family $\{E_\lambda\}_{\lambda \in \mathbb{R}}$.*

(i) $\lambda_0 \in \sigma(A)$ if and only if $E_{\lambda_0} \neq E_{\lambda_0+\varepsilon}$ for all $\varepsilon > 0$.

(ii) λ_0 is an eigenvalue of A if and only if $E_{\lambda_0} \neq E_{\lambda_0+0} = \lim_{\varepsilon \to 0} E_{\lambda_0+\varepsilon}$; the corresponding eigensubspace is then given by $(E_{\lambda_0+0} - E_{\lambda_0})(\mathcal{X})$.

We consider now two special cases. If A is a strictly positive selfadjoint operator satisfying, with $\gamma > 0$,

$$\langle Ax, x \rangle \geq \gamma \|x\|^2 \quad \text{for all } x \in \mathcal{D}(A),$$

then for all $f \in \mathcal{M}_0$

$$\int_{-\infty}^{+\infty} f(\lambda)\, dE_\lambda x = \int_{\gamma}^{+\infty} f(\lambda)\, dE_\lambda x.$$

Hence, the function f may be restricted to the interval $[\gamma, \infty)$.

If $T: \mathcal{X} \to \mathcal{Y}$ is a linear bounded operator and $A := T^*T$, then for all $f \in \mathcal{M}_0$

$$\int_{-\infty}^{+\infty} f(\lambda)\, dE_\lambda x = \int_0^{\|T\|^2+} f(\lambda)\, dE_\lambda x = \lim_{\varepsilon \to 0} \int_0^{\|T\|^2+\varepsilon} f(\lambda)\, dE_\lambda x.$$

Hence, the function f may be restricted to the interval $[0, \|T\|^2 + \varepsilon]$ for some $\varepsilon > 0$. It follows from above that (2.43) and (2.45) – (2.48) remain valid if K^*K is replaced by T^*T. Moreover, it is an immediate consequence of Proposition 2.16 (i) and the Hölder inequality that the *interpolation inequality*

$$\|(T^*T)^r x\| \leq \|(T^*T)^q x\|^{\frac{r}{q}} \|x\|^{1-\frac{r}{q}}, \qquad (2.49)$$

holds for all $q > r \geq 0$.

48   2. Ill-Posed Linear Operator Equations

Finally, we show that the subspaces $\mathcal{R}(T^*)$ and $\mathcal{R}((T^*T)^{\frac{1}{2}})$ are equal, which will be needed later.

Proposition 2.18. *Let $T : \mathcal{X} \to \mathcal{Y}$ be a linear bounded operator. Then*

$$\mathcal{R}(T^*) = \mathcal{R}((T^*T)^{\frac{1}{2}})$$

Proof. (2.43) and (2.46) applied to the functions

$$f_\rho(\lambda) := \begin{cases} \frac{1}{\lambda}, & \lambda \geq \rho, \\ 0, & \lambda < \rho, \end{cases}$$

for $\rho > 0$, yields

$$\begin{aligned}
\int_\rho^\infty \frac{1}{\lambda^2} d\|F_\lambda Tx\|^2 &= \|f_\rho(TT^*)Tx\|^2 = \langle Tf_\rho(T^*T)x, Tf_\rho(T^*T)x \rangle \\
&= \langle f_\rho(T^*T)x, T^*Tf_\rho(T^*T)x \rangle \\
&= \langle (T^*T)^{\frac{1}{2}} f_\rho(T^*T)x, (T^*T)^{\frac{1}{2}} f_\rho(T^*T)x \rangle \\
&= \|(T^*T)^{\frac{1}{2}} f_\rho(T^*T)x\|^2 = \int_\rho^\infty \frac{1}{\lambda} d\|E_\lambda x\|^2.
\end{aligned}$$

This together with Definition 2.15 implies that

$$\begin{aligned}
x \in \mathcal{R}(T^*) &\iff Tx \in \mathcal{R}(TT^*) \wedge x \in \mathcal{N}(T)^\perp \\
&\iff Tx \in \mathcal{D}((TT^*)^\dagger) \wedge x \in \mathcal{N}(T)^\perp \\
&\iff \int_0^\infty \frac{1}{\lambda^2} d\|F_\lambda Tx\|^2 < \infty \wedge x \in \mathcal{N}(T)^\perp \\
&\iff \int_0^\infty \frac{1}{\lambda} d\|E_\lambda x\|^2 < \infty \wedge x \in \mathcal{N}(T)^\perp \\
&\iff x \in \mathcal{D}((T^*T|_{\mathcal{N}(T)^\perp})^{-\frac{1}{2}}) \wedge x \in \mathcal{N}(T)^\perp \\
&\iff x \in \mathcal{R}((T^*T)^{\frac{1}{2}})
\end{aligned}$$

and hence the assertion is shown. ∎

3. Regularization Operators

3.1. Definition and Basic Results

In general terms, *regularization* is the approximation of an ill-posed problem by a family of neighbouring well-posed problems. We motivate the definition of a *regularization operator* and of a *regularization method*:

We want to approximate the best-approximate solution $x^\dagger = T^\dagger y$ of

$$Tx = y \qquad (3.1)$$

for a specific right-hand side y in the situation that the "exact data" y are not known precisely, but that only an approximation y^δ with

$$\|y^\delta - y\| \leq \delta \qquad (3.2)$$

is available; we will call y^δ the "noisy data" and δ the "noise level".

In the ill-posed case, $T^\dagger y^\delta$ is certainly not a good approximation of $T^\dagger y$ due to the unboundedness of $T^\dagger$ even if it exists (which will, in general, also not be the case, since $\mathcal{D}(T^\dagger)$ is then a proper subset of $\mathcal{Y}$). We are looking for some approximation, say x_α^δ, of $x^\dagger$ which does, on the one hand, depend continuously on the (noisy) data y^δ, so that it can be computed in a stable way, and has, on the other hand, the property that as the noise level δ decreases to zero and the *regularization parameter* α is chosen appropriately (whatever this means), then x_α^δ tends to $x^\dagger$.

The construction of x_α^δ will in general involve the operator T. Although this seems to be a trivial remark, there are situations where this is not necessarily the case: if $\|y^\delta\| < \delta$, i.e., if the noise level is larger than the signal, one might be best off just to take $x_\alpha^\delta := 0$ independent of the operator T, since in such a situation, the noisy right-hand side contains no information at all anyway. But except in this case, the operator T certainly has to play some role in the construction of x_α^δ. But then, it makes more sense to look not only at the equation (3.1) for one specific right-hand side y, but to consider (3.1) as a collection of equations for every $y \in \mathcal{R}(T)$ or every $y \in \mathcal{D}(T^\dagger)$. In this situation, we are not only talking about regularizing a specific equation, but this collection of equations or, in other words, about regularizing the solution operator $T^\dagger$ (on $\mathcal{R}(T)$ if we assume attainability, or on $\mathcal{D}(T^\dagger)$ otherwise). Intuitively, a regularization of $T^\dagger$ should then be the replacement of the unbounded operator $T^\dagger$ by a parameter-dependent family $\{R_\alpha\}$ of continuous operators. As approximation of $x^\dagger$, we then take $x_\alpha^\delta := R_\alpha y^\delta$, which can then be computed in a stable way (at least in principle, since R_α is assumed to be continuous). A requirement for α is that, if the noise level δ tends to 0, then the regularized solution x_α^δ should tend to $x^\dagger$. Therefore, the regularization parameter α has to be somehow linked with δ and/or y^δ and maybe with other information about T or about y. We will see that such *parameter choice rules* should be linked with either y^δ or some a-priori information about the exact data y, so that they depend on the specific equation. Therefore, we will define regularization operators for the whole collection of equations

$$Tx = y, \quad y \in \mathcal{D}(T^\dagger), \qquad (3.3)$$

or, equivalenty, for the operator T, but parameter choice rules for a specific equation out of this collection. Both together then form a regularization method for solving one specific equation. Together with the way the parameter choice rule depends on y and/or y^δ, a regularization method can then also be seen in relation with the collection of equations (3.3).

A regularization method also depends on the concept of solution we want to consider: in the case of multiple solutions of (3.1), all we said so far (and all we will do in Chapters 3 and 4) refers to approximating the minimum-norm (least-squares) solution $x^\dagger = T^\dagger y$. We will consider regularization methods for approximating other solutions in Chapter 8.

These considerations lead to the following definition:

Definition 3.1. *Let $T : \mathcal{X} \to \mathcal{Y}$ be a bounded linear operator between the Hilbert spaces $\mathcal{X}$ and $\mathcal{Y}$, $\alpha_0 \in (0, +\infty]$. For every $\alpha \in (0, \alpha_0)$, let*

$$R_\alpha : \mathcal{Y} \to \mathcal{X}$$

be a continuous (not necessarily linear) operator. The family $\{R_\alpha\}$ is called a regularization or a regularization operator (for $T^\dagger$), if, for all $y \in \mathcal{D}(T^\dagger)$, there exists a parameter choice rule $\alpha = \alpha(\delta, y^\delta)$ such that

$$\lim_{\delta \to 0} \sup\{\|R_{\alpha(\delta, y^\delta)} y^\delta - T^\dagger y\| \mid y^\delta \in \mathcal{Y}, \|y^\delta - y\| \leq \delta\} = 0 \qquad (3.4)$$

holds. Here,

$$\alpha : \mathbb{R}^+ \times \mathcal{Y} \to (0, \alpha_0) \qquad (3.5)$$

is such that

$$\lim_{\delta \to 0} \sup\{\alpha(\delta, y^\delta) \mid y^\delta \in \mathcal{Y}, \|y^\delta - y\| \leq \delta\} = 0. \qquad (3.6)$$

For a specific $y \in \mathcal{D}(T^\dagger)$, a pair (R_α, α) is called a (convergent) regularization method (for solving $Tx = y$) if (3.4) and (3.6) hold.

Thus, a regularization method consists of a regularization operator and a parameter choice rule which is convergent in the sense that, if the regularization parameter is chosen according to that rule, then the regularized solutions converge (in the norm) as the noise level tends to zero; this is assured for any collection of noisy right-hand sides compatible with the noise level and thus is a "worst-case" concept of convergence with respect to the right-hand side.

We mention that it is possible to extend Definition 3.1 to include besides right-hand side perturbations the case of perturbations in the operator ("modelling errors"). In this case, one assumes that instead of T only some approximation T_η is

known with
$$\|T - T_\eta\| \le \eta,$$
and the parameter choice rule would depend on δ, η, y^δ, and T_η. The natural requirement for a convergent regularization method would then be that the lim sup condition in (3.4) holds for $\delta, \eta \to 0$. Although many results in this book can be extended to this situation, we restrict ourselves for the ease of exposition to the case when the operator is known precisely. For the general case, we refer to, e.g., [194, 278].

Usually, a parameter choice rule will only be applied to pairs (δ, y^δ) where (3.2) holds. Thus, one could define a parameter choice rule α in (3.5) only as a function
$$\alpha : \{(\delta, y^\delta) \mid \delta > 0, \|y^\delta - y\| \le \delta\} \to (0, \alpha_0).$$
In [235], the question is studied what happens if the noisy data do not fulfill (3.2), i.e., if a parameter choice rule is applied to pairs (δ, y^δ) where possibly $\|y - y^\delta\| > \delta$.

One could also consider convergence of the regularized solutions in the weak topology: see [68] for results about (non)convergence of regularization methods in the weak topology.

Note that we did not require the regularization operator $\{R_\alpha\}$ to be a family of *linear* operators. If the R_α are linear, then we call the corresponding method a linear regularization method, and the family $\{R_\alpha\}$ a linear regularization operator. However, it also makes sense to consider nonlinear regularization methods for solving linear problems, like the method of conjugate gradients which will be considered in Chapter 7. We chose to define regularization methods for *linear* problems only so far in order to keep the definition simple; in Chapter 10, we turn to nonlinear problems.

The parameter choice rule $\alpha = \alpha(\delta, y^\delta)$ depends (so far) explicitly on the noise level δ and on the actual perturbed data y^δ. Also, remember that we defined it for every specific $y \in \mathcal{D}(T^\dagger)$, so that α depends also on the exact data y. Since y is not known, this dependence can only be on some qualitative a-priori knowledge about y like smoothness properties. Finally, α depends also on the operator T, which we do not indicate explicitly in the notation.

We distinguish between two types of parameter choice rules:

Definition 3.2. *Let α be a parameter choice rule according to Definition* 3.1. *If α does not depend on y^δ, but only on δ, then we call α an* a-priori *parameter choice rule and write $\alpha = \alpha(\delta)$. Otherwise, we call α an* a-posteriori *parameter choice rule.*

Thus, an a-priori parameter choice rule depends only on the noise level, not on the actual data and, hence, not on results obtained during the actual computation like the residual $\|Tx_\alpha^\delta - y^\delta\|$, where $x_\alpha^\delta = R_\alpha y^\delta$ is the regularized solution. Such a rule may be devised before the actual calculation, hence the name a-priori parameter choice rule.

One could also think of parameter choice rules that depend *only* on y^δ, not on the noise level δ. The following result due to Bakushinskii [18] shows that such rules

52   3. Regularization Operators

cannot be part of a convergent regularization method in the sense of Definition 3.1 for an ill-posed problem:

Theorem 3.3.   *Let $T : \mathcal{X} \to \mathcal{Y}$ be a bounded linear operator and assume that there is a regularization $\{R_\alpha\}$ for $T^\dagger$ with a parameter choice rule α which depends on y^δ only (and not on δ) such that the regularization method (R_α, α) is convergent for every $y \in \mathcal{D}(T^\dagger)$. Then $T^\dagger$ is bounded.*

Proof. If α is independent of δ, say, $\alpha = \alpha(y^\delta)$, then it follows from (3.4) that

$$\limsup_{\delta \to 0} \{\|R_{\alpha(y^\delta)} y^\delta - T^\dagger y\| \mid y^\delta \in \mathcal{Y}, \|y^\delta - y\| \leq \delta\} = 0, \tag{3.7}$$

so that especially $R_{\alpha(y)} y = T^\dagger y$ for all $y \in \mathcal{D}(T^\dagger)$. Hence, by (3.7), for any sequence $\{y_n\}$ in $\mathcal{D}(T^\dagger)$ which converges to a $y \in \mathcal{D}(T^\dagger)$, $T^\dagger y_n = R_{\alpha(y_n)} y_n \to T^\dagger y$, so that $T^\dagger$ is continuous on $\mathcal{D}(T^\dagger)$. But then, Proposition 2.4 implies that $\mathcal{D}(T^\dagger) = \mathcal{Y}$. ∎

Thus, if $T^\dagger$ is unbounded, no *error-free* parameter choice strategy can yield a convergent regularization method. This asymptotic result should just be taken for what it is; it does not say that error-free parameter choice rules cannot behave well for finite noise levels δ. Such heuristic parameter choice rules will be considered in Section 4.5.

The following questions now arise:

(i) How can one construct regularization operators?

(ii) How can one construct parameter choice rules that give rise to convergent regularization methods?

(iii) How can these steps be performed in some "optimal" way?

For linear operators T, the following preliminary answer to the first question can be given:

Proposition 3.4.   *Let, for all $\alpha > 0$, R_α be a continuous (possibly nonlinear) operator. Then, the family $\{R_\alpha\}$ is a regularization for $T^\dagger$ if*

$$R_\alpha \to T^\dagger \quad \text{pointwise on } \mathcal{D}(T^\dagger) \text{ as } \alpha \to 0. \tag{3.8}$$

In this case, there exists, for every $y \in \mathcal{D}(T^\dagger)$, an a-priori parameter choice rule α such that (R_α, α) is a convergent regularization method for solving $Tx = y$.

Proof. Let $y \in \mathcal{D}(T^\dagger)$ be arbitrary, but fixed. By assumption, there is a monotone function $\sigma : \mathbb{R}^+ \to \mathbb{R}^+$ with $\lim_{\varepsilon \to 0} \sigma(\varepsilon) = 0$ such that for every $\varepsilon > 0$,

$$\|R_{\sigma(\varepsilon)} y - T^\dagger y\| \leq \frac{\varepsilon}{2}.$$

Since each $R_{\sigma(\varepsilon)}$ is continuous, for every $\varepsilon > 0$, there exists a $\rho(\varepsilon)$ such that, if $\|z - y\| \leq \rho(\varepsilon)$, then
$$\|R_{\sigma(\varepsilon)}z - R_{\sigma(\varepsilon)}y\| \leq \frac{\varepsilon}{2}.$$
This defines a function $\rho : \mathbb{R}^+ \to \mathbb{R}^+$ which we can assume, without loss of generality, to be strictly monotone, continuous and have the property that $\lim_{\varepsilon \to 0} \rho(\varepsilon) = 0$. Hence, the inverse function ρ^{-1} exists on the range of ρ, is strictly monotone and continuous, and has the property that $\lim_{\delta \to 0} \rho^{-1}(\delta) = 0$. We can extend ρ^{-1} to all of $\mathbb{R}^+$ and thus define
$$\alpha : \mathbb{R}^+ \to \mathbb{R}^+$$
$$\delta \mapsto \sigma(\rho^{-1}(\delta)).$$
The function α is monotone and has the property that $\lim_{\delta \to 0} \alpha(\delta) = 0$. Furthermore, for all $\varepsilon > 0$, there is a $\delta > 0$, namely $\delta := \rho(\varepsilon)$, such that, if $\|y^\delta - y\| \leq \delta$, then
$$\|R_{\alpha(\delta)}y^\delta - T^\dagger y\| \leq \|R_{\alpha(\delta)}y^\delta - R_{\alpha(\delta)}y\| + \|R_{\alpha(\delta)}y - T^\dagger y\| \leq \frac{\varepsilon}{2} + \frac{\varepsilon}{2} = \varepsilon,$$
since $\alpha(\delta) = \sigma(\varepsilon)$. Thus, for the method (R_α, α), (3.4) and (3.6) hold. The function α defines an a-priori parameter choice rule. ∎

Remark 3.5. The converse of Proposition 3.4 holds in the following sense: if (R_α, α) is a convergent regularization method, then it follows from (3.6) that
$$\lim_{\delta \to 0} R_{\alpha(\delta, y)} y = T^\dagger y$$
holds for all $y \in \mathcal{D}(T^\dagger)$. If α is continuous in δ, this implies that
$$\lim_{\sigma \to 0} R_\sigma y = T^\dagger y;$$
otherwise, this holds only over the set of σ-values which are in the range of the parameter choice strategy α.

Thus, regularizations are pointwise approximations of the Moore-Penrose inverse of T. If $\{R_\alpha\}$ is uniformly bounded and linear and if $\mathcal{R}(T)$ is non-closed, the convergence in (3.8) cannot be in the operator norm, since then, $T^\dagger$ would have to be bounded. Also, due to the Banach-Steinhaus Theorem,
$$\|R_\alpha\| \to +\infty \quad \text{as} \quad \alpha \to 0 \tag{3.9}$$
if $\mathcal{R}(T)$ is non-closed. By the Uniform Boundedness Principle, (3.9) implies that there must exist $y \in \mathcal{Y}$ such that
$$\|R_\alpha y\| \to +\infty \quad \text{as} \quad \alpha \to 0. \tag{3.10}$$

In fact, (3.10) holds for all $\mathcal{Y}\backslash\mathcal{M}$, where $\mathcal{M}$ is a set of first Baire category. On the other hand,
$$x_\alpha := R_\alpha y \tag{3.11}$$
converges to $T^\dagger y$ on the dense set $\mathcal{D}(T^\dagger)$ due to (3.8). It turns out that under a (reasonable) additional condition, the set where (3.10) holds is precisely the complement of $\mathcal{D}(T^\dagger)$:

Proposition 3.6. *Let $\{R_\alpha\}$ be a linear regularization, x_α be defined by (3.11) for all $y \in \mathcal{Y}$. Then*

$$\{x_\alpha\} \quad \text{converges to} \quad T^\dagger y \quad \text{as} \quad \alpha \to 0 \quad \text{for} \quad y \in \mathcal{D}(T^\dagger) \tag{3.12}$$

and if

$$\sup\{\|TR_\alpha\| \mid \alpha > 0\} < \infty, \tag{3.13}$$

then

$$\|x_\alpha\| \to +\infty \quad \text{as} \quad \alpha \to 0 \quad \text{for} \quad y \notin \mathcal{D}(T^\dagger). \tag{3.14}$$

Here, the limits in (3.12) and (3.14) have to be understood as explained in Remark 3.5.

Proof. (3.12) follows as in Remark 3.5. Also, from there we conclude that (in the sense described there), $R_\alpha \to T^\dagger$ pointwise on $\mathcal{D}(T^\dagger)$, so that by (2.13), $TR_\alpha \to Q$ pointwise on the dense set $\mathcal{D}(T^\dagger)$. Since, by assumption, $\|TR_\alpha\|$ is uniformly bounded, $TR_\alpha \to Q$ pointwise on all of $\mathcal{Y}$.

Now, assume that there is a sequence $\{\alpha_n\} \to 0$ such that $\{\|x_{\alpha_n}\|\}$ is bounded, so that $\{x_{\alpha_n}\}$ has a subsequence (denoted again by $\{x_{\alpha_n}\}$) which converges weakly to an $x \in \mathcal{X}$. Since T is also weakly sequentially continuous, $Tx_{\alpha_n} \to Tx$. On the other hand, $Tx_{\alpha_n} = TR_{\alpha_n}y \to Qy$, so that $Tx = Qy$. Hence, $y \in \mathcal{D}(T^\dagger)$. Thus, if $y \notin \mathcal{D}(T^\dagger)$, no bounded sequence $\|x_{\alpha_n}\|$ (where α_n is of the form $\alpha_n(\delta_n, y) \to 0$) can exist; hence, (3.14) holds. ∎

In Proposition 3.4 we have seen that, if (3.8) holds, then there exists an a-priori parameter choice rule α such that (R_α, α) is a convergent regularization method. Such parameter choice rules can be characterized as follows:

Proposition 3.7. *Let $\{R_\alpha\}$ be a linear regularization; for every $y \in \mathcal{D}(T^\dagger)$, let $\alpha : \mathbb{R}^+ \to \mathbb{R}^+$ be an a-priori parameter choice rule. Then (R_α, α) is a convergent regularization method if and only if*

$$\lim_{\delta \to 0} \alpha(\delta) = 0 \tag{3.15}$$

and

$$\lim_{\delta \to 0} \delta \|R_{\alpha(\delta)}\| = 0 \tag{3.16}$$

hold.

Proof. Assume that (3.15) and (3.16) hold. With the notation (3.11), we have that for any $y^\delta \in \mathcal{Y}$ with $\|y^\delta - y\| \leq \delta$,

$$\|R_{\alpha(\delta)} y^\delta - T^\dagger y\| \leq \|x_{\alpha(\delta)} - T^\dagger y\| + \|x_{\alpha(\delta)} - R_{\alpha(\delta)} y^\delta\|$$
$$\leq \|x_{\alpha(\delta)} - T^\dagger y\| + \|R_{\alpha(\delta)}\| \delta.$$

Because of (3.12), (3.15) and (3.16), this implies (3.4).

We now show the converse: since α is assumed to be an a-priori parameter choice rule, (3.15) holds by definition. Thus, we assume that (3.16) does not hold, so that there is a sequence $\delta_n \to 0$ such that $\|\delta_n R_{\alpha(\delta_n)}\| \geq C > 0$. Hence, there is a sequence $\{z_n\}$ in $\mathcal{Y}$ with $\|z_n\| = 1$ such that $\|\delta_n R_{\alpha(\delta_n)} z_n\| \geq C/2$. Then, for any $y \in \mathcal{D}(T^\dagger)$ and $y_n := y + \delta_n z_n$, $R_{\alpha(\delta_n)} y_n - T^\dagger y = (R_{\alpha(\delta_n)} y - T^\dagger y) + \delta_n R_{\alpha(\delta_n)} z_n$ does not converge to 0, since the first term converges to 0 and the second does not. ∎

Remark 3.8. If (3.16) is replaced by

$$\limsup_{\delta \to 0} \delta \|R_{\alpha(\delta)}\| < +\infty, \tag{3.17}$$

then (R_α, α) is weakly convergent, i.e., for all sequences $\{\delta_k\} \to 0$ and $y_k \in \mathcal{Y}$ with $\|y_k - y\| \leq \delta_k$, $\{R_{\alpha(\delta_k)} y_k\}$ converges weakly to $T^\dagger y$. Conversely, (3.17) is necessary for weak convergence in the sense that, if (3.17) does not hold, then there exist sequences $\{\delta_k\}$ and y_k as above such that $\{R_{\alpha(\delta_k)} y_k\}$ diverges in the weak topology (and is even unbounded). This can be proven as in [68].

3.2. Order Optimality

When talking about the rate of convergence of a regularization method (R_α, α) one can think of the rate with which

$$\|x_\alpha - x^\dagger\| \to 0 \quad \text{as} \quad \alpha \to 0, \tag{3.18}$$

where x_α is defined by (3.11) and $x^\dagger := T^\dagger y$, or of the rate with which

$$\|x^\delta_{\alpha(\delta, y^\delta)} - x^\dagger\| \to 0 \quad \text{as} \quad \delta \to 0, \tag{3.19}$$

where

$$x^\delta_{\alpha(\delta, y^\delta)} := R_{\alpha(\delta, y^\delta)} y^\delta$$

and (3.2) holds. Only the rate in (3.19) depends on the parameter choice rule, while the rate in (3.18) depends on the regularization operator (and, as we will see, on the exact data y). Since

$$\|x^\delta_{\alpha(\delta, y^\delta)} - x^\dagger\| \leq \|x_{\alpha(\delta, y^\delta)} - x^\dagger\| + \|x_{\alpha(\delta, y^\delta)} - x^\delta_{\alpha(\delta, y^\delta)}\|,$$

56   3. Regularization Operators

both rates are connected.

We first consider the convergence rate in (3.18), i.e., the rate with which a regularization method converges with the exact data as right-hand side. If $\{R_\alpha\}$ would converge to $T^\dagger$ in the operator norm, then, since $\|x_\alpha - x^\dagger\| \leq \|R_\alpha - T^\dagger\| \|y\|$, a rate which is uniform in y could be found in (3.18). It is in principle impossible to give a uniform convergence rate for ill-posed problems, the rate in (3.18) can be arbitrarily slow, as we will see in Proposition 3.11 below.

As we will see, (best-possible) rates for regularization methods under additional assumptions on the solution (or, equivalently, on the exact data) are closely related to a "modulus of continuity" for $T^\dagger$ on subsets defined by such a-priori assumptions; we will first define this modulus of continuity:

Definition 3.9. For $\mathcal{M} \subseteq \mathcal{X}, \delta > 0$, let

$$\Omega(\delta, \mathcal{M}) := \sup\{\|x\| \mid x \in \mathcal{M}, \|Tx\| \leq \delta\}. \tag{3.20}$$

In general, $\Omega(\delta, \mathcal{M})$ will be infinite, e.g., if $\mathcal{M} \cap \mathcal{N}(T) \neq \{0\}$ and $\mathcal{M}$ is unbounded. If $\mathcal{M} \cap \mathcal{N}(T) = \{0\}$, then $\Omega(\delta, \mathcal{M})$ is finite if and only if $T^\dagger$ is continuous on $T\mathcal{M}$. E.g., if $\mathcal{M}$ is compact and $\mathcal{M} \cap \mathcal{N}(T) = \{0\}$, then $\Omega(\delta, \mathcal{M}) < \infty$.

Any method for solving (3.1) has the form $x = Ry$, where R is a (not necessarily linear) mapping from $\mathcal{Y}$ into $\mathcal{X}$. If this method is used for perturbed data y^δ, then the worst-case error under the information that $\|y^\delta - y\| \leq \delta$ and under the a-priori assumption that

$$x^\dagger \in \mathcal{M} \tag{3.21}$$

is given by

$$\Delta(\delta, \mathcal{M}, R) := \sup\{\|Ry^\delta - x\| \mid x \in \mathcal{M}, y^\delta \in \mathcal{Y}, \|Tx - y^\delta\| \leq \delta\}. \tag{3.22}$$

Again, $\Delta(\delta, \mathcal{M}, R) = +\infty$ if $\mathcal{M} \cap \mathcal{N}(T) \neq \{0\}$ and $\mathcal{M}$ is unbounded.

An "optimal method" R_0 in a class of methods $\mathcal{R}$ would be one for which

$$\Delta(\delta, \mathcal{M}, R_0) = \inf\{\Delta(\delta, \mathcal{M}, R) \mid R \in \mathcal{R}\} \tag{3.23}$$

holds, where "optimality" is to be understood with respect to the a-priori information (3.21) and the class of methods considered. We will now see that the accuracy of any method is bounded by $\Omega(\delta, \mathcal{M})$:

Proposition 3.10. Let $\mathcal{M} \subseteq \mathcal{X}$, $\delta > 0$, $R : \mathcal{Y} \to \mathcal{X}$ be an arbitrary map with $R(0) = 0$. Then

$$\Delta(\delta, \mathcal{M}, R) \geq \Omega(\delta, \mathcal{M}). \tag{3.24}$$

Proof. Let $x \in \mathcal{M}$ with $\|Tx\| \leq \delta$ be arbitrary. From (3.22) we obtain (with $y^\delta = 0$) that $\Delta(\delta, \mathcal{M}, R) \geq \|R(0) - x\| = \|x\|$. By taking the supremum over all $x \in \mathcal{M}$ with $\|Tx\| \leq \delta$, we obtain (3.24). ∎

3.2. Order Optimality 57

Proposition 3.10 can be used to show that there can be no uniform convergence rate for a regularization method (or any method) for solving (3.1) if $\mathcal{R}(T)$ is non-closed (cf. also [247]), i.e., the convergence is *arbitrarily slow*:

Proposition 3.11. *Let $\mathcal{R}(T)$ be non-closed, $\{R_\alpha\}$ be a regularization operator for $T^\dagger$ with $R_\alpha(0) = 0$, $\alpha = \alpha(\delta, y^\delta)$ be a parameter choice rule. Then, there can be no function $f: \mathbb{R}^+ \to \mathbb{R}^+$ with $\lim_{\delta \to 0} f(\delta) = 0$ such that*

$$\|R_{\alpha(\delta,y^\delta)} y^\delta - T^\dagger y\| \leq f(\delta) \tag{3.25}$$

holds for all $y \in \mathcal{D}(T^\dagger)$ with $\|y\| \leq 1$ and all $\delta > 0$.

Proof. Assume that a function f with $\lim_{\delta \to 0} f(\delta) = 0$ exists such that (3.25) holds. For $\delta > 0$, let $R^\delta: \mathcal{Y} \to \mathcal{X}$ be defined by $R^\delta y^\delta := R_{\alpha(\delta,y^\delta)} y^\delta$. If (3.25) holds for all $y \in \mathcal{D}(T^\dagger)$ with $\|y\| \leq 1$, then, since $\mathcal{R}(T^\dagger) = \mathcal{N}(T)^\perp$,

$$\Delta(\delta, \mathcal{N}(T)^\perp \cap T^{-1}(\mathcal{B}_1(0)), R^\delta) \leq f(\delta).$$

(Here, $\mathcal{B}_1(0) := \{y \in \mathcal{Y} \mid \|y\| \leq 1\}$.) By Proposition 3.10, this implies that

$$\Omega(\delta, \mathcal{N}(T)^\perp \cap T^{-1}(\mathcal{B}_1(0))) \leq f(\delta)$$

holds for all $\delta > 0$. Especially,

$$\lim_{\delta \to 0} \Omega(\delta, \mathcal{N}(T)^\perp \cap T^{-1}(\mathcal{B}_1(0))) = 0. \tag{3.26}$$

Let $\{y_k\}$ be any sequence in $\mathcal{B}_1(0) \cap \mathcal{R}(T)$ converging to a $y \in \mathcal{B}_1(0) \cap \mathcal{R}(T)$, $x_k := T^\dagger y_k$, $x := T^\dagger y$. If $k \in \mathbb{N}$ is such that $\|y_k - y\| \leq \delta$, then, by Definition 3.9, $\|x_k - x\| \leq \Omega(\delta, \mathcal{N}(T)^\perp \cap T^{-1}(\mathcal{B}_1(0)))$. Together with (3.26) and the convergence of $\{y_k\}$ to y, this implies that $\{x_k\}$ converges to x. Hence, $T^\dagger$ is continuous on $\mathcal{B}_1(0) \cap \mathcal{R}(T)$, i.e., $T^\dagger$ is bounded, which contradicts the non-closedness of $\mathcal{R}(T)$. ∎

Remark 3.12. Proposition 3.11 holds (with the same proof) for any method, not just for regularization methods in the sense of Definition 3.1. Also, the definitions and results of this subsection do not depend in an essential way on the linearity of T and could easily be reformulated for nonlinear problems; the definition (3.20) of Ω would have to be replaced by

$$\sup\{\|x_1 - x_2\| \mid x_1, x_2 \in \mathcal{M}, \|Tx_1 - Tx_2\| \leq \delta\}.$$

Convergence rates can thus be given only on subsets of $\mathcal{D}(T^\dagger)$ (or of $\mathcal{X}$), i.e., under a-priori assumptions on the exact data (or, equivalently, on the exact solution). We will formulate such a-priori assumptions in terms of the exact solutions by considering subsets of $\mathcal{X}$ of the form

$$\{x \in \mathcal{X} \mid x = Bw, \|w\| \leq \rho\}, \tag{3.27}$$

where B is a bounded linear operator from some Hilbert space into $\mathcal{X}$. While in Section 8.5, we will use another choice of B, we will consider the choice

$$B = (T^*T)^\mu$$

for some $\mu > 0$ here; we then denote the set in (3.27) by

$$\mathcal{X}_{\mu,\rho} := \{x \in \mathcal{X} \mid x = (T^*T)^\mu w, \|w\| \leq \rho\} \tag{3.28}$$

and use the further notation

$$\mathcal{X}_\mu := \bigcup_{\rho > 0} \mathcal{X}_{\mu,\rho} = \mathcal{R}((T^*T)^\mu). \tag{3.29}$$

These sets are usually called *source sets*, $x \in \mathcal{X}_{\mu,\rho}$ is said to have a *source representation*. Since for ill-posed problems T is usually a smoothing operator, the requirement for an element to be in $\mathcal{X}_{\mu,\rho}$ can be considered as an (abstract) smoothness condition.

For compact operators, $\mathcal{X}_\mu$ can be characterized via the singular values as follows:

Proposition 3.13. *Let K be compact with singular system $(\sigma_n; v_n, u_n)$. Then, for $\mu > 0$,*

$$K^\dagger y \in \mathcal{R}((K^*K)^\mu) \tag{3.30}$$

if and only if

$$\sum_{n=1}^\infty \frac{|\langle y, u_n \rangle|^2}{\sigma_n^{2+4\mu}} < \infty. \tag{3.31}$$

Proof. (3.30) holds if and only if there is a $w \in \mathcal{X}$ with

$$K^\dagger y = (K^*K)^\mu w = \sum_{n=1}^\infty \sigma_n^{2\mu} \langle w, v_n \rangle v_n.$$

Because of (2.24), this is equivalent to

$$\frac{\langle y, u_n \rangle}{\sigma_n} = \sigma_n^{2\mu} \langle w, v_n \rangle \quad \text{for all} \quad n \in \mathbb{N}.$$

Since $w \in \mathcal{X}$ if and only if $\{\langle w, v_n \rangle\} \in l^2$, this is in turn equivalent to (3.31). ∎

Thus, (3.30) can be seen as a scale of conditions starting (for $\mu = 0$) from the Picard criterion (Theorem 2.8 (i)). Like the Picard criterion, (3.30) can be seen as a condition on the decay rate of $\{\langle y, u_n \rangle\}$, which becomes more severe as μ becomes larger. Also, the condition becomes more severe the "more ill-posed" the problem is, i.e., the faster the singular values σ_n decay.

Proposition 3.10 has provided us with a lower bound for the accuracy of any method for solving (3.1) (and, hence, also with a bound for the convergence rate of any regularization method) in terms of the modulus of continuity of $T^\dagger$ on a set

3.2. Order Optimality

$\mathcal{M}$ describing an a-priori assumption. Now, we make this estimate concrete for $\mathcal{M} = \mathcal{X}_{\mu,\rho}$ as defined in (3.28):

Proposition 3.14. *For any $\mu, \rho > 0$, let $\mathcal{X}_{\mu,\rho}$ be defined by (3.28). Then, for any $\delta > 0$,*

$$\Omega(\delta, \mathcal{X}_{\mu,\rho}) \leq \delta^{\frac{2\mu}{2\mu+1}} \rho^{\frac{1}{2\mu+1}} \tag{3.32}$$

holds.

Proof. Let $x \in \mathcal{X}_{\mu,\rho}, w \in \mathcal{X}$ with $\|w\| \leq \rho$ and $x = (T^*T)^\mu w$. Let $\{E_\lambda\}$ be the spectral family for T^*T. Then, by the interpolation inequality (2.49) (with $r = \mu$ and $q = \mu + 1/2$)

$$\begin{aligned}
\|x\| &= \|(T^*T)^\mu w\| \leq \|(T^*T)^{\mu+\frac{1}{2}} w\|^{\frac{2\mu}{2\mu+1}} \|w\|^{\frac{1}{2\mu+1}} \\
&= \|(T^*T)^{\frac{1}{2}} x\|^{\frac{2\mu}{2\mu+1}} \|w\|^{\frac{1}{2\mu+1}} = \|Tx\|^{\frac{2\mu}{2\mu+1}} \|w\|^{\frac{1}{2\mu+1}}.
\end{aligned}$$

Thus, if $\|Tx\| \leq \delta$, $\|x\| \leq \delta^{\frac{2\mu}{2\mu+1}} \rho^{\frac{1}{2\mu+1}}$. Taking the supremum over all $x \in \mathcal{X}_{\mu,\rho}$ with $\|Tx\| \leq \delta$, we obtain (3.32). ∎

Thus, $\Omega(\delta, \mathcal{X}_{\mu,\rho})$ goes at least as fast to 0 as $O(\delta^{\frac{2\mu}{2\mu+1}})$ as $\delta \to 0$. We now show that this estimate is sharp at least for one sequence $\{\delta_k\}$ converging to 0 if $\mathcal{R}(T)$ is non-closed. We give the proof only for the case of a compact operator:

Proposition 3.15. *Let K be compact with non-closed range. Then, for any $\mu, \rho > 0$, there is a sequence $\{\delta_k\}$ converging to 0 such that*

$$\Omega(\delta_k, \mathcal{X}_{\mu,\rho}) = \delta_k^{\frac{2\mu}{2\mu+1}} \rho^{\frac{1}{2\mu+1}}. \tag{3.33}$$

Proof. By assumption, the singular values $\{\sigma_k\}$ of K converge to 0. Let $\delta_k := \rho \sigma_k^{2\mu+1}$. Then, $(\delta_k/\rho)^{\frac{2}{2\mu+1}} = \sigma_k^2$ is an eigenvalue of K^*K (with eigenvector $v_k, \|v_k\| = 1$). Let $x_k := \rho(K^*K)^\mu v_k$, so that $x_k \in \mathcal{X}_{\mu,\rho}$. By definition, $x_k = \rho \sigma_k^{2\mu} v_k = \delta_k^{\frac{2\mu}{2\mu+1}} \rho^{\frac{1}{2\mu+1}} v_k$, i.e., $\|x_k\| = \delta_k^{\frac{2\mu}{2\mu+1}} \rho^{\frac{1}{2\mu+1}}$.

We have $K^*Kx_k = \delta_k^{\frac{2\mu}{2\mu+1}} \rho^{\frac{1}{2\mu+1}} \sigma_k^2 v_k = \delta_k^{\frac{2\mu+2}{2\mu+1}} \rho^{-\frac{1}{2\mu+1}} v_k$, so that $\|Kx_k\|^2 = \langle K^*Kx_k, x_k \rangle = \delta_k^2$.

Hence, $\Omega(\delta_k, \mathcal{X}_{\mu,\rho}) \geq \|x_k\| = \delta_k^{\frac{2\mu}{2\mu+1}} \rho^{\frac{1}{2\mu+1}}$. Together with Proposition 3.14, this implies (3.33). ∎

Remark 3.16. If T is not necessarily compact and has a non-closed range, one can proceed as in the proof of Proposition 2.3 in [68], using approximate eigenvectors, to prove that there is a sequence $\{\delta_k\}$ converging to 0 such that $\Omega(\delta_k, \mathcal{X}_{\mu,\rho})$ does not go to 0 faster than $\delta_k^{\frac{2\mu}{2\mu+1}}$.

60   3. Regularization Operators

From Propositions 3.10, 3.14 and 3.15, we can conclude that, if $\mathcal{R}(T)$ is non-closed, a regularization algorithm cannot converge to 0 faster than $\delta^{\frac{2\mu}{2\mu+1}} \rho^{\frac{1}{2\mu+1}}$ as $\delta \to 0$ under the a-priori assumption

$$x^\dagger \in \mathcal{X}_{\mu,\rho}, \qquad (3.34)$$

or, if we are concerned with the rate only, not faster than $O(\delta^{\frac{2\mu}{2\mu+1}})$ under the a-priori assumption

$$x^\dagger \in \mathcal{X}_\mu. \qquad (3.35)$$

This also means that under all rates of the form δ^s, $s = 2\mu/(2\mu+1)$ is best possible. Note, however, that (3.33) holds in general only for a special sequence $\{\delta_k\}$ converging to 0, for other values of δ, $\Omega(\delta, \mathcal{M})$ may well be smaller than $\delta^{\frac{2\mu}{2\mu+1}} \rho^{\frac{1}{2\mu+1}}$.

These considerations lead to different possible notions of optimal regularization algorithms (in $\mathcal{X}_{\mu,\rho}$ or in $\mathcal{X}_\mu$): one might call an algorithm (R_α, α) "optimal" if $\Delta(\delta, \mathcal{M}, R_\alpha)$ is minimal (for all $\delta > 0$) under all possible regularization methods. This is certainly the strictest possible requirement. But consider that due to the definition of Δ, it refers to the worst-case situation as far as y^δ compatible with $\|y^\delta - y\| \leq \delta$ is concerned.

We will use the following weaker notion of optimality:

Definition 3.17. *Let $\mathcal{R}(T)$ be non-closed, $\{R_\alpha\}$ be a regularization operator for $T^\dagger$. For $\mu, \rho > 0$ and $y \in T\mathcal{X}_{\mu,\rho}$, let α be a parameter choice rule for solving (3.1). We call (R_α, α) optimal in $\mathcal{X}_{\mu,\rho}$ if*

$$\Delta(\delta, \mathcal{X}_{\mu,\rho}, R_\alpha) = \delta^{\frac{2\mu}{2\mu+1}} \rho^{\frac{1}{2\mu+1}} \qquad (3.36)$$

holds for all $\delta > 0$. We call (R_α, α) of optimal order in $\mathcal{X}_{\mu,\rho}$ if there exists a constant $c \geq 1$ such that

$$\Delta(\delta, \mathcal{X}_{\mu,\rho}, R_\alpha) \leq c\delta^{\frac{2\mu}{2\mu+1}} \rho^{\frac{1}{2\mu+1}} \qquad (3.37)$$

holds for all $\delta > 0$.

This notion of optimality refers to the source sets $\mathcal{X}_{\mu,\rho}$ only. As $\mu \to 0$, these source sets become larger, the optimal convergence rate becomes slower. As can be seen from comparing Theorem 2.8 and Proposition 3.13, in general, $\bigcup_{\mu>0} T(\mathcal{X}_\mu) \neq \mathcal{R}(T)$. Nevertheless, a result due to Plato [227] shows under very weak assumptions, that a regularization method which is optimal on a specific set $\mathcal{X}_\mu$ is convergent for all $y \in \mathcal{R}(T)$. In fact, all that is required as assumption on the regularization method is that the resulting approximation (or better, the parameter choice rule) depends on the right-hand side y^δ, and on a slightly larger bound than δ for the noise level.

To be precise, assume that we have a regularization operator $\{R_\alpha\}$ and a parameter choice rule $\alpha(\delta, y^\delta)$. Then, defining

$$\alpha_\tau(y^\delta, \delta) := \alpha(y^\delta, \tau\delta), \qquad \tau > 1, \qquad (3.38)$$

we have the following result:

3.2. Order Optimality

Theorem 3.18. *If, for all $\tau > \tau_0 \geq 1$, the regularization method (R_α, α_τ) is of optimal order in $\mathcal{X}_{\mu,\rho}$ for some $\mu > 0$ and all $\rho > 0$, then all regularization methods (R_α, α_τ) with $\tau > \tau_0 \geq 1$ are convergent for $y \in \mathcal{R}(T)$, and they are of optimal order for all $\mathcal{X}_{\nu,\rho}$ with $0 < \nu \leq \mu$ and $\rho > 0$.*

Proof. The idea of the proof is to consider the right-hand side y^δ as a perturbation of $(I - F_\varepsilon)y$ with a clever choice of $\varepsilon = \varepsilon(\delta)$, where $\{F_\varepsilon\}$ is the spectral family of TT^* (cf. (2.44) for the compact case). Note that

$$(I - F_\varepsilon)y = T x_\varepsilon, \quad x_\varepsilon = (I - E_\varepsilon)x^\dagger,$$

i.e., x_ε is the truncated spectral expansion of $x^\dagger$ that will be considered again in Example 4.8. Obviously, $x_\varepsilon \in \mathcal{R}((T^*T)^\mu)$, in fact, this holds for all $\mu > 0$. More precisely, $x_\varepsilon \in \mathcal{X}_{\mu,\tilde{\rho}}$ with

$$\tilde{\rho}^2 = \int_\varepsilon^\infty \lambda^{-2\mu} d\|E_\lambda x^\dagger\|^2. \tag{3.39}$$

We now fix $\tau > \tau_0$, and let

$$\tilde{\tau} := \frac{\tau + \tau_0}{2}, \quad \tilde{\delta} := \frac{2\tau}{\tau + \tau_0}\delta \in (\delta, 2\delta).$$

Next, we define $\varepsilon = \varepsilon(\delta)$ via

$$\varepsilon(\delta) := \inf\{\tilde{\varepsilon} > 0 \mid \|F_{\tilde{\varepsilon}} y\| \geq \frac{\tau - \tau_0}{\tau + \tau_0}\delta\}.$$

It follows that

$$\|F_\varepsilon y\| \leq \frac{\tau - \tau_0}{\tau + \tau_0}\delta, \quad \|F_{2\varepsilon} y\| \geq \frac{\tau - \tau_0}{\tau + \tau_0}\delta. \tag{3.40}$$

Moreover, y^δ is a perturbation of $(I - F_\varepsilon)y$ with

$$\|(I - F_\varepsilon)y - y^\delta\| \leq \|y - y^\delta\| + \|F_\varepsilon y\| \leq \left(1 + \frac{\tau - \tau_0}{\tau + \tau_0}\right)\delta = \tilde{\delta}.$$

Since we have $\tau\delta = \tilde{\tau}\tilde{\delta}$ and $\tilde{\tau} > \tau_0$, there holds

$$\alpha_\tau(y^\delta, \delta) = \alpha(y^\delta, \tau\delta) = \alpha(y^\delta, \tilde{\tau}\tilde{\delta}) = \alpha_{\tilde{\tau}}(y^\delta, \tilde{\delta}).$$

In other words, the parameter choice strategy α_τ gives the same regularization parameter α as the strategy $\alpha_{\tilde{\tau}}$ with bound $\tilde{\delta}$ for the perturbation. In view of this interpretation, the order-optimality of $(R_\alpha, \alpha_{\tilde{\tau}})$ for $\mathcal{X}_{\mu,\tilde{\rho}}$ yields

$$\|x_\alpha^\delta - x_\varepsilon\| \leq c_{\tilde{\tau}} \tilde{\delta}^{\frac{2\mu}{2\mu+1}} \tilde{\rho}^{\frac{1}{2\mu+1}} \leq 2^{\frac{2\mu}{2\mu+1}} c_{\tilde{\tau}} \delta^{\frac{2\mu}{2\mu+1}} \tilde{\rho}^{\frac{1}{2\mu+1}}. \tag{3.41}$$

We first establish the order-optimality of the rule α_τ for $\mathcal{X}_{\nu,\rho}$. Let $x^\dagger \in \mathcal{X}_{\nu,\rho}$ for some $0 < \nu < \mu$, i.e., $x^\dagger = (T^*T)^\nu z$ with $\|z\| \leq \rho$. Then we obtain from (3.39) that

$$\tilde{\rho}^2 = \int_\varepsilon^\infty \lambda^{2(\nu-\mu)} d\|E_\lambda z\|^2 \leq \varepsilon^{2(\nu-\mu)}\rho^2,$$

and hence, from (3.41),

$$\|x_\alpha^\delta - x_\varepsilon\| \leq c\left(\frac{\delta}{\rho}\right)^{\frac{2\mu}{2\mu+1}} \varepsilon^{\frac{\nu-\mu}{2\mu+1}} \rho. \tag{3.42}$$

By virtue of (3.40) we have

$$\delta \leq \frac{\tau+\tau_0}{\tau-\tau_0}\|F_{2\varepsilon}T(T^*T)^\nu z\| \leq \frac{\tau+\tau_0}{\tau-\tau_0}(2\varepsilon)^{\nu+\frac{1}{2}}\rho.$$

Thus, we can bound ε from below:

$$\varepsilon \geq c\left(\frac{\delta}{\rho}\right)^{\frac{2}{2\nu+1}}.$$

Inserting this into (3.42) we obtain, note that $\nu < \mu$,

$$\|x_\alpha^\delta - x_\varepsilon\| \leq c\left(\frac{\delta}{\rho}\right)^{\frac{2\mu}{2\mu+1}}\left(\frac{\delta}{\rho}\right)^{\frac{2\nu-2\mu}{(2\nu+1)(2\mu+1)}}\rho = c\delta^{\frac{2\nu}{2\nu+1}}\rho^{\frac{1}{2\nu+1}}.$$

On the other hand, by the interpolation inequality and (3.40),

$$\|x_\varepsilon - x^\dagger\| = \|E_\varepsilon x^\dagger\| \leq \|F_\varepsilon y\|^{\frac{2\nu}{2\nu+1}}\|E_\varepsilon z\|^{\frac{1}{2\nu+1}} \leq \left(\frac{\tau-\tau_0}{\tau+\tau_0}\right)^{\frac{2\nu}{2\nu+1}}\delta^{\frac{2\nu}{2\nu+1}}\rho^{\frac{1}{2\nu+1}}.$$

Since $\|x_\alpha^\delta - x^\dagger\| \leq \|x_\alpha^\delta - x_\varepsilon\| + \|x_\varepsilon - x^\dagger\|$, this proves the order-optimality of $\mathcal{X}_{\nu,\rho}$ for $0 < \nu < \mu$ and all $\rho > 0$.

Next, we establish convergence of (R_α, α_τ) for any $y \in \mathcal{R}(T)$. From (3.39) we have $\tilde\rho \leq \varepsilon^{-\mu}\|x^\dagger\|$, so that (3.41) becomes

$$\|x_\alpha^\delta - x_\varepsilon\| \leq c\left(\frac{\delta^2}{\varepsilon}\right)^{\frac{\mu}{2\mu+1}}\|x^\dagger\|^{\frac{1}{2\mu+1}}.$$

Consequently,

$$\begin{aligned}\|x_\alpha^\delta - x^\dagger\| &\leq \|x_\alpha^\delta - x_\varepsilon\| + \|x_\varepsilon - x^\dagger\| \\ &\leq c\left(\frac{\delta^2}{\varepsilon}\right)^{\frac{\mu}{2\mu+1}}\|x^\dagger\|^{\frac{1}{2\mu+1}} + \|E_\varepsilon x^\dagger\|.\end{aligned}$$

It remains to be shown that $\delta^2/\varepsilon(\delta) \to 0$ and $E_{\varepsilon(\delta)}x^\dagger \to 0$ as $\delta \to 0$. Since $\varepsilon(\delta)$ is monotone, two different cases have to be considered. First, if $\varepsilon(\delta) \to \varepsilon_0 > 0$ as $\delta \to 0$, then, obviously, $\delta^2/\varepsilon(\delta) \to 0$. To see that $E_{\varepsilon(\delta)}x^\dagger \to 0$, we observe from (3.40) that $F_{\varepsilon(\delta)}y \to F_{\varepsilon_0}y = 0$. This implies that

$$0 = E_{\varepsilon_0}x^\dagger = \lim_{\delta\to 0}E_{\varepsilon(\delta)}x^\dagger.$$

In the second case we have $\varepsilon(\delta) \to 0$ as $\delta \to 0$. Thus, we see at once that $E_\varepsilon x^\dagger \to 0$, since $x^\dagger \in \mathcal{N}(T)^\perp$. From (3.40) we, furthermore, conclude that

$$\frac{\tau - \tau_0}{\tau + \tau_0}\delta \leq \|F_{2\varepsilon}y\| = \|F_{2\varepsilon}Tx^\dagger\| \leq \sqrt{2\varepsilon}\|E_{2\varepsilon}x^\dagger\|,$$

which implies that

$$\frac{\delta^2}{\varepsilon} \leq 2\left(\frac{\tau + \tau_0}{\tau - \tau_0}\right)^2 \|E_{2\varepsilon}x^\dagger\|^2 \to 0 \quad \text{as} \quad \delta \to 0.$$

Therefore, $x_\alpha^\delta \to x^\dagger$ as $\varepsilon \to 0$, and the proof is complete. ∎

As we will see the assumptions of Theorem 3.18 are fulfilled for all parameter choice strategies that will be considered in this book. Later on (cf. Sections 4.1 and 4.2) we will see that the supremum of all possible μ for which a regularization operator can be order-optimal is termed the *qualification* of the method.

In Chapter 4 we will construct regularization methods which are order-optimal on $\mathcal{X}_{\mu,\rho}$ or on $\mathcal{X}_\mu$, first with a-priori, then with a-posteriori parameter choice rules.

3.3. Regularization by Projection

We have shown in the sections above that a stable solution of ill-posed problems $Tx = y$ can be obtained via regularization methods. For numerical calculations, we have to look for an implementable method, i.e., for one which can be realized in finite-dimensional spaces. One approach in this direction is *regularization by projection*, where the regularization is achieved by a finite-dimensional approximation alone, e.g., through discretization (cf. Section 1.1 on numerical differentiation), collocation, Galerkin or Ritz approximation. A general mathematical framework of this class of methods was developed by Natterer [205]. The combination of regularization as treated before and projection will be considered in Section 5.3.

A natural practical approach to approximating $x^\dagger$ is *least-squares projection*, i.e., to find the minimum-norm solution of $Tx = y$ in a finite-dimensional subspace of $\mathcal{X}$. That is, given a sequence

$$\mathcal{X}_1 \subset \mathcal{X}_2 \subset \mathcal{X}_3 \subset \ldots$$

of finite-dimensional subspaces of $\mathcal{X}$ whose union is dense in $\mathcal{X}$, let x_n be the least-squares solution of minimal norm in the space $\mathcal{X}_n$. Obviously,

$$x_n = T_n^\dagger y,$$

where $T_n := TP_n$ and P_n is the orthogonal projector onto $\mathcal{X}_n$. Note that, since T_n has closed range, the operator $T_n^\dagger$ is bounded (see Proposition 2.4). Thus, x_n is a stable approximation of $x^\dagger$.

3. Regularization Operators

Note that without additional assumptions it cannot be guaranteed that x_n converges to $x^\dagger$, as the following example due to Seidman [252] shows.

Example 3.19. Let $\{e_n\}$ be any orthonormal basis of a infinite-dimensional Hilbert space $\mathcal{X}$ and let $\mathcal{X}_n := \operatorname{span}\{e_1, \ldots, e_n\}$ for all $n \in \mathbb{N}$; then the union of the n-dimensional subspaces $\mathcal{X}_n$ is dense in $\mathcal{X}$.

We define an operator $T : \mathcal{X} \to \mathcal{Y} := \mathcal{X}$ as follows:

$$T : \sum_{i=1}^{\infty} \xi_i e_i \mapsto \sum_{i=1}^{\infty} (a_i \xi_i + b_i \xi_1) e_i$$

with

$$b_i := \begin{cases} 0, & i = 1, \\ i^{-1}, & i > 1, \end{cases} \qquad a_i := \begin{cases} i^{-1}, & i \text{ odd}, \\ i^{-\frac{5}{2}}, & i \text{ even}. \end{cases}$$

This operator is linear, injective and compact.

If x_n denotes the best-approximate solution of $Tx = y := Tx^\dagger$ in $\mathcal{X}_n$ with

$$x^\dagger := \sum_{i=1}^{\infty} i^{-1} e_i \quad \text{and} \quad x_n := \sum_{i=1}^{n} \xi_{i,n} e_i,$$

then $(\xi_{1,n}, \ldots, \xi_{n,n})$ solve the minimization problem

$$\sum_{i=1}^{n} (a_i(\xi_i - i^{-1}) + b_i(\xi_1 - 1))^2 + \sum_{i=n+1}^{\infty} i^{-2}(1 + a_i - \xi_1)^2 \to \min.$$

From the first order necessary conditions for a minimum (which are also sufficient in this case) we obtain the solution:

$$\xi_{1,n} = 1 + \Big(\sum_{i=n+1}^{\infty} a_i i^{-2}\Big)\Big(1 + \sum_{i=n+1}^{\infty} i^{-2}\Big)^{-1},$$

$$\xi_{i,n} = i^{-1} + (a_i i)^{-1}(\xi_{1,n} - 1), \quad 2 \leq i \leq n.$$

Thus,

$$\|x_n - P_n x^\dagger\|^2 = \Big(\sum_{i=1}^{n}(a_i i)^{-2}\Big)\Big(\sum_{i=n+1}^{\infty} a_i i^{-2}\Big)^2 \Big(1 + \sum_{i=n+1}^{\infty} i^{-2}\Big)^{-2}.$$

It follows easily that

$$\|x_n - P_n x^\dagger\| \sim n,$$

i.e., x_n does not converge.

We will now present some results on convergence and convergence rates for x_n from [108].

Theorem 3.20. *Let $y \in \mathcal{D}(T^\dagger)$ and let x_n be as above.*

(i) $x_n \rightharpoonup x^\dagger$ if and only if $\{\|x_n\|\}$ is bounded.

3.3. Regularization by Projection

(ii) $x_n \to x^\dagger$ *if and only if* $\limsup_{n\to\infty} \|x_n\| \leq \|x^\dagger\|$.

Proof. (i) If x_n converges weakly to $x^\dagger$, then $\{\|x_n\|\}$ is bounded by the uniform boundedness principle. Suppose now that $\{\|x_n\|\}$ is bounded. Then any subsequence $\{x_k\}$ of $\{x_n\}$ has a subsequence, again denoted by $\{x_k\}$, converging weakly to some $u \in \mathcal{X}$. Since T is bounded, we then have that

$$Tx_k \rightharpoonup Tu. \tag{3.43}$$

By the definition of x_k, $\|Tx_k - Tx^\dagger\| \leq \|T(P_k - I)x^\dagger\| \leq \|T\|\|(I - P_k)x^\dagger\|$. Due to the choice of the subspaces $\mathcal{X}_k$, this implies that $Tx_k \to Tx^\dagger$ and hence together with (3.43) that

$$u - x^\dagger \in \mathcal{N}(T). \tag{3.44}$$

Since

$$x_k \in \mathcal{N}(T_k)^\perp = \mathcal{N}(TP_k)^\perp = \mathcal{X}_k \cap (\mathcal{N}(T) \cap \mathcal{X}_k)^\perp$$

and

$$(\mathcal{N}(T) \cap \mathcal{X}_k)^\perp \supset (\mathcal{N}(T) \cap \mathcal{X}_{k+1})^\perp \supset \ldots \supset \mathcal{N}(T)^\perp,$$

it follows that $x_k \in (\mathcal{N}(T) \cap \mathcal{X}_m)^\perp$ for all $k \geq m$. Since $(\mathcal{N}(T) \cap \mathcal{X}_m)^\perp$ is weakly closed, this implies that $u \in (\mathcal{N}(T) \cap \mathcal{X}_m)^\perp$ for all $m \in \mathbb{N}$. However,

$$\overline{\bigcup_{m=1}^\infty (\mathcal{N}(T) \cap \mathcal{X}_m)} = \mathcal{N}(T)$$

and hence $u \in \mathcal{N}(T)^\perp$. But $x^\dagger \in \mathcal{N}(T)^\perp$ and finally, due to (3.44), $u = x^\dagger$. Therefore, every subsequence of $\{x_n\}$ has a subsequence converging weakly to $x^\dagger$ and hence $x_n \rightharpoonup x^\dagger$.

(ii) The result follows with (i) and the weak lower semicontinuity of the norm on the Hilbert space $\mathcal{X}$. ∎

The characterization of strong convergence in Theorem 3.20 (ii) is not very useful, as it requires knowledge of the norm of the true solution. In [186] it is shown that the condition

$$\limsup_{n\to\infty} \|(T_n^\dagger)^* x_n\| = \limsup_{n\to\infty} \|(T_n^*)^\dagger x_n\| < \infty \tag{3.45}$$

is a sufficient condition for the convergence of $\{x_n\}$ to $x^\dagger$:

Proposition 3.21. *Let* $y \in \mathcal{D}(T^\dagger)$ *and let* x_n *be as above. If condition (3.45) holds, then* $x_n \to x^\dagger$.

Proof. Let $z_n := (T_n^*)^\dagger x_n$. Since $x_n \in \mathcal{R}(T_n^\dagger) = \mathcal{R}(T_n^*)$, we have that $T_n^* z_n = x_n$ and hence, by the definition of x_n,

$$\begin{aligned}
\langle x_n - x^\dagger, x_n \rangle &= \langle x_n - x^\dagger, T_n^* z_n \rangle \\
&= \langle Tx_n - Tx^\dagger, z_n \rangle + \langle T(I - P_n)x^\dagger, z_n \rangle \\
&\leq (\|Tx_n - Tx^\dagger\| + \|T(I - P_n)x^\dagger\|)\|z_n\| \tag{3.46} \\
&\leq 2\|T\|\|(I - P_n)x^\dagger\|\|z_n\| =: r_n.
\end{aligned}$$

66   3. Regularization Operators

Thus, we obtain the estimate
$$\|x_n\|^2 \leq \|x^\dagger\| \|x_n\| + r_n$$
and hence
$$\|x_n\| \leq \|x^\dagger\| + r_n^{\frac{1}{2}}.$$
This together with (3.45) and the fact that $\|(I - P_n)x^\dagger\| \to 0$ implies that $\limsup_{n\to\infty} \|x_n\| \leq \|x^\dagger\|$. Now the assertion follows with Theorem 3.20 (ii). ∎

We now investigate some consequences of condition (3.45) for compact operators. First we show that (3.45) imposes a certain type of regularity on the solution $x^\dagger$:

Proposition 3.22. *If T is compact and (3.45) is satisfied, then $x^\dagger \in \mathcal{R}(T^*)$.*

Proof. Let $z_n := (T_n^*)^\dagger x_n$. As a consequence of (3.45), there is a subsequence $\{z_k\}$ with $z_k \rightharpoonup v$ for some $v \in \mathcal{Y}$. Since T is compact, so is T^* and hence
$$T^* z_k \to T^* v. \tag{3.47}$$
However,
$$T^* z_k = P_k T^* z_k + (I - P_k) T^* z_k = x_k + (I - P_k) T^* z_k.$$
This together with (3.47), Proposition 3.21 and the fact that $\|(I - P_k)T^*\| \to 0$ for a compact operator T implies that $T^* v = x^\dagger$. Therefore, $x^\dagger \in \mathcal{R}(T^*)$. ∎

Concerning the situation when (3.45) is violated, in [108] compact operator equations have been constructed showing that it is possible, on the one hand, that $x^\dagger \in \mathcal{R}(T^*)$ but $\{x_n\}$ is not even weakly convergent to $x^\dagger$ and that, on the other hand, $\{x_n\}$ may converge strongly to $x^\dagger \notin \mathcal{R}(T^*)$. This shows that condition (3.45) is too strong merely to guarantee convergence. We now show that in fact it implies a convergence rate:

Theorem 3.23. *Let $y \in \mathcal{D}(T^\dagger)$ and let x_n be as above. If T is compact and condition (3.45) holds, then*
$$\|x_n - x^\dagger\| = O(\|(I - P_n)T^*\|).$$

Proof. Due to Proposition 3.22, there is a $v \in \mathcal{Y}$ with $x^\dagger = T^* v$. Hence, we have
$$\langle x_n - x^\dagger, x^\dagger \rangle \leq \|Tx_n - Tx^\dagger\| \|v\| \leq \|T(P_n - I)x^\dagger\| \|v\|.$$
This together with (3.45) and (3.46) implies that
$$\begin{aligned}\|x_n - x^\dagger\|^2 &= O(\|T(I - P_n)x^\dagger\|) \\ &= O(\|T(I - P_n)\| \|(I - P_n)T^* v\|) \\ &= O(\|(I - P_n)T^*\|^2) \end{aligned}$$ ∎

3.3. Regularization by Projection

Note that, if (3.45) holds, for non-compact operators T we obtain at least the rate

$$\|x_n - x^\dagger\| = O(\|(I - P_n)x^\dagger\|^{\frac{1}{2}})$$

if we assume that $x^\dagger \in \mathcal{R}(T^*)$. Condition (3.45) is of course only an a-posteriori criterion for the convergence of x_n.

We will present now a different projection method, called *dual least-squares method* in [205], for which convergence is always guaranteed. That is, given a sequence

$$\mathcal{Y}_1 \subset \mathcal{Y}_2 \subset \mathcal{Y}_3 \subset \ldots$$

of finite-dimensional subspaces of $\overline{\mathcal{R}(T)} = \mathcal{N}(T^*)^\perp \subset \mathcal{Y}$ whose union is dense in $\mathcal{N}(T^*)^\perp$. x_n is now defined as the least-squares solution of minimal norm of the equation

$$T_n x = y_n, \quad T_n := Q_n T, \quad y_n := Q_n y, \tag{3.48}$$

where Q_n is the orthogonal projector onto $\mathcal{Y}_n$. Again x_n is a stable approximation of $x^\dagger$. In the next theorem we show that x_n has a special characterization:

Theorem 3.24. *Let $y \in \mathcal{D}(T^\dagger)$ and let x_n be as above. Then $x_n = P_n x^\dagger$, where P_n is the orthogonal projector onto $\mathcal{X}_n := T^* \mathcal{Y}_n$. Moreover,*

$$x_n \to x^\dagger \quad as \quad n \to \infty.$$

Proof. Note that $\mathcal{N}(T_n) = \mathcal{R}(T^* Q_n)^\perp = (T^* \mathcal{Y}_n)^\perp = \mathcal{X}_n^\perp$. Hence,

$$Q_n T(I - P_n) = 0. \tag{3.49}$$

By the definition of the spaces $\mathcal{Y}_n$, $Q_n y = Q_n Q y = Q_n T x^\dagger$. This together with (3.49) implies that

$$\|T_n x - Q_n y\| = \|Q_n T(x - x^\dagger)\| = \|Q_n T P_n(x - P_n x^\dagger)\|.$$

Thus, $\{P_n x^\dagger\} + \mathcal{X}_n^\perp$ is the set of least-squares solutions of the equation in (3.48) and $x_n = P_n x^\dagger$. Since $\mathcal{X}_n \subset \mathcal{X}_{n+1}$ for all $n \in \mathbb{N}$, $\bigcup_{n \in \mathbb{N}} \mathcal{X}_n$ is dense in $\mathcal{N}(T)^\perp$, and $x^\dagger \in \mathcal{N}(T)^\perp$, we now obtain that $x_n \to x^\dagger$. ∎

This theorem shows that $\{T_n^\dagger\}$, with T_n defined as in (3.48), is a regularization operator (cf. Proposition 3.4).

Before we turn to the convergence analysis of the dual least-squares method for noisy data, we show that *least-squares collocation*, also called *moment discretization* (cf. [69, 204], see also [165, chapter 17]), fits into this framework.

Example 3.25. Let $K : \mathcal{L}^2[0, 1] \to \mathcal{L}^2[0, 1]$ be an integral operator with continuous kernel k (cf. (2.19)). For the numerical solution of $Kx = y$ by least-squares

68   3. Regularization Operators

collocation we choose n collocation points $s_j \in [0,1], j = 1, \ldots, n$ and approximate $x^\dagger$ by the solution of minimal $\mathcal{L}^2$-norm satisfying the integral equation at the collocation points, i.e.,

$$\int_0^1 k(s_j, t) x(t)\, dt = y(s_j), \quad j = 1, \ldots, n. \tag{3.50}$$

If we define $\mathcal{Y}$ as the space $\mathcal{R}(K)$ equipped with the inner product

$$\langle y, z \rangle_\mathcal{Y} := \langle K^\dagger y, K^\dagger z \rangle_{\mathcal{L}^2},$$

then $\mathcal{Y}$ is a reproducing kernel Hilbert space with reproducing kernel

$$M(s, t) := \int_0^1 k(s, u) k(t, u)\, du \tag{3.51}$$

(cf., e.g., [165, Theorem 17.10]). Therefore, equation (3.50) is equivalent to

$$\langle Kx, M_{s_j} \rangle_\mathcal{Y} = \langle y, M_{s_j} \rangle_\mathcal{Y},$$

where $M_{s_j} := M(s_j, \cdot)$, and hence equivalent to equation (3.48) with $\mathcal{Y}_n$ defined by $\mathrm{span}\{M_{s_1}, \ldots, M_{s_n}\}$. Due to Theorem 3.24, applied to K as an operator from $\mathcal{L}^2[0,1]$ in $\mathcal{Y}$, we now know that the solution of minimal norm of (3.50) is given by $P_n x^\dagger$, where P_n is the orthogonal projector onto $\mathcal{X}_n = K^* \mathcal{Y}_n = K^\dagger \mathcal{Y}_n = \mathrm{span}\{k_{s_1}, \ldots, k_{s_n}\}$ with $k_{s_j} := k(s_j, \cdot)$. Assuming that the kernel functions $k_{s_1}, \ldots, k_{s_n}$ are linearly independent, we thus have that

$$x_n = P_n x^\dagger = \sum_{i=1}^n \xi_i k_{s_i},$$

where the ξ_i, $i = 1, \ldots, n$, solve the linear system

$$\sum_{i=1}^n M(s_i, s_j) \xi_i = y(s_j), \quad j = 1, \ldots, n.$$

For the results above, about convergence of x_n towards $x^\dagger$, to hold one needs that P_n converges pointwise towards I. This is the case if the sequence of collocation points, $\{s_n\}$, is dense in $[0, 1]$.

For projection methods no regularization parameter is introduced explicitly. We will see in the next Theorem that there is a *hidden* regularization parameter, namely the smallest singular value of the operator T_n. For a stability analysis we assume as in the Section 3.1 that only perturbed data y^δ are available with

$$\|Q_n(y - y^\delta)\| \leq \delta. \tag{3.52}$$

The least-squares solution of (3.48) with the data y replaced by y^δ is denoted by x_n^δ.

3.3. Regularization by Projection

Theorem 3.26. *Let $y \in \mathcal{D}(T^\dagger)$ and let (3.52) hold. If $\delta/\mu_n \to 0$ as $\delta \to 0$ and $n \to \infty$, where μ_n is the smallest singular value of T_n, then*

$$x_n^\delta \to x^\dagger \quad as \quad \delta \to 0, \ n \to \infty.$$

Proof. The assertion follows with Theorem 3.24 and the estimate

$$\begin{aligned} \|x_n^\delta - x^\dagger\| &\leq \|x_n - x^\dagger\| + \|T_n^\dagger Q_n(y - y^\delta)\| \\ &\leq \|x_n - x^\dagger\| + \|T_n^\dagger\|\delta \\ &\leq \|x_n - x^\dagger\| + \frac{\delta}{\mu_n} \qquad \blacksquare \end{aligned} \qquad (3.53)$$

Remark 3.27. As we have remarked earlier, the projection method (3.48) is a regularization in the sense of Definition 3.1. Theorem 3.26 is therefore the analogue of Proposition 3.7, since $\|T_n^\dagger\| = 1/\mu_n$; $1/n$ plays the role of $\alpha(\delta)$.

If $\mathcal{Y}_n$ has fixed dimension n, we can ask ourselves how to choose $\mathcal{Y}_n$ to maximize μ_n. In view of Theorem 3.26, this would be an optimal choice in the sense that we then obtain minimal error due to noise for fixed n. For compact operators T, we can give an answer to this question.

Proposition 3.28. *Let T be compact with singular system $(\sigma_n; v_n, u_n)$ and let $\mathcal{Y}_n$ be such that $\dim(\mathcal{Y}_n) = n$. Then*

$$\mu_n \leq \sigma_n,$$

with μ_n defined as in Theorem 3.26.

Proof. Let $\mathcal{U}_{n-1} := \operatorname{span}\{u_1, \ldots, u_{n-1}\}$. Since $\dim(\mathcal{Y}_n) = n$, there is an element $\bar{y} \in \mathcal{U}_{n-1}^\perp \cap \mathcal{Y}_n$ with $\|\bar{y}\| = 1$. Thus,

$$\begin{aligned} \mu_n^2 &= \inf\{\langle TT^*y, y\rangle \mid \|y\| = 1 \ \wedge \ y \in \mathcal{Y}_n\} \leq \langle TT^*\bar{y}, \bar{y}\rangle \\ &\leq \sup\{\langle TT^*y, y\rangle \mid \|y\| = 1 \ \wedge \ y \in \mathcal{U}_{n-1}^\perp\} = \sigma_n^2 \end{aligned}$$

due to the Courant-Fischer inequalities. $\blacksquare$

Obviously, equality holds in the Proposition above for the special choice of $\mathcal{Y}_n = \mathcal{U}_n := \operatorname{span}\{u_1, \ldots, u_n\}$. The resulting method is the truncated singular value expansion (to be discussed in Example 4.8).

If T is compact and $x^\dagger \in \mathcal{R}(T^*)$, it follows immediately from (3.53) and Theorem 3.24 that

$$\|x_n^\delta - x^\dagger\| = O(\|(I - P_n)T^*\| + \frac{\delta}{\mu_n}).$$

We show in the next proposition that the choice $\mathcal{Y}_n = \mathcal{U}_n$ is also optimal with resepect to the approximation error term.

3. Regularization Operators

Proposition 3.29. *Let T be compact with singular system $(\sigma_n; v_n, u_n)$ and let $\mathcal{Y}_n$ be such that $\dim(\mathcal{Y}_n) = n$. Then*

$$\|(I - P_n)T^*\| \geq \sigma_{n+1}.$$

If $\mathcal{Y}_n = \mathcal{U}_n$, then equality holds.

Proof. By the Courant-Fischer characterization of eigenvalues,

$$\sigma_{n+1}^2 = \inf\sup\{\langle T^*Tx, x\rangle \mid \|x\| = 1 \,\wedge\, x \in \mathcal{M}_n^\perp\}$$

holds, where the infimum is taken over all n-dimensional subspaces $\mathcal{M}_n$. Thus,

$$\begin{aligned}
\sigma_{n+1}^2 &\leq \sup\{\|Tx\|^2 \mid \|x\| = 1 \,\wedge\, x \in \mathcal{X}_n^\perp\} \\
&= \sup\{\|T(I - P_n)x\|^2 \mid \|x\| = 1\} \\
&= \|T(I - P_n)\|^2 = \|(I - P_n)T^*\|^2.
\end{aligned}$$

If $\mathcal{Y}_n = \mathcal{U}_n$, equality follows from the fact that then $\mathcal{X}_n = \text{span}\{v_1, \ldots, v_n\}$ and that for this set the infimum is attained in the characterization of σ_{n+1} above. ∎

Thus, the convergence rate

$$\|x_n^\delta - x^\dagger\| = O(\sigma_{n+1} + \frac{\delta}{\sigma_n})$$

is the best possible rate one can expect for compact operators T and $x^\dagger \in \mathcal{R}(T^*)$; it is achieved for the truncated singular value expansion.

4. Continuous Regularization Methods

4.1. A-priori Parameter Choice Rules

After the general considerations of the last chapters, we now consider a class of linear regularization methods based on spectral theory for selfadjoint linear operators (see Section 2.3 for basic definitions and results about spectral theory).

The basic idea is the following one: let $\{E_\lambda\}$ be a spectral family for T^*T (cf. Section 2.3). If T^*T is continuously invertible, then $(T^*T)^{-1} = \int \frac{1}{\lambda} dE_\lambda$. Since the best-approximate solution $x^\dagger = T^\dagger y$ can be characterized by the Gaußian normal equation (2.18),

$$x^\dagger = \int \frac{1}{\lambda} dE_\lambda T^* y \qquad (4.1)$$

holds in this case. Now, if $\mathcal{R}(T)$ is non-closed and $y \notin \mathcal{D}(T^\dagger)$, i.e., if

$$Tx = y \qquad (4.2)$$

is ill-posed, then the integral in (4.1) does not exist, since the integrand $1/\lambda$ has a pole in 0, which belongs to the spectrum of T^*T. The idea is now to replace the integrand $1/\lambda$ by a parameter-dependent family of functions $g_\alpha(\lambda)$ which are at least piecewise continuous on $[0, \|T\|^2]$ (i.e., on a set containing the spectrum of T^*T) and, for convenience, continuous from the right in points of discontinuity and to replace (4.1) by

$$x_\alpha := \int g_\alpha(\lambda) \, dE_\lambda T^* y \, . \qquad (4.3)$$

By construction, the operator on the right-hand side of (4.3) acting on y is continuous, so that, for noisy data y^δ with $\|y - y^\delta\| \leq \delta$, one can bound the error between x_α and

$$x_\alpha^\delta := \int g_\alpha(\lambda) \, dE_\lambda T^* y^\delta, \qquad (4.4)$$

see (4.16), i.e., x_α^δ can be computed in a stable way.

In order to obtain convergence as $\alpha \to 0$, it is certainly necessary to choose g_α such that it approximates $1/\lambda$, i.e., such that, $\lim_{\alpha \to 0} g_\alpha(\lambda) = 1/\lambda$ holds for all $\lambda \in (0, \|T\|^2]$.

Before studying concrete choices of g_α, i.e., concrete methods, we study the question under which condition the family $\{R_\alpha\}$ with

$$R_\alpha := \int g_\alpha(\lambda) \, dE_\lambda T^* \qquad (4.5)$$

is a regularization operator for $T^\dagger$ and then, how one can construct parameter choice rules.

If x_α is defined by (4.3), then, by (2.18), the residual has the representation

$$x^\dagger - x_\alpha = x^\dagger - g_\alpha(T^*T)T^*y = (I - g_\alpha(T^*T)T^*T)x^\dagger = \int (1 - \lambda g_\alpha(\lambda))\, dE_\lambda x^\dagger.$$

Hence, if we define, for all (α, λ) for which $g_\alpha(\lambda)$ is defined,

$$r_\alpha(\lambda) := 1 - \lambda g_\alpha(\lambda), \tag{4.6}$$

and hence

$$r_\alpha(0) = 1, \tag{4.7}$$

then

$$x^\dagger - x_\alpha = r_\alpha(T^*T)x^\dagger. \tag{4.8}$$

Theorem 4.1. *Let, for all $\alpha > 0$, $g_\alpha : [0, \|T\|^2] \to \mathbb{R}$ fulfill the following assumptions: g_α is piecewise continuous, and there is a $C > 0$ such that*

$$|\lambda g_\alpha(\lambda)| \leq C, \tag{4.9}$$

and

$$\lim_{\alpha \to 0} g_\alpha(\lambda) = \frac{1}{\lambda} \tag{4.10}$$

for all $\lambda \in (0, \|T\|^2]$. Then, for all $y \in \mathcal{D}(T^\dagger)$,

$$\lim_{\alpha \to 0} g_\alpha(T^*T)T^*y = x^\dagger \tag{4.11}$$

holds with $x^\dagger = T^\dagger y$. If $y \notin \mathcal{D}(T^\dagger)$, then $\lim_{\alpha \to 0} \|g_\alpha(T^*T)T^*y\| = +\infty$.

Proof. Because of (4.3) and (4.8),

$$\|x^\dagger - x_\alpha\|^2 = \int_0^{\|T\|^2+} r_\alpha^2(\lambda)\, d\|E_\lambda x^\dagger\|^2. \tag{4.12}$$

The integrand in (4.12) is bounded by the constant $(C+1)^2$, which is integrable with respect to the measure $d\|E_\lambda x^\dagger\|^2$. Hence, by the Dominated Convergence Theorem,

$$\lim_{\alpha \to 0} \int_0^{\|T\|^2+} r_\alpha^2(\lambda)\, d\|E_\lambda x^\dagger\|^2 = \int_0^{\|T\|^2+} \lim_{\alpha \to 0} r_\alpha^2(\lambda)\, d\|E_\lambda x^\dagger\|^2 \tag{4.13}$$

holds. Since, by (4.10) and (4.7), $\lim_{\alpha \to 0} r_\alpha^2(\lambda) = 0$ for $\lambda > 0$ and $\lim_{\alpha \to 0} r_\alpha^2(0) = 1$, the integral on the right-hand side of (4.13) equals the "jump" of $\lambda \mapsto \|E_\lambda x^\dagger\|^2$ at $\lambda = 0$, i.e., $\lim_{\lambda \to 0+} \|E_\lambda x^\dagger\|^2 - \|E_0 x^\dagger\|^2 = \|Px^\dagger\|^2$, where P is the orthogonal projector onto $\mathcal{N}(T)$. Since $x^\dagger \in \mathcal{N}(T)^\perp$, $Px^\dagger = 0$, so that we finally obtain from (4.12) and (4.13) that,

$$\lim_{\alpha \to 0} \|x^\dagger - x_\alpha\|^2 = 0,$$

i.e., (4.11) holds. The fact that $\|x_\alpha\| \to +\infty$ as $\alpha \to 0$ if $y \notin \mathcal{D}(T^\dagger)$, follows from Proposition 3.6. Note that (4.9) implies, for $R_\alpha := g_\alpha(T^*T)T^*$, that $\|TR_\alpha\| = \|TT^*g_\alpha(TT^*)\| \leq C$, so that (3.13) holds. ∎

4.1. A-priori Parameter Choice Rules

We remark in passing that, when the operator T is already selfadjoint and positive semidefinite, one will not use $R_\alpha := g_\alpha(T^*T)T^*$, but apply the theory of regularization methods for equations with selfadjoint operators, where R_α would be just $g_\alpha(T)$ (see, e.g., [233, 248, 278]).

While Theorem 4.1 was about convergence of regularized solutions with exact data, the next theorem is concerned with the stability question, i.e., with the propagation of data noise:

Theorem 4.2. *Let g_α and C be as in Theorem 4.1, x_α and x_α^δ be defined by (4.3) and (4.4), respectively. For $\alpha > 0$, let*

$$G_\alpha := \sup\{|g_\alpha(\lambda)| \mid \lambda \in [0, \|T\|^2]\}. \tag{4.14}$$

Then,

$$\|Tx_\alpha - Tx_\alpha^\delta\| \leq C\delta \tag{4.15}$$

and

$$\|x_\alpha - x_\alpha^\delta\| \leq \delta\sqrt{CG_\alpha} \tag{4.16}$$

hold.

Proof. We first prove (4.15). Since

$$g_\alpha(T^*T)T^* = T^*g_\alpha(TT^*), \tag{4.17}$$

we can estimate

$$\|Tx_\alpha - Tx_\alpha^\delta\| = \|Tg_\alpha(T^*T)T^*(y - y^\delta)\| \leq \|TT^*g_\alpha(TT^*)\| \|y - y^\delta\|, \tag{4.18}$$

so that

$$\|Tx_\alpha - Tx_\alpha^\delta\| \leq \delta \|TT^*g_\alpha(TT^*)\|. \tag{4.19}$$

Let $\{F_\lambda\}$ be the spectral family of TT^*. Then, for all $y \in \mathcal{Y}$ with $\|y\| = 1$,

$$\begin{aligned}
\|TT^*g_\alpha(TT^*)y\|^2 &= \int_0^{\|T\|^2+} (\lambda g_\alpha(\lambda))^2 \, d\|F_\lambda y\|^2 \\
&\leq \int_0^{\|T\|^2+} C^2 \, d\|F_\lambda y\|^2 = C^2 \|y\|^2.
\end{aligned}$$

Hence,

$$\|TT^*g_\alpha(TT^*)\|^2 \leq C^2 \tag{4.20}$$

holds. Now, (4.19) and (4.20) imply (4.15).

In the same way as (4.20), one proves that

$$\|g_\alpha(TT^*)\| \leq G_\alpha. \tag{4.21}$$

Now, (4.17) implies that

$$\begin{aligned}
\|x_\alpha - x_\alpha^\delta\|^2 &= \langle x_\alpha - x_\alpha^\delta, T^*g_\alpha(TT^*)(y - y^\delta) \rangle \\
&= \langle Tx_\alpha - Tx_\alpha^\delta, g_\alpha(TT^*)(y - y^\delta) \rangle \\
&\leq \|Tx_\alpha - Tx_\alpha^\delta\| \|g_\alpha(TT^*)\| \delta
\end{aligned}$$

74   4. Continuous Regularization Methods

which, with (4.15) and (4.21), implies (4.16). ∎

Thus, for the total error we have the estimate

$$\|x_\alpha^\delta - x^\dagger\| \leq \|x_\alpha - x^\dagger\| + \delta\sqrt{CG_\alpha}. \tag{4.22}$$

By Theorem 4.1, the first term in this estimate goes to 0 if $y \in \mathcal{D}(T^\dagger)$. However, since $g_\alpha(\lambda) \to 1/\lambda$ as $\alpha \to 0$,

$$\lim_{\alpha \to 0} G_\alpha = +\infty, \tag{4.23}$$

so that for fixed $\delta > 0$, the second term in the estimate (4.22) explodes. Unless $\dim \mathcal{R}(T) < \infty$, the estimate (4.16) is sharp in a worst-case sense, so that also $\|x_\alpha - x_\alpha^\delta\|$ generally explodes (cf. [103]).

We now give an estimate for the convergence rate in $\|x_\alpha - x^\dagger\|$ in terms of an estimate for the function r_α defined by (4.6):

Theorem 4.3. *Let g_α fulfill the assumptions of Theorem 4.1, r_α be defined by (4.6), $\mu, \rho > 0$. Let $\omega_\mu : (0, \alpha_0) \to \mathbb{R}$ be such that for all $\alpha \in (0, \alpha_0)$ and $\lambda \in [0, \|T\|^2]$,*

$$\lambda^\mu |r_\alpha(\lambda)| \leq \omega_\mu(\alpha) \tag{4.24}$$

holds. Then, for $x^\dagger \in \mathcal{X}_{\mu,\rho}$ (cf. (3.28)),

$$\|x_\alpha - x^\dagger\| \leq \omega_\mu(\alpha)\rho \quad \text{and} \quad \|Tx_\alpha - Tx^\dagger\| \leq \omega_{\mu+\frac{1}{2}}(\alpha)\rho \tag{4.25}$$

holds, where x_α is defined by (4.3) and $x^\dagger := T^\dagger y$.

Proof. Let $w \in \mathcal{X}$ be such that $x^\dagger = (T^*T)^\mu w$, $\|w\| \leq \rho$. Then, by (4.8), $x^\dagger - x_\alpha = r_\alpha(T^*T)(T^*T)^\mu w$, and $Tx^\dagger - Tx_\alpha = Tr_\alpha(T^*T)(T^*T)^\mu w$. Thus, by spectral theory, (4.25) follows, where for the estimation of the residual norm $\|Tx^\dagger - Tx_\alpha\|$ we have used the identity

$$\|Tz\|^2 = \langle Tz, Tz \rangle = \langle T^*Tz, z \rangle = \langle (T^*T)^{\frac{1}{2}}z, (T^*T)^{\frac{1}{2}}z \rangle = \|(T^*T)^{\frac{1}{2}}z\|^2, \tag{4.26}$$

which holds for any $z \in \mathcal{X}$. ∎

Corollary 4.4. *Let the assumptions of Theorem 4.3 hold with*

$$\omega_\mu(\alpha) = c\alpha^\mu \tag{4.27}$$

for some $c > 0$, and assume that G_α as defined in (4.14) fulfills

$$G_\alpha = O\left(\frac{1}{\alpha}\right) \quad \text{as} \quad \alpha \to 0. \tag{4.28}$$

Then, with the parameter choice rule

$$\alpha \sim \left(\frac{\delta}{\rho}\right)^{\frac{2}{2\mu+1}}, \tag{4.29}$$

4.1. A-priori Parameter Choice Rules

the regularization method (R_α, α) is of optimal order in $\mathcal{X}_{\mu,\rho}$.

Proof. By Theorems 4.2 and 4.3,

$$\|x_\alpha^\delta - x^\dagger\| \leq c\alpha^\mu \rho + \delta\sqrt{\frac{C}{\alpha}}.$$

By (4.29), the right-hand side of this estimate can be bounded by $\delta^{\frac{2\mu}{2\mu+1}} \rho^{\frac{1}{2\mu+1}}$ multiplied by a constant. Thus, according to Definition 3.17, (R_α, α) is of optimal order in $\mathcal{X}_{\mu,\rho}$. ■

Remark 4.5. In Corollary 4.4, we have found an a-priori parameter choice rule which yields optimal order in $\mathcal{X}_{\mu,\rho}$ for those regularization methods for which the assumptions (especially (4.27) and (4.28)) are fulfilled; see below for some examples. Note that (4.29) means that with suitable constants $c_1, c_2 > 0$,

$$c_1 \left(\frac{\delta}{\rho}\right)^{\frac{2}{2\mu+1}} \leq \alpha(\delta) \leq c_2 \left(\frac{\delta}{\rho}\right)^{\frac{2}{2\mu+1}}.$$

One could also try to choose $\alpha(\delta) = \bar{c}(\delta/\rho)^{\frac{2}{2\mu+1}}$ and compute $\bar{c}$ such that the constant c in the error estimate (3.37) is minimized. If, as will be usually the case, ρ is not known, then a choice of $\alpha \sim \delta^{\frac{2}{2\mu+1}}$ will lead to an error estimate which is at least optimal with respect to the power of δ, i.e., of optimal order in $\mathcal{X}_\mu$: $\|x_\alpha^\delta - x^\dagger\| = O(\delta^{\frac{2\mu}{2\mu+1}})$. If, as will also usually be the case, also μ is unknown, then an a-priori parameter choice rule is not suitable, and one will have to use a-posteriori rules to be discussed below.

It is much more difficult to prove results about optimality (instead of just optimal order). In [276], such results have been proven for regularization operators of the form (4.5) under special assumptions on the function g_α (see also [183]). For the sharp estimates needed there, the "Melkman-Micchelli formula" (cf. [191])

$$\sup\{\|z\| \mid z \in \mathcal{Z}, \|z\|_0 \leq 1, \|z\|_1 \leq 1\} = \min_{0 \leq t \leq 1} \sup\{\|z\| \mid z \in \mathcal{Z}, (1-t)\|z\|_0^2 + t\|z\|_1^2 \leq 1\} \quad (4.30)$$

is used; here, $\|\cdot\|$, $\|\cdot\|_0$, and $\|\cdot\|_1$ are seminorms on a vector space $\mathcal{Z}$. For two different detailed proofs of (4.30) see [100].

Remark 4.6. Sometimes, one is not interested in the convergence rate of $x_\alpha^\delta - x^\dagger$, but in the rate with which some linear functional $\langle f, x_\alpha^\delta - x^\dagger \rangle$ converges. Now, if

$$f \in \mathcal{X}_{\nu,\eta},$$

i.e., $f = (T^*T)^\nu z$ with $\|z\| \leq \eta$, one sees as in the proof of Theorem 4.3 that

$$|\langle f, x_\alpha - x^\dagger \rangle| \leq \omega_{\mu+\nu}(\alpha)\rho\eta$$

holds, i.e., if g_α is such that (4.27) holds for $\mu + \nu$, then

$$|\langle f, x_\alpha - x^\dagger \rangle| = O(\alpha^{\mu+\nu}) \quad \text{as} \quad \alpha \to 0. \tag{4.31}$$

In the same way, one sees that

$$\|(T^*T)^\nu (x_\alpha - x^\dagger)\| \le \omega_{\mu+\nu}(\alpha)\rho \tag{4.32}$$

and especially (4.25) holds, i.e., if (4.27) holds for $\mu + \nu$, then

$$\|(T^*T)^\nu (x_\alpha - x^\dagger)\| = O(\alpha^{\mu+\nu}) \quad \text{as} \quad \alpha \to 0 \tag{4.33}$$

and

$$\|T(x_\alpha - x^\dagger)\| = O(\alpha^{\mu+\frac{1}{2}}) \quad \text{as} \quad \alpha \to 0. \tag{4.34}$$

But note that the rate with which $\omega_\mu(\alpha)$ decreases does not necessarily improve as μ increases, and that (4.27) may only hold for $\mu \in (0, \mu_0]$ and not for $\mu > \mu_0$; such an index μ_0 is sometimes called the *qualification* of the regularization operator $\{R_\alpha\}$. If (4.27) holds for all $\mu > 0$, we set $\mu_0 := +\infty$. If $\mu_0 < \infty$, then the parameter choice (4.29) will never yield a better rate than $O(\delta^{\frac{2\mu_0}{2\mu_0+1}})$, while, if (4.27) holds for all $\mu > 0$, the rate may be arbitrarily close to $O(\delta)$ (but never reaches this best-possible conceivable rate) if $x^\dagger \in \mathcal{X}_\mu$ with μ large enough. This is closely related to the effect of *saturation* to be discussed in Section 4.2. Note that the above definition of the term qualification is in good agreement with the way we have already used this term in the discussion of Theorem 3.18.

We will study the most widely-used regularization method, Tikhonov regularization, in Chapter 5. Thus, we give two other examples of the form (4.5) here:

Example 4.7. We consider the initial value problem

$$\begin{aligned} u'_\delta(t) + T^*T u_\delta(t) &= T^* y^\delta, \quad t \in \mathbb{R}_0^+, \\ u_\delta(0) &= 0. \end{aligned} \tag{4.35}$$

Here, $u_\delta : \mathbb{R}_0^+ \to \mathcal{X}$. We denote by

$$x_\alpha^\delta := u_\delta \left(\frac{1}{\alpha}\right) \tag{4.36}$$

(assuming the existence and uniqueness of u for the moment, which will follow from the considerations below). We omit δ, i.e., write u and x_α if we replace y^δ by y in (4.35).

For $t > 0$, let

$$v(t) := \int \gamma(t, \lambda) \, dE_\lambda T^* y^\delta,$$

where $\{E_\lambda\}$ is the spectral family of T^*T and $\gamma(t, \lambda) = (1 - e^{-\lambda t})/\lambda$. Then v solves (4.35), since

$$v'(t) = \int e^{-\lambda t} \, dE_\lambda T^* y^\delta$$

and
$$T^*Tv(t) = \int \lambda\gamma(t,\lambda)\,dE_\lambda T^* y^\delta = \int (1-e^{-\lambda t})\,dE_\lambda T^* y^\delta\,.$$
Hence, x_α^δ can be written as
$$x_\alpha^\delta = g_\alpha(T^*T)T^*y^\delta$$
with
$$g_\alpha(\lambda) = \frac{1}{\lambda}(1-e^{-\frac{\lambda}{\alpha}}) = \int_0^{\frac{1}{\alpha}} e^{-\lambda s}\,ds\,.$$
Hence, this method is of the form (4.5), and
$$r_\alpha(\lambda) = e^{-\frac{\lambda}{\alpha}}.$$

The assumptions of Theorems 4.1, 4.2, and 4.3 hold with $C = 1$, $G_\alpha = 1/\alpha$, $\omega_\mu(\alpha) = \mu^\mu e^{-\mu}\alpha^\mu$ for all $\mu > 0$. Thus, also the assumptions of Corollary 4.4 hold. Hence, we can conclude that with the choice (4.29), this method is of optimal order in $\mathcal{X}_{\mu,\rho}$ for all $\mu > 0$. Also, (4.31), (4.33), and (4.34) hold for all $\mu, \nu > 0$ and $f \in \mathcal{X}_\nu$. The optimality of this method is analyzed in [276].

It is quite interesting that, if $\mathcal{R}(T)$ is non-closed, the solution of (4.35) depends continuously on y^δ for all $t > 0$, but not its limit for $t \to +\infty$; $\lim_{t\to+\infty} u(t)$ exists and equals $T^\dagger y$ if $y \in \mathcal{D}(T^\dagger)$, otherwise, $\lim_{t\to+\infty} \|u(t)\| = +\infty$. This follows from Theorem 4.1 and Proposition 3.6, respectively. Another way of writing this down is the "Showalter formula"
$$\int_0^{+\infty} e^{-sT^*T}\,ds\,T^*y = T^\dagger y, \tag{4.37}$$
which holds precisely for $y \in \mathcal{D}(T^\dagger)$; otherwise, the integral in (4.37) is divergent.

The method (4.35) is known under the names *Showalter's method* or *asymptotic regularization*. The "regularization" in using (4.35) is in integrating this initial value problem not as far as possible (theoretically to $+\infty$), but only up to an abscissa $1/\alpha$, where α is the regularization parameter. As α becomes smaller, i.e., the abscissa up to which we integrate becomes larger, the accuracy $\|x_\alpha - x^\dagger\|$ improves, but eventually, the stability $\|x_\alpha^\delta - x_\alpha\|$ worsens. We will see an analogous behaviour for iterative regularization methods in Chapters 6 and 7. In fact, (4.35) can be seen as a "continuous analogue" of an iterative regularization method: if we would solve (4.35) by a forward Euler method with stepsize β, then
$$u(t+\beta) \approx u(t) + \beta(T^*y^\delta - T^*Tu(t)) = (I - \beta T^*T)u(t) + \beta T^*y^\delta\,,$$
which is quite reminiscent of the Landweber method to be discussed in Section 6.1.

Example 4.8. Let, for $\alpha \in (0, \alpha_0), \lambda \in [0, \|T\|^2]$,
$$g_\alpha(\lambda) := \begin{cases} \dfrac{1}{\lambda}, & \lambda \geq \alpha, \\ 0, & \lambda < \alpha. \end{cases}$$

78   4. Continuous Regularization Methods

Then, the assumptions of Theorems 4.1, 4.2, and 4.3 hold with $C = 1$, $G_\alpha = 1/\alpha$, $\omega_\mu(\alpha) = \alpha^\mu$ for all $\mu > 0$, so that also the assumptions of Corollary 4.4 hold. Thus, with x_α^δ defined by

$$x_\alpha^\delta := g_\alpha(T^*T)T^*y^\delta = \int_\alpha^{\|T\|^2+1} \frac{1}{\lambda} dE_\lambda T^*y^\delta, \qquad (4.38)$$

the estimates (4.31), (4.33), and (4.34) hold for all $\mu, \nu > 0$ and $f \in \mathcal{X}_\nu$. With the parameter choice rule (4.29), the method is of optimal order in $\mathcal{X}_{\mu,\rho}$ for all $\mu, \rho > 0$.

If T is replaced by a compact operator K with singular system $(\sigma_n; v_n, u_n)$, then (4.38) can be computed as follows:

$$x_\alpha^\delta = \sum_{\substack{n=1 \\ \sigma_n^2 \geq \alpha}}^\infty \frac{1}{\sigma_n^2} \langle K^*y^\delta, v_n \rangle v_n = \sum_{\substack{n=1 \\ \sigma_n^2 \geq \alpha}}^\infty \frac{1}{\sigma_n^2} \langle y^\delta, Kv_n \rangle v_n = \sum_{\substack{n=1 \\ \sigma_n^2 \geq \alpha}}^\infty \frac{1}{\sigma_n^2} \langle y^\delta, \sigma_n u_n \rangle v_n,$$

and hence

$$x_\alpha^\delta = \sum_{\substack{n=1 \\ \sigma_n^2 \geq \alpha}}^\infty \frac{\langle y^\delta, u_n \rangle}{\sigma_n} v_n, \qquad (4.39)$$

which can be seen as a truncated version of the singular value expansion (2.24) for $K^\dagger y$. Thus, this method is called *truncated singular value expansion*. It can also be interpreted as replacing K by the finite-rank operator K_α defined by

$$K_\alpha x := \sum_{\substack{n=1 \\ \sigma_n^2 \geq \alpha}}^\infty \sigma_n \langle x, v_n \rangle u_n$$

and by computing $x_\alpha^\delta = K_\alpha^\dagger y^\delta$, i.e., as a projection method. We have seen in Section 3.3 that this is in some sense an optimal projection method.

The truncation level α, which decides which singular values are replaced by 0, acts as regularization parameter.

The truncated singular value expansion is, for a compact operator with infinite-dimensional range, not optimal in any $\mathcal{X}_{\mu,\rho}$ for an a-priori parameter choice rule (cf. [183]).

We close this section by making a few remarks about uniform convergence of regularization methods. So far, "convergence" was always meant in the Hilbert space sense. Let, until the end of this section, $\mathcal{X} := \mathcal{L}^2(\mathcal{I})$, where $\mathcal{I} \subset \mathbb{R}$ is a compact interval. Let g_α fulfill the assumptions of Theorem 4.1, and let $x_\alpha, x_\alpha^\delta$ and R_α be defined by (4.3), (4.4), and (4.5), respectively.

The key to proving results about uniform convergence, i.e., convergence in the space $\mathcal{C}(\mathcal{I})$ of continuous functions on $\mathcal{I}$ with the supremum norm, is the fact that

$$R_\alpha TT^* = T^*g_\alpha(TT^*)TT^*.$$

This can be used to prove convergence of $R_\alpha TT^*w$ in $\mathcal{C}(\mathcal{I})$ from convergence of $g_\alpha(TT^*)TT^*w$ (which can be considered as a regularization method for the adjoint

4.1. A-priori Parameter Choice Rules

equation $T^*w = v$) in $\mathcal{Y}$, provided that T^* maps $\mathcal{Y}$ continuously into $\mathcal{C}(\mathcal{I})$. For the latter, a simple sufficient condition can be given:

Theorem 4.9. *Assume that $T : \mathcal{L}^2(I) \to \mathcal{Y}$ is bounded, with*

$$\mathcal{R}(T^*) \subseteq \mathcal{C}(\mathcal{I}), \qquad (4.40)$$

and that

$$x^\dagger \in \mathcal{R}(T^*). \qquad (4.41)$$

Then x_α converges to $x^\dagger$ in $\mathcal{C}(\mathcal{I})$, i.e., uniformly on $\mathcal{I}$.

Proof. Let (y_n, T^*y_n) be a sequence in the graph of T^*, $gr(T^*)$, converging to (y, z) in the topology of $\mathcal{Y} \times \mathcal{C}(\mathcal{I})$ (and hence also in $\mathcal{Y} \times \mathcal{L}^2(\mathcal{I})$). Since $T^* : \mathcal{Y} \to \mathcal{L}^2(\mathcal{I})$ is bounded, $(y, z) \in gr(T^*)$ by the Closed Graph Theorem. By (4.40), it follows, again from the Closed Graph Theorem, that $T^* : \mathcal{Y} \to \mathcal{C}(\mathcal{I})$ is bounded.

Let $w \in \mathcal{Y}$ (without loss of generality $w \in \mathcal{N}(T^*)^\perp$) be such that

$$T^*w = x^\dagger. \qquad (4.42)$$

From Theorem 4.1, applied to (4.42) (considered as an equation for w), we obtain that $g_\alpha(TT^*)Tx^\dagger$ converges to w in $\mathcal{Y}$. Since

$$x_\alpha - x^\dagger = g_\alpha(T^*T)T^*Tx^\dagger - x^\dagger = g_\alpha(T^*T)T^*Tx^\dagger - T^*w = T^*(g_\alpha(TT^*)Tx^\dagger - w),$$

uniform convergence of x_α to $x^\dagger$ follows. ∎

Remark 4.10. The principle of this proof carries over also to proving convergence rates: any result about convergence rates for the equation (4.42) translates into a result about a convergence rate in $\mathcal{C}(\mathcal{I})$ for the original equation, provided that (4.40) and (4.41) hold. For results about convergence rates for (4.42), one needs source conditions of the type

$$w \in \mathcal{R}((TT^*)^\mu),$$

which translates into the condition

$$x^\dagger \in \mathcal{R}(T^*(TT^*)^\mu) = \mathcal{R}((T^*T)^\mu T^*) = \mathcal{R}((T^*T)^{\mu+\frac{1}{2}}) = \mathcal{X}_{\mu+\frac{1}{2}} \qquad (4.43)$$

for $x^\dagger$. On the other hand, the stability bound (4.16), applied to (4.42), is multiplied by $\|T^*\|_{\mathcal{Y},\mathcal{C}(\mathcal{I})}$ for the original equation if (4.40) holds.

Thus, we can, e.g., conclude from Theorem 4.3 that, with ω_μ as in that theorem,

$$\|x_\alpha - x^\dagger\|_{\mathcal{C}(\mathcal{I})} = O(\omega_\mu(\alpha)) \qquad (4.44)$$

provided that (4.40), (4.41), and (4.43) hold. Note that (4.43) is stronger than the requirement that $x^\dagger \in \mathcal{X}_\mu$ needed for the $\mathcal{L}^2$-convergence rate provided by Theorem 4.3. Under the same assumptions,

$$\|x_\alpha^\delta - x^\dagger\|_{\mathcal{C}(\mathcal{I})} = O(\omega_\mu(\alpha)) + O(G_\alpha) \qquad (4.45)$$

80   4. Continuous Regularization Methods

holds (with G_α as in (4.14)).

An important case where (4.40) is fulfilled is that T is an integral operator with a continuous kernel. For this case, uniform convergence results for regularization methods have been proven in [104].

Results on regularization methods for linear ill-posed problems in Banach spaces can be found in [228].

4.2. Saturation and Converse Results

We now touch on the subject of what is called *converse results* and *saturation* in regularization theory:

In Theorem 4.3, we have seen that for regularization methods for which (4.27) holds,

$$\|x_\alpha - x^\dagger\| = O(\alpha^\mu), \qquad (4.46)$$

if

$$x^\dagger \in \mathcal{X}_\mu. \qquad (4.47)$$

The statement that (4.47) or a similar condition is not only sufficient for (4.46), but also necessary, is called a converse result. The same word is used for results about (4.47) being necessary for the convergence rate

$$\|x_\alpha^\delta - x^\dagger\| = O(\delta^{\frac{2\mu}{2\mu+1}}) \qquad (4.48)$$

when noisy data are used. Note that the rate in (4.48) has been recognized as optimal in $\mathcal{X}_\mu$ in Propositon 3.15, but in a slightly different sense.

The term "saturation" is used to describe the behaviour of some (not all!) regularization methods for which (4.48) does not hold for all $\mu > 0$, but only up to a finite value μ_0, the qualification of the regularization method.

Recall that the qualification μ_0 was defined as the largest value such that, with ω_μ as defined in (4.24), (4.27) holds for $0 < \mu \leq \mu_0$. Equivalently, μ_0 is the largest value such that

$$\lambda^\mu |r_\alpha(\lambda)| = O(\alpha^\mu) \qquad (4.49)$$

holds for $0 < \mu \leq \mu_0$.

For converse results, indices μ where (4.49) is sharp in the sense that there exist constants $c, \gamma > 0$ (independent of α) such that

$$\lambda^\mu |r_\alpha(\lambda)| \geq \gamma \alpha^\mu, \qquad \lambda \in [c\alpha, \|T\|^2], \qquad (4.50)$$

are of special importance (cf., e.g., Schock [250]). Usually (but, strictly speaking, not necessarily), the qualification μ_0 is such an index. For $\mu < \mu_0$, (4.50) cannot hold. For such indices, we have the following simple converse result:

4.2. Saturation and Converse Results

Theorem 4.11. *Let x_α be defined via (4.3), where g_α fulfills the assumptions of Theorem 4.1. Assume that μ is such that (4.50) holds. Then $\|x_\alpha - x^\dagger\| = O(\alpha^\mu)$ implies that $x^\dagger \in \mathcal{X}_\mu$.*

Proof. By (4.12) and (4.50),

$$\|x_\alpha - x^\dagger\|^2 \geq \int_{c\alpha}^{\|T\|^2+} r_\alpha^2(\lambda)\, d\|E_\lambda x^\dagger\|^2 \geq \gamma^2 \alpha^{2\mu} \int_{c\alpha}^{\|T\|^2+} \lambda^{-2\mu}\, d\|E_\lambda x^\dagger\|^2.$$

Hence, by (4.46),

$$\int_{c\alpha}^{\|T\|^2+} \lambda^{-2\mu}\, d\|E_\lambda x^\dagger\|^2 = O(1) \quad \text{as} \quad \alpha \to 0,$$

so that

$$\int_0^{\|T\|^2+} \lambda^{-2\mu}\, d\|E_\lambda x^\dagger\|^2 < +\infty. \tag{4.51}$$

Thus, we can define

$$w := \int_0^{\|T\|^2+} \lambda^{-\mu}\, dE_\lambda x^\dagger.$$

It follows from spectral theory that $(T^*T)^\mu w = x^\dagger$, so that (4.47) holds. ∎

It follows as in the proof of Theorem 4.11 that (4.47) and (4.51) are actually equivalent. From (4.51), we obtain, since then

$$\lim_{t \to 0} \int_0^t \lambda^{-2\mu}\, d\|E_\lambda x^\dagger\|^2 = 0,$$

that

$$\int_0^t 1\, d\|E_\lambda x^\dagger\|^2 = \int_0^t \lambda^{2\mu} \lambda^{-2\mu}\, d\|E_\lambda x^\dagger\|^2 = o(t^{2\mu})$$

as $t \to 0$. Hence, (4.47) implies the weaker estimate

$$\|E_t x^\dagger\|^2 = \int_0^t 1\, d\|E_\lambda x^\dagger\|^2 = O(t^{2\mu}). \tag{4.52}$$

The converse is in general not true, but the following holds.

Lemma 4.12. *If (4.52) holds, then*

$$x^\dagger \in \bigcap_{\nu < \mu} \mathcal{X}_\nu. \tag{4.53}$$

Proof. Let $0 < \nu < \mu$ and $\varepsilon > 0$. Then we obtain by integration by parts that

$$\int_\varepsilon^{\|T\|^2+} \lambda^{-2\nu}\, d\|E_\lambda x^\dagger\|^2 = \|T\|^{-4\nu} \|x^\dagger\|^2 - \varepsilon^{-2\nu} \|E_\varepsilon x^\dagger\|^2 + 2\nu \int_\varepsilon^{\|T\|^2+} \lambda^{-2\nu-1} \|E_\lambda x^\dagger\|^2\, d\lambda,$$

and hence, because of (4.52) there is a constant $c > 0$ (independent of ε) with

$$\int_\varepsilon^{\|T\|^2+} \lambda^{-2\nu}\, d\|E_\lambda x^\dagger\|^2 \leq c\left(1 + \int_\varepsilon^{\|T\|^2+} \lambda^{2\mu-2\nu-1}\, d\lambda\right).$$

82   4. Continuous Regularization Methods

Since $2\mu - 2\nu - 1 > -1$ by definition, we obtain with $\varepsilon \to 0$ that

$$\int_0^{\|T\|^2+} \lambda^{-2\nu} \, d\|E_\lambda x^\dagger\|^2 < \infty,$$

i.e., $x^\dagger \in \mathcal{X}_\nu$. ∎

The condition (4.52) (which also implies the convergence rate (4.46), cf. [218]) is actually suitable for a converse result if (4.50) does not hold:

Proposition 4.13. *Let g_α fulfill the assumptions of Theorem 4.1. Assume, in addition, that*

$$G_\alpha \leq \frac{\hat{c}}{\alpha}, \quad \alpha > 0, \tag{4.54}$$

holds with a suitable constant $\hat{c} > 0$, where G_α is defined by (4.14). Then $\|x_\alpha - x^\dagger\| = O(\alpha^\mu)$ implies (4.52) and (4.53).

Proof. Let $c := 1/(2\hat{c})$. Then, for $\lambda \in [0, c\alpha]$, we obtain from (4.54) that

$$|r_\alpha(\lambda)| \geq 1 - \lambda|g_\alpha(\lambda)| \geq 1 - \frac{\lambda \hat{c}}{\alpha} \geq \frac{1}{2}.$$

Hence,

$$\|x_\alpha - x^\dagger\|^2 = \int_0^{\|T\|^2+} r_\alpha^2(\lambda) \, d\|E_\lambda x^\dagger\|^2 \geq \int_0^{c\alpha} r_\alpha^2(\lambda) \, d\|E_\lambda x^\dagger\|^2 \geq \frac{1}{4} \int_0^{c\alpha} 1 \, d\|E_\lambda x^\dagger\|^2.$$

Together with (4.46), this implies (4.52), and hence (4.53) follows from Lemma 4.12. ∎

Remark 4.14. Proposition 4.13 is applicable to the methods discussed in Examples 4.7 and 4.8, respectively, since (4.54) holds there. For a compact operator with singular system $(\sigma_n; v_n, u_n)$, (4.52) reduces to

$$\sum_{n=k}^{\infty} |\langle x^\dagger, v_n \rangle|^2 = O(\sigma_k^{4\mu}), \tag{4.55}$$

i.e., to a decay condition for the Fourier components of $x^\dagger$ with respect to the singular vectors in relation to the singular values.

For converse results involving noisy data, we refer to [218]: under some additional assumptions on g_α, the equivalence of the worst-case convergence rate

$$\sup\{\inf_{\alpha > 0} \|x_\alpha^\delta - x^\dagger\| \mid \|Q(y - y^\delta)\| \leq \delta\} = O(\delta^{\frac{2\mu}{2\mu+1}})$$

with (4.52) (and even with (4.46) if (4.50) holds) is shown there. Under the same technical conditions on g_α, the following saturation result is also shown there:

If $\mathcal{R}(T)$ is non-closed and μ is such that (4.50) holds, and if

$$\sup\{\|x_\alpha^\delta - x^\dagger\| \mid \|Q(y - y^\delta)\| \le \delta\} = o(\delta^{\frac{2\mu}{2\mu+1}}),$$

then $x^\dagger = 0$, i.e., a convergence rate $o(\delta^{\frac{2\mu}{2\mu+1}})$ cannot hold (for any parameter choice rule) except in the trivial case that $x^\dagger = 0$.

As we have seen in Examples 4.7 and 4.8, there is no saturation for Showalter's method and for truncated singular value expansion, respectively. Thus, for these methods, (4.50) cannot hold for any μ (which can of course be checked directly), so that the converse result from Theorem 4.11 does not apply, but only the weaker converse result from Proposition 4.13.

Example 4.15. To show that methods with finite qualification μ_0 exist at all, we anticipate at this point Tikhonov regularization, which will be introduced in more detail in Chapter 5. Here we only mention that it can be defined via (4.3) with

$$g_\alpha(\lambda) = \frac{1}{\lambda + \alpha}, \qquad r_\alpha(\lambda) = \frac{\alpha}{\lambda + \alpha}.$$

To determine the qualification μ_0 of Tikhonov regularization we have to estimate the function

$$h_\mu(\lambda) = \lambda^\mu \frac{\alpha}{\lambda + \alpha}.$$

For $\mu < 1$ this function assumes its maximum for $\lambda = \mu\alpha/(1-\mu)$, so that $h_\mu(\lambda) \le \mu^\mu(1-\mu)^{1-\mu}\alpha^\mu$. For $\mu \ge 1$, the function h_μ is strictly increasing, and thus assumes its largest value in $[0, \|T\|^2]$ at the right end of this interval. Hence, for ω_μ as defined by (4.24) we can take

$$\omega_\mu(\alpha) = \begin{cases} \alpha^\mu, & \mu \le 1, \\ c\alpha, & \mu > 1, \end{cases}$$

with $c = \|T\|^{2\mu-2}$. It follows that (4.27) holds for $\mu \in (0,1]$, i.e., the qualification of Tikhonov regularization is $\mu_0 = 1$.

Furthermore, for $\mu = \mu_0 = 1$, (4.50) holds, e.g., with $c = 1$ and $\gamma = 1/2$. Hence, the results of this section apply with $\mu = 1$; cf. Section 5.1 for more details.

4.3. The Discrepancy Principle

In general, a $\mu > 0$ such that $x^\dagger \in \mathcal{X}_\mu$ will not be known. But, such a μ has to be known if one wants to construct an a-priori parameter choice rule which is at least of optimal order. We now turn our attention to a-posteriori parameter choice rules. We start with the widely-used "discrepancy principle" due to Morozov [193]; there are several variants of this principle in the literature (cf., e.g., [274, 275]). The version we use here is the following one:

4. Continuous Regularization Methods

Let g_α be as in Theorem 4.1, and let r_α be defined by (4.6). Furthermore, let

$$\tau > \sup\{|r_\alpha(\lambda)| \mid \alpha > 0, \lambda \in [0, \|T\|^2]\}. \tag{4.56}$$

The regularization parameter defined via the *discrepancy principle* is

$$\alpha(\delta, y^\delta) := \sup\{\alpha > 0 \mid \|Tx_\alpha^\delta - y^\delta\| \leq \tau\delta\}. \tag{4.57}$$

We assume that for each $\lambda > 0$, $\alpha \mapsto g_\alpha(\lambda)$ is continuous from the left, so that also the functional

$$\alpha \mapsto \|Tx_\alpha^\delta - y^\delta\|$$

is continuous from the left. Then, the supremum in (4.57) is attained, so that also

$$\|Tx_{\alpha(\delta,y^\delta)}^\delta - y^\delta\| \leq \tau\delta. \tag{4.58}$$

If $\|Tx_\alpha^\delta - y^\delta\| \leq \tau\delta$ for all $\alpha > 0$, then $\alpha(\delta, y^\delta) = +\infty$, and $x_{\alpha(\delta,y^\delta)}^\delta$ has then to be understood in the sense of a limit as $\alpha \to +\infty$. Note that under the assumption (4.54),

$$x_\infty^\delta := \lim_{\alpha \to \infty} x_\alpha^\delta = 0. \tag{4.59}$$

Thus, the regularization parameter is chosen via a comparison between the residual (or *discrepancy*) $\|Tx_\alpha^\delta - y^\delta\|$ and the assumed bound δ for the noise level. A heuristic motivation for this is the following:

Remark 4.16. We want to solve $Tx = y$, but instead of y, we have only noisy data y^δ and know that $\|y - y^\delta\| \leq \delta$. Thus, it does not make sense to ask for an approximate solution $\tilde{x}$ with a discrepancy $\|T\tilde{x} - y^\delta\| < \delta$, a residual in the order of δ is the best we should ask for. Since, on the other hand, a smaller regularization parameter means less stability, one should take the largest possible regularization parameter which results in a discrepancy in the order of δ. This is what is done in (4.57); note that, due to (4.7), (4.56) implies that $\tau > 1$.

After this heuristic motivation, we study the convergence properties of the regularization method (R_α, α) with R_α defined by (4.5) and the a-posteriori parameter choice rule α defined by (4.57). We assume that y is attainable, i.e.,

$$y \in \mathcal{R}(T). \tag{4.60}$$

Otherwise, the discrepancy $\|Tx_\alpha - y^\delta\|$ is never smaller than $\|y - Qy\| - \delta$, so that the set in (4.57) might be empty. Formally, the non-attainable case may be reduced to the attainable case by considering the equation

$$Tx = Qy \tag{4.61}$$

or the normal equation

$$T^*Tx = T^*y,$$

which are always solvable if $y \in \mathcal{D}(T^\dagger)$. At first sight, it seems not to be practical to use (4.61), since in general, Q will not be easily computable. However, this problem

4.3. The Discrepancy Principle

will disappear when (4.61) is approximated by a finite-dimensional equation in an appropriate way, as will be discussed in Section 5.2.

In addition to the assumptions made about g_α in Theorem 4.1, we assume that G_α as in (4.14) satisfies (4.54) and that, with

$$\mu_0 > \frac{1}{2} \tag{4.62}$$

and ω_μ defined in (4.24),

$$\omega_\mu \sim \alpha^\mu \quad \text{for} \quad 0 < \mu \leq \mu_0 \tag{4.63}$$

holds. By Theorem 4.3, this implies that with the a-priori parameter choice rule given there, (R_α, α) is of optimal order in $\mathcal{X}_{\mu,\rho}$ for $\mu \in (0, \mu_0]$.

As mentioned before, the largest value μ_0 for which (4.63) holds is called the qualification of the regularization $\{R_\alpha\}$ if it is finite. By Corollary 4.4, a regularization $\{R_\alpha\}$ can be made an order-optimal regularization method in $\mathcal{X}_{\mu,\rho}$ by an *a-priori* parameter choice rule for μ up to the qualification. Under our assumptions, the discrepancy principle has the following convergence properties:

Theorem 4.17. *The regularization method (R_α, α), where α is defined via the discrepancy principle (4.57), is convergent for all $y \in \mathcal{R}(T)$, and of optimal order in $\mathcal{X}_{\mu,\rho}$ for $\mu \in (0, \mu_0 - 1/2]$.*

Proof. If there is a sequence $\delta_n \to 0$ such that with corresponding y^{δ_n}, $\alpha(\delta_n, y^{\delta_n}) = +\infty$ for all n, then, because of (4.59), $\|y^{\delta_n}\| \leq \tau \delta_n$. Together with $\|y^{\delta_n} - y\| \leq \delta$, this yields $y = 0$, and hence

$$x^\dagger = 0 = x^{\delta_n}_{\alpha(\delta_n, y^{\delta_n})}.$$

Thus, we can assume that for δ sufficiently small, $\alpha(\delta, y^\delta) < +\infty$ for all y^δ with $\|y^\delta - y\| \leq \delta$.

We will make repeated use of the identity

$$\|Tx_\beta - y\| = \|r_\beta(TT^*)y\| = \|(T^*T)^{\frac{1}{2}} r_\beta(T^*T) x^\dagger\|, \quad \beta > 0, \tag{4.64}$$

where r_β is defined as in (4.6), and which follows from (4.8), (2.43), and (4.26).

Now let $\mu \in (0, \mu_0 - 1/2]$ and $x^\dagger \in \mathcal{X}_{\mu,\rho}$, i.e., $x^\dagger = (T^*T)^\mu w$ with $\|w\| \leq \rho$, and assume that $\|y^\delta - y\| \leq \delta$. We set $\alpha := \alpha(\delta, y^\delta)$. Then we have as in the proof of Theorem 4.3 that

$$\|x_\alpha - x^\dagger\| = \|(T^*T)^\mu r_\alpha(T^*T) w\|. \tag{4.65}$$

By the interpolation inequality (2.49) applied to $z := r_\alpha(T^*T)w$, we obtain together with (4.65) that

$$\|x_\alpha - x^\dagger\| \leq \|r_\alpha(T^*T)w\|^{\frac{1}{2\mu+1}} \|(T^*T)^{\frac{1}{2}} r_\alpha(T^*T)(T^*T)^\mu w\|^{\frac{2\mu}{2\mu+1}}.$$

86   4. Continuous Regularization Methods

By definition of w and (4.64), this implies that

$$\|x_\alpha - x^\dagger\| \leq \|r_\alpha(T^*T)w\|^{\frac{1}{2\mu+1}} \|Tx_\alpha - y\|^{\frac{2\mu}{2\mu+1}},$$

and hence,

$$\|x_\alpha - x^\dagger\| \leq (\gamma\rho)^{\frac{1}{2\mu+1}} \|Tx_\alpha - y\|^{\frac{2\mu}{2\mu+1}}, \qquad (4.66)$$

where γ is the supremum in (4.56), i.e.,

$$\gamma := \sup\{|r_\alpha(\lambda)| \mid \alpha > 0, \lambda \in [0, \|T\|^2]\}. \qquad (4.67)$$

Now, since $x_\alpha - x_\alpha^\delta = g_\alpha(T^*T)T^*(y - y^\delta)$, we have that

$$\begin{aligned}
\|T(x_\alpha - x_\alpha^\delta) - (y - y^\delta)\| &= \|(I - TT^*g_\alpha(TT^*))(y^\delta - y)\| \\
&= \|r_\alpha(TT^*)(y - y^\delta)\| \leq \gamma\delta.
\end{aligned} \qquad (4.68)$$

Therefore, it follows from (4.58) that

$$\|Tx_\alpha - y\| \leq \|Tx_\alpha^\delta - y^\delta\| + \gamma\delta \leq (\tau + \gamma)\delta.$$

Hence, inserting this into (4.66) we obtain that

$$\|x_\alpha - x^\dagger\| = O(\rho^{\frac{1}{2\mu+1}} \delta^{\frac{2\mu}{2\mu+1}}). \qquad (4.69)$$

Since, by (4.54) and Theorem 4.2,

$$\|x_\alpha - x_\alpha^\delta\| \leq c\delta\alpha^{-\frac{1}{2}}, \qquad (4.70)$$

it remains to be shown that

$$\alpha \geq c\delta^{\frac{2}{2\mu+1}} \rho^{-\frac{2}{2\mu+1}} \qquad (4.71)$$

holds with a generic constant $c > 0$.

By definition of α and (4.64), we have that

$$\|Tx_{2\alpha}^\delta - y^\delta\| = \|r_{2\alpha}(TT^*)y^\delta\| > \tau\delta.$$

Using (4.68) and the triangle inequality, this implies that

$$\|Tx_{2\alpha} - y\| \geq \|Tx_{2\alpha}^\delta - y^\delta\| - \gamma\delta \geq (\tau - \gamma)\delta,$$

where $\tau - \gamma > 0$ by virtue of (4.56) and (4.67). Together with (4.63), (4.64), and the definition of w, we now obtain that

$$\delta \leq c\|Tx_{2\alpha} - y\| = c\|(T^*T)^{\mu+\frac{1}{2}} r_{2\alpha}(T^*T)w\| \leq c\omega_{\mu+\frac{1}{2}}(2\alpha)\rho \leq c\alpha^{\mu+\frac{1}{2}}\rho.$$

This implies (4.71), and thus completes the proof of the order optimality of the discrepancy principle for $0 < \mu \leq \mu_0 - 1/2$.

Finally, to prove the convergence of the regularization method we use Theorem 3.18; note that the assumptions of this theorem are fulfilled for $\mu = \mu_0 - 1/2$

4.3. The Discrepancy Principle

and $\tau_0 = \gamma$, with γ as in (4.67). The parameter choice rule α_τ defined in (3.38) corresponds to the version of the discrepancy principle (4.57) with the respective parameter τ. ∎

Remark 4.18. One can observe from the proof of Theorem 4.17 that the assertions already hold for any parameter choice $\alpha = \alpha(\delta, y^\delta)$ which satisfies

$$\|Tx_\alpha^\delta - y^\delta\| \leq \tau\delta \leq \|Tx_\beta^\delta - y^\delta\|$$

for some β with $\alpha \leq \beta \leq 2\alpha$, and with τ constrained by (4.56). This is an important point in practical computations, because it allows a certain "sloppyness" in the determination of $\alpha(\delta, y^\delta)$ from (4.57).

Remark 4.19. In the proof of Theorem 4.17, we have applied Theorem 3.18 to handle the case $\mu = 0$, i.e., convergence of the regularization method. The arguments used for the case $\mu > 0$, i.e., order optimality in $\mathcal{X}_{\mu,\rho}$, when applied to $\mu = 0$ would only yield boundedness of the terms in (4.69) and (4.70) instead of convergence to 0. Similar arguments as used in the proof of Theorem 3.18 can also be applied to replace the O-estimates by o-estimates in (4.69) and (4.70) for any $0 < \mu < \mu_0 - 1/2$. In this way, we obtain that the method (R_α, α), where $\alpha = \alpha(\delta, y^\delta)$ is defined via the discrepancy principle (4.57), has the convergence property

$$\|x^\delta_{\alpha(\delta,y^\delta)} - x^\dagger\| = o(\delta^{\frac{2\mu}{2\mu+1}}) \qquad (4.72)$$

for $x^\dagger \in \mathcal{X}_\mu$ and $\mu \in [0, \mu_0 - 1/2)$.

We mention that in general, a regularization method (R_α, α) with α defined via the discrepancy principle (4.57) need not be of optimal order in $\mathcal{X}_{\mu,\rho}$ for $\mu > \mu_0 - 1/2$, as the following result for Tikhonov regularization (cf. Example 4.15 and Section 5.1) due to Groetsch [102] implies.

Proposition 4.20. Let K be compact, $R_\alpha := (K^*K + \alpha I)^{-1}K^*$, $\alpha(\delta, y^\delta)$ be defined via (4.57). If

$$\|x^\delta_{\alpha(\delta,y^\delta)} - x^\dagger\| = o(\sqrt{\delta}) \qquad (4.73)$$

holds for all $y \in \mathcal{R}(K)$ and $y^\delta \in \mathcal{Y}$ fulfilling $\|y^\delta - y\| \leq \delta$, then $\mathcal{R}(K)$ is finite-dimensional.

Proof. Let $(\sigma_n; v_n, u_n)$ be a singular system for the operator K and assume that $\dim \mathcal{R}(K) = \infty$, so that there are infinitely many σ_n and $\lim_{n \to \infty} \sigma_n = 0$. Let, with $\delta_n \to 0$, $y := u_1$, $y^{\delta_n} := y + \delta_n u_n$, $\alpha_n := \alpha(\delta_n, y^{\delta_n})$. Then $x^\dagger = v_1/\sigma_1$ and $\|y - y^{\delta_n}\| = \delta_n < \|y^{\delta_n}\|$ (at least for sufficiently large n; we consider only those n from now on).

4. Continuous Regularization Methods

Furthermore, by definition of R_α,

$$x_{\alpha_n}^{\delta_n} - x^\dagger = (K^*K + \alpha_n I)^{-1} K^* u_1 + \delta_n (K^*K + \alpha_n I)^{-1} K^* u_n - \frac{v_1}{\sigma_1}$$

$$= \frac{\sigma_1}{\sigma_1^2 + \alpha_n} v_1 + \frac{\sigma_n \delta_n}{\sigma_n^2 + \alpha_n} v_n - \frac{v_1}{\sigma_1}.$$

Hence, if we choose $\delta_n = \sigma_n^2$, which is possible by assumption, for $n \geq 2$,

$$\|x_{\alpha_n}^{\delta_n} - x^\dagger\|^2 \geq \frac{\delta_n^3}{(\alpha_n + \delta_n)^2} = \left(\frac{\sqrt{\delta_n}}{1 + \frac{\alpha_n}{\delta_n}}\right)^2$$

holds. By (4.73), this implies

$$\frac{\sqrt{\delta_n}}{1 + \frac{\alpha_n}{\delta_n}} = o(\sqrt{\delta_n})$$

and hence

$$\lim_{n \to \infty} \frac{\alpha_n}{\delta_n} = +\infty. \tag{4.74}$$

On the other hand, by (4.58),

$$\|y^{\delta_n}\| - \tau\delta_n \leq \|y^{\delta_n}\| - \|Kx_{\alpha_n}^{\delta_n} - y^{\delta_n}\| \leq \|Kx_{\alpha_n}^{\delta_n}\|$$

and, since

$$Kx_{\alpha_n}^{\delta_n} = \frac{1}{\alpha_n} K[\alpha_n (K^*K + \alpha_n I)^{-1} K^* y^{\delta_n}]$$

$$= \frac{1}{\alpha_n} K[(\alpha_n I + K^*K) - K^*K](K^*K + \alpha_n I)^{-1} K^* y^{\delta_n}$$

$$= \frac{1}{\alpha_n} K[K^* y^{\delta_n} - K^* K x_{\alpha_n}^{\delta_n}] = \frac{1}{\alpha_n} KK^*[y^{\delta_n} - Kx_{\alpha_n}^{\delta_n}]$$

and hence

$$\|Kx_{\alpha_n}^{\delta_n}\| \leq \frac{\|K\|^2}{\alpha_n} \|Kx_{\alpha_n}^{\delta_n} - y^{\delta_n}\|$$

holds, we have

$$\|y^{\delta_n}\| - \tau\delta_n \leq \frac{\|K\|^2}{\alpha_n} \|Kx_{\alpha_n}^{\delta_n} - y^{\delta_n}\|.$$

By (4.58), this implies

$$\alpha_n \leq \frac{\|K\|^2 \tau \delta_n}{\|y^{\delta_n}\| - \tau\delta_n},$$

which, for n sufficiently large, contradicts (4.74). Hence, our assumption that $\dim \mathcal{R}(K) = \infty$ was wrong, which implies the statement of the proposition. ∎

Remark 4.21. As we have seen in Example 4.15, we have, for R_α as in Proposition 4.20, that (4.63) holds for $\mu_0 = 1$. Since, by Proposition 4.20, for (R_α, α) with

α chosen according to the discrepancy principle, $\|x^\delta_{\alpha(\delta,y^\delta)} - x^\dagger\|$ does not converge faster than $O(\sqrt{\delta})$ in general, unless $\mathcal{R}(K)$ is finite-dimensional (i.e., the problem of determining $x^\dagger$ is well-posed), (R_α, α) is *not* of optimal order in $\mathcal{X}_{\mu,\rho}$ for $\mu \in (1/2, 1]$. It is therefore of interest to look for another a-posteriori parameter choice strategy which is of optimal order in $\mathcal{X}_{\mu,\rho}$ for μ up to μ_0 (instead of $\mu_0 - 1/2$). Before turning to this question, we remark that for the regularization methods considered in Examples 4.7 and 4.8, (4.63) holds for all $\mu \in (0, +\infty)$, so that by Theorem 4.17, the discrepancy principle results in a regularization method that is of optimal order in $\mathcal{X}_{\mu,\rho}$ for all $\mu, \rho > 0$.

4.4. Improved A-posteriori Rules

We now present other a-posteriori parameter choice strategies which will turn out to be always of optimal order. We first give a heuristic motivation:

We start from the estimate (4.22) and square it to obtain

$$\|x^\delta_\alpha - x^\dagger\|^2 \leq 2(\|r_\alpha(T^*T)x^\dagger\|^2 + C\delta^2 G_\alpha). \tag{4.75}$$

Here, x^δ_α is defined by (4.4), C is as in Theorem 4.1, G_α is defined by (4.14) and r_α is as in (4.6). Minimizing the right-hand side of (4.75) should certainly yield a good choice for α. A necessary condition for such a minimum is

$$\frac{\partial}{\partial \alpha}(\|r_\alpha(T^*T)x^\dagger\|^2 + C\delta^2 G_\alpha) = 0. \tag{4.76}$$

We compute this derivative using the auxiliary function

$$f(\alpha, w) := 2\left(\frac{\partial G_\alpha}{\partial \alpha}\right)^{-1} \langle \frac{\partial g_\alpha}{\partial \alpha}(TT^*)r_\alpha(TT^*)Qw, Qw \rangle \tag{4.77}$$

defined for $\alpha \in \mathbb{R}^+$ and $w \in \mathcal{Y}$.

Lemma 4.22. *In addition to the assumptions made about g_α in Theorem 4.1, we assume that g_α and G_α are continuously differentiable with respect to α and that $\frac{\partial G_\alpha}{\partial \alpha} \neq 0$. Then, for $w \in \mathcal{D}(T^\dagger)$,*

$$\frac{\partial}{\partial \alpha}\|r_\alpha(T^*T)T^\dagger w\|^2 = -\frac{\partial G_\alpha}{\partial \alpha} f(\alpha, w).$$

Proof. Since differentiation with respect to α and integration with respect to $d\|E_\lambda T^\dagger w\|^2$ can be interchanged, the integrand being piecewise continuously differentiable, we have (since $\frac{\partial r_\alpha}{\partial \alpha} = -\lambda \frac{\partial g_\alpha}{\partial \alpha}$)

90   4. Continuous Regularization Methods

$$\frac{\partial}{\partial \alpha}\|r_\alpha(T^*T)T^\dagger w\|^2 = 2\langle r_\alpha(T^*T)T^\dagger w, \frac{\partial r_\alpha}{\partial \alpha}(T^*T)T^\dagger w\rangle$$

$$= 2\langle r_\alpha(T^*T)T^\dagger w, -T^*T\frac{\partial g_\alpha}{\partial \alpha}(T^*T)T^\dagger w\rangle$$

$$= -2\langle r_\alpha(TT^*)TT^\dagger w, \frac{\partial g_\alpha}{\partial \alpha}(TT^*)TT^\dagger w\rangle$$

$$= -\frac{\partial G_\alpha}{\partial \alpha} f(\alpha, w)$$

because of (2.13). ∎

Because of Lemma 4.22, the necessary condition (4.76) is equivalent to

$$f(\alpha, y) = C\delta^2. \tag{4.78}$$

If we could find a solution $\alpha = \alpha(\delta, y)$ of (4.78), this would (because of the argument leading to (4.76)) yield (at least under conditions on g_α that make (4.76) also a sufficient condition for an extremum) a good parameter choice strategy. However, (4.78) cannot be solved, because this equation cannot even be written down since it involves the unknown exact data y. The idea is now to replace y by y^δ in (a slightly changed version of) (4.78). We will analyze this strategy for regularization methods based on functions g_α fulfilling the following assumptions:

Assumption 4.23. Let g_α fulfill the assumptions made in Theorem 4.1 and assume, in addition, that $\alpha \mapsto g_\alpha$ and $\alpha \mapsto G_\alpha$ is continuously differentiable, that there exists a constant $K > 0$ such that

$$\left|\frac{\partial g_\alpha}{\partial \alpha}(\lambda)\left(\frac{\partial G_\alpha}{\partial \alpha}\right)^{-1}\right| \leq K \tag{4.79}$$

holds for all $\alpha > 0, \lambda \geq 0$, and that the function

$$\alpha \mapsto \left(\frac{\partial G_\alpha}{\partial \alpha}\right)^{-1}\frac{\partial g_\alpha}{\partial \alpha}(\lambda)r_\alpha(\lambda) \tag{4.80}$$

is strictly increasing for all $\lambda > 0$.

Lemma 4.24. *Under Assumption 4.23, the function f defined by (4.77) is, for $Qw \neq 0$, continuous and strictly increasing in α. Furthermore,*

$$\lim_{\alpha \to 0} f(\alpha, w) = 0 \tag{4.81}$$

and

$$\lim_{\alpha \to +\infty} f(\alpha, w) = h(w) \tag{4.82}$$

4.4. Improved A-posteriori Rules

with

$$h(w) := 2 \int \lim_{\alpha \to +\infty} \left[\left(\frac{\partial G_\alpha}{\partial \alpha}\right)^{-1} \frac{\partial g_\alpha}{\partial \alpha}(\lambda) r_\alpha(\lambda) \right] d\|F_\lambda Q w\|^2, \qquad (4.83)$$

where $\{F_\lambda\}$ is the spectral family generated by TT^*.

Proof. By (4.77),

$$f(\alpha, w) = 2 \int \left(\frac{\partial G_\alpha}{\partial \alpha}\right)^{-1} \frac{\partial g_\alpha}{\partial \alpha}(\lambda) r_\alpha(\lambda) \, d\|F_\lambda Q w\|^2, \qquad (4.84)$$

so that, by assumption, $f(\cdot, w)$ is continuous and strictly increasing. Since the integrand in (4.84) is bounded by $K(1 + C)$, (4.81) and (4.82) follow from the Dominated Convergence Theorem: note that since $\lim_{\alpha \to 0} r_\alpha(\lambda) = 0$ for $\lambda \neq 0$ and $\lim_{\alpha \to 0} r_\alpha(0) = 1$, the integral on the right-hand side of (4.84) tends to a multiple of $\|P_{\mathcal{N}(TT^*)} Q w\|^2$ as $\alpha \to 0$, where $P_{\mathcal{N}(TT^*)}$ is the orthogonal projector onto $\mathcal{N}(TT^*)$. Since $Qw \in \overline{\mathcal{R}(T)} = \mathcal{N}(TT^*)^\perp$, this expression vanishes. ∎

Proposition 4.25. *Under Assumption 4.23, for any $\delta > 0$ and $y^\delta \in \mathcal{Y}$ with $Q y^\delta \neq 0$, there is a unique $\alpha = \alpha(\delta, y^\delta) > 0$ such that*

$$f(\alpha, y^\delta) = \tau \delta^2 \qquad (4.85)$$

holds, provided that

$$\tau \in (0, h(y^\delta) \delta^{-2}) \qquad (4.86)$$

with h defined by (4.83).

Proof. The assertion follows from Lemma 4.24 with the Intermediate Value Theorem. ∎

Instead of the (non-implementable) parameter choice strategy (4.78), we choose the parameter according to the a-posteriori rule (4.85). Thus, the parameter choice rule is

$$2 \left(\frac{\partial G_\alpha}{\partial \alpha}\right)^{-1} \langle \frac{\partial g_\alpha}{\partial \alpha}(TT^*) r_\alpha(TT^*) Q y^\delta, Q y^\delta \rangle = \tau \delta^2$$

with τ as in (4.86). We show that this replacement of the unknown exact data y by the actual data y^δ does not change the asymptotic behaviour in the following sense:

Proposition 4.26. *With the assumptions and notations of Proposition 4.25, let*

$$L := 2 \sup\{ \left| \left[\frac{\partial G_\alpha}{\partial \alpha}\right]^{-1} \frac{\partial g_\alpha}{\partial \alpha}(\lambda) r_\alpha(\lambda) \right| \mid \alpha > 0, \lambda \geq 0 \}, \qquad (4.87)$$

and let $\tau > L, Q y \neq 0$. For $\delta > 0$, let $y^\delta \in \mathcal{Y}$ be such that $\|y^\delta - y\| \leq \delta$. Furthermore, let

$$\tau_1 := (\sqrt{\tau} - \sqrt{L})^2, \quad \tau_2 := (\sqrt{\tau} + \sqrt{L})^2.$$

92   4. Continuous Regularization Methods

If $Qy^\delta \neq 0$, (4.86) holds and $\alpha(\delta, y^\delta)$ is defined via (4.85), then there is a $\tau = \tau(\delta, y^\delta) \in [\tau_1, \tau_2]$ such that

$$f(\alpha(\delta, y^\delta), y) = \tau(\delta, y^\delta)\delta^2 \tag{4.88}$$

holds. (4.86) and $Qy^\delta \neq 0$ hold especially if

$$\delta^2 < \frac{h(y)}{\tau_2}. \tag{4.89}$$

Proof. By Assumption 4.23, the function in (4.80) is positive for all $\alpha > 0$ and $\lambda \geq 0$, so that the operator

$$H_\alpha := \left[2\left(\frac{\partial G_\alpha}{\partial \alpha}\right)^{-1} \frac{\partial g_\alpha}{\partial \alpha}(TT^*)r_\alpha(TT^*)\right]^{\frac{1}{2}}$$

is well-defined for all $\alpha > 0$. By definition of L,

$$\|H_\alpha\| \leq \sqrt{L} \tag{4.90}$$

holds. By (4.77), (2.45) and (2.46),

$$\sqrt{f(\alpha, w)} = \|H_\alpha Q w\| \tag{4.91}$$

holds. Hence, for all $\alpha > 0$,

$$\begin{aligned}
\sqrt{f(\alpha, y^\delta)} &= \|H_\alpha Q y - H_\alpha(Qy - Qy^\delta)\| \\
&\geq \|H_\alpha Q y\| - \|H_\alpha Q(y - y^\delta)\| \\
&\geq \sqrt{f(\alpha, y)} - \sqrt{L}\delta.
\end{aligned}$$

Because of (4.82), this implies

$$\sqrt{h(y^\delta)} > \sqrt{h(y)} - \sqrt{L}\delta. \tag{4.92}$$

Now, if (4.89) holds, (4.92) implies

$$\sqrt{h(y^\delta)} > \delta(\sqrt{\tau_2} - \sqrt{L}) = \delta\sqrt{\tau},$$

which implies (4.86) and also $Qy^\delta \neq 0$.

Now, assume that (4.86) holds, that $Qy^\delta \neq 0$, and that $\alpha = \alpha(\delta, y^\delta)$ is defined by (4.85). Then, by (4.90) and (4.91),

$$\begin{aligned}
\sqrt{f(\alpha, y)} &= \|H_\alpha Q y^\delta + H_\alpha(Qy - Qy^\delta)\| \leq \|H_\alpha Q y^\delta\| + \sqrt{L}\delta \\
&= \sqrt{f(\alpha, y^\delta)} + \sqrt{L}\delta = \sqrt{\tau}\delta + \sqrt{L}\delta = \sqrt{\tau_2}\delta
\end{aligned}$$

4.4. Improved A-posteriori Rules

and

$$\sqrt{f(\alpha,y)} = \|H_\alpha Qy^\delta - H_\alpha(Qy^\delta - Qy)\| \geq \|H_\alpha Qy^\delta\| - \sqrt{L}\delta$$
$$= \sqrt{f(\alpha,y^\delta)} - \sqrt{L}\delta = \sqrt{\tau}\delta - \sqrt{L}\delta = \sqrt{\tau_1}\delta,$$

which implies (4.88) with $\tau(\delta, y^\delta) \in [\tau_1, \tau_2]$. ∎

Comparing (4.88) and (4.78), we see that the solution of (4.85) can be bounded by the solutions of (4.78) with C replaced by τ_1 and τ_2, respectively. This leads us to expect that the regularized solution obtained for $\alpha(\delta, y^\delta)$ according to (4.85) has (up to constants) the same (optimal) convergence properties as that with α determined by (4.78), and that it should minimize the right-hand sides of (4.22) and of (4.75) with C replaced by $\tau(\delta, y^\delta)$:

Theorem 4.27. *Let Assumption 4.23 hold, let $y \in \mathcal{R}(T)$, $y \neq 0$, let $\tau > L$ (with L as in (4.87)), and let, for $\delta > 0$, $y^\delta \in \mathcal{Y}$ fulfill $\|y^\delta - y\| \leq \delta$ and (4.86). Then, if $\alpha = \alpha(\delta, y^\delta)$ is determined as the unique solution of (4.85), there is a constant η (independent of y, δ and y^δ) such that with x_α and x_α^δ defined by (4.3) and (4.4), respectively,*

$$\|x^\delta_{\alpha(\delta,y^\delta)} - T^\dagger y\| \leq \eta \inf\{\|x_\alpha - x^\dagger\| + \delta\sqrt{CG_\alpha} \mid \alpha > 0\} \tag{4.93}$$

holds.

Proof. Due to Propositions 4.25 and 4.26, (4.85) has a unique solution $\alpha(\delta, y^\delta)$. Now, it follows from Lemma 4.22 that (4.85) is the first-order necessary condition for minimizing the functional

$$\alpha \mapsto \|r_\alpha(T^*T)x^\dagger\|^2 + \tau(\delta, y^\delta)\delta^2 G_\alpha. \tag{4.94}$$

By the strict monotonicity of $f(\cdot, y)$, the functional in (4.94) is strictly decreasing (increasing) to the left (right) of $\alpha(\delta, y^\delta)$, so that it has $\alpha(\delta, y^\delta)$ as unique minimizer.

Thus, with $a := \max\{1, C/\tau(\delta, y^\delta)\}$ and $b := \max\{1, \tau(\delta, y^\delta)/C\}$, we obtain, using (4.75) and (4.85):

$$\frac{1}{2}\|x^\delta_{\alpha(\delta,y^\delta)} - x^\dagger\|^2 \leq \|r_{\alpha(\delta,y^\delta)}(T^*T)x^\dagger\|^2 + C\delta^2 G_{\alpha(\delta,y^\delta)}$$
$$\leq a[\|r_{\alpha(\delta,y^\delta)}(T^*T)x^\dagger\|^2 + \tau(\delta, y^\delta)\delta^2 G_{\alpha(\delta,y^\delta)}]$$
$$= a\min\{\|r_\alpha(T^*T)x^\dagger\|^2 + \tau(\delta, y^\delta)\delta^2 G_\alpha \mid \alpha > 0\}$$
$$\leq ab\inf\{\|r_\alpha(T^*T)x^\dagger\|^2 + C\delta^2 G_\alpha \mid \alpha > 0\}$$
$$\leq ab\inf\{\|r_\alpha(T^*T)x^\dagger\| + \delta\sqrt{CG_\alpha} \mid \alpha > 0\}^2,$$

so that, by (4.8), (4.93) holds (since $\tau(\delta, y^\delta) \in [\tau_1, \tau_2]$) with

$$\eta^2 := 2\max\{\frac{C}{\tau_1}, \frac{\tau_2}{C}\}. \quad \blacksquare$$

94   4. Continuous Regularization Methods

Remark 4.28. In (4.93), the actual error with the a-posteriori parameter choice rule (4.85) is compared to the estimate (4.22) for the error obtained with the parameter choice which minimizes this error estimate. However, since this error estimate is used to show that certain a-priori parameter choice strategies are of optimal order, we can use this comparison to show the same for the new a-posteriori strategy; however, this strategy is so far only defined for y^δ fulfilling (4.86). We define formally

$$\alpha(\delta, y^\delta) := +\infty, \quad x^\delta_{\alpha(\delta,y^\delta)} := 0 \quad \text{if} \quad h(y^\delta) \leq \tau\delta^2, \tag{4.95}$$

so that our new parameter choice strategy is now always well-defined.

Corollary 4.29. *Let the assumptions of Corollary 4.4 and Assumption 4.23 be fulfilled. Furthermore, let $\tau > L$ (where L is as in (4.87)), and for $\delta > 0$, $y^\delta \in \mathcal{Y}$ fulfill $\|y^\delta - y\| \leq \delta$. Let $\alpha = \alpha(\delta, y^\delta)$ be determined by (4.85) and (4.95). Then, the regularization method $(R_\alpha, \alpha(\delta, y^\delta))$ is of optimal order in $\mathcal{X}_{\mu,\rho}$, and convergent for all $y \in \mathcal{R}(T)$.*

Proof. It follows from the properties of f that $h(y) = 0$ if and only if $Qy = 0$. Hence, by Proposition 4.26, (4.86) holds for sufficiently small $\delta > 0$ as soon as $Qy \neq 0$. Thus, $\alpha(\delta, y^\delta)$ is not defined by (4.85), but by (4.95), also for sufficiently small $\delta > 0$ only if $Qy = 0$; but then, $x^\dagger = x_{\alpha(\delta, y^\delta)} = 0$. Since we want to prove an asymptotic statement, we can therefore assume that $\alpha(\delta, y^\delta)$ is always determined by (4.85), so that Theorem 4.27 is applicable.

For $x^\dagger \in \mathcal{X}_{\mu,\rho}$, let

$$\tilde{\alpha} \sim \left(\frac{\delta}{\rho}\right)^{\frac{2}{2\mu+1}}.$$

It follows from Theorems 4.2, 4.3 and Corollary 4.4 that $(R_\alpha, \tilde{\alpha})$ is of optimal order in $\mathcal{X}_{\mu,\rho}$ and also that

$$\|r_{\tilde{\alpha}}(T^*T)x^\dagger\| + \delta\sqrt{CG_{\tilde{\alpha}}} \sim \delta^{\frac{2\mu}{2\mu+1}} \rho^{\frac{1}{2\mu+1}}.$$

Hence, the right-hand side of (4.93) can be bounded by $\delta^{\frac{2\mu}{2\mu+1}} \rho^{\frac{1}{2\mu+1}}$, multiplied by a constant, which, due to (4.93), implies that

$$\|x^\delta_{\alpha(\delta,y^\delta)} - T^\dagger y\| = O(\delta^{\frac{2\mu}{2\mu+1}} \rho^{\frac{1}{2\mu+1}}),$$

i.e., according to Definition 3.17, $(R_\alpha, \alpha(\delta, y^\delta))$ is of optimal order in $\mathcal{X}_{\mu,\rho}$.

For arbitrary $y \in \mathcal{R}(T)$, convergence follows from Theorem 3.18. ∎

Remark 4.30. Thus, the parameter choice strategy (4.85) is of optimal order in $\mathcal{X}_{\mu,\rho}$ for all μ for which (4.63) holds. I.e., under the conditions of Theorem 4.17, where we could show optimal order in $\mathcal{X}_{\mu,\rho}$ for the discrepancy principle only for

$\mu \in (0, \mu_0 - 1/2]$, which cannot be improved in general due to Proposition 4.20, the parameter choice strategy (4.85) gives optimal order up to the qualification μ_0.

The conditions (4.86) and (4.89), together with the requirement that $\tau > L$, can be interpreted as "signal-to-noise ratio conditions", as will become clearer when we consider this parameter choice strategy for specific regularization methods.

As mentioned in Remark 4.28, the estimate (4.93) compares the actual error for $(R_\alpha, \alpha(\delta, y^\delta))$ with the best-possible estimate of the error. However, the following stronger result shows that the parameter choice strategy $\alpha(\delta, y^\delta)$ is (up to a multiplicative constant) asymptotically at least as good (always in the worst-case sense) as any other conceivable parameter choice strategy:

Theorem 4.31. *Let the assumptions of Theorem 4.27 hold and assume the following in addition: there are $\overline{\lambda} > 0, \overline{\alpha} > 0$ such that for $\lambda \in (0, \overline{\lambda}]$, $\alpha \mapsto g_\alpha(\lambda)$ is positive and decreasing, and that for $\alpha \in [0, \overline{\alpha}]$, g_α is decreasing on $(0, \overline{\lambda}]$. Furthermore, there are $C_1 > 0$ and a continuous increasing function $\overline{\alpha} : [0, \overline{\lambda}] \to [0, +\infty)$ with $\overline{\alpha}(0) = 0$ such that, for all $\lambda \in (0, \overline{\lambda}]$,*

$$\lambda g_{\overline{\alpha}(\lambda)}^2(\lambda) \geq C_1 G_{\overline{\alpha}(\lambda)}$$

holds. Then, if $\mathcal{R}(T)$ is non-closed and $Qy \neq 0$, the following statements hold:

(i) There is a sequence $\{\delta_k\} \to 0$ such that

$$\sup\{\|x^{\delta_k}_{\alpha(\delta_k, y_k)} - T^\dagger y\| \mid \|y_k - y\| \leq \delta_k\} =$$
$$O[\sup\{\inf_{\alpha > 0} \|x^{\delta_k}_\alpha - T^\dagger y\| \mid \|y_k - y\| \leq \delta_k\}].$$

(ii) If there is a sequence $\{\lambda_k\}$ in the spectrum of TT^* decreasing to 0 with

$$\sup\left\{\frac{\lambda_k}{\lambda_{k+1}} \mid k \in \mathbb{N}\right\} < \infty,$$

then

$$\sup\{\|x^\delta_{\alpha(\delta, y^\delta)} - T^\dagger y\| \mid \|y^\delta - y\| \leq \delta\} =$$
$$O[\sup\{\inf_{\alpha > 0} \|x^\delta_\alpha - T^\dagger y\| \mid \|y^\delta - y\| \leq \delta\}]$$

as $\delta \to 0$.

Proof. [70]. ∎

In concrete examples, we will also verify the conditions of Theorem 4.31. The condition on the spectrum in (ii) requires that the spectrum does not decay too fast, i.e., that the problem is not too ill-posed. It is, e.g., fulfilled if K is compact with singular values σ_n that do not decay faster than e^{-n}, but not if $\sigma_n \sim e^{-n^2}$.

96   4. Continuous Regularization Methods

In the next chapter, we will apply the new parameter choice strategy to Tikhonov regularization. We give another example here:

Example 4.32. We consider the method of Example 4.7. Let u_δ and x_α^δ be defined by (4.35) and (4.36), respectively. Since f as defined by (4.77) involves TT^* instead of T^*T, it is advisable to calculate u_δ equivalently via

$$\begin{aligned} v_\delta'(t) + TT^* v_\delta(t) &= Qy^\delta, \quad t \in \mathbb{R}_0^+, \\ v_\delta(0) &= 0, \\ u_\delta(t) &:= T^* v_\delta(t). \end{aligned} \qquad (4.96)$$

The assumptions of Theorems 4.27 and 4.31 are, as easy calculations show, fulfilled with $L = 2$, $K = 1$, $h(w) = 2\|Qw\|^2$; f as defined by (4.77) turns out to be

$$f(\alpha, w) = 2\|\exp\left(-\frac{TT^*}{\alpha}\right) Qw\|^2. \qquad (4.97)$$

Thus, the parameter choice strategy (4.85) is

$$2\|\exp\left(-\frac{TT^*}{\alpha}\right) Qy^\delta\|^2 = \tau \delta^2$$

and can be performed for any $\tau > 2$, provided that also (4.86), i.e., $\tau < 2\|Qy^\delta\|^2/\delta^2$ holds. Both conditions on τ can be fulfilled simultaneously if the signal-to-noise ratio condition

$$\delta^2 < \|Qy^\delta\|^2$$

is fulfilled, the range of admissible values for τ is then $(2, 2\|Qy^\delta\|^2/\delta^2)$.

It follows from the considerations in Example 4.7 that $v_\delta'(t) = \int e^{-\lambda t} dF_\lambda Qy^\delta$, where F_λ is the spectral family of TT^*, hence

$$\|\exp\left(-\frac{TT^*}{\alpha}\right) Qy^\delta\| = \|v_\delta'\left(\frac{1}{\alpha}\right)\|.$$

Thus, the parameter choice strategy (4.97) can also be written as

$$\|v_\delta'\left(\frac{1}{\alpha}\right)\|^2 = \frac{\tau}{2}\delta^2,$$

which, because of (4.96), is equivalent to $\|Tu_\delta(1/\alpha) - Qy^\delta\|^2 = \tau\delta^2/2$, i.e., by (4.36), to the discrepancy principle

$$\|Tx_\alpha^\delta - Qy^\delta\| = \sqrt{\frac{\tau}{2}}\delta$$

with $\tau > 2$.

If $Qy^\delta = y^\delta$ then the parameter choice rule that we have analyzed in Theorem 4.27 defines $\alpha(\delta, y^\delta)$ as a solution of the nonlinear equation

$$\eta_\alpha := \langle y^\delta, s_\alpha(TT^*)y^\delta \rangle = \tau\delta^2 \qquad (4.98)$$

(or $\alpha := +\infty$ if (4.98) has no solution), where

$$s_\alpha(\lambda) = \left(\frac{\partial G_\alpha}{\partial \alpha}\right)^{-1} \frac{\partial g_\alpha}{\partial \alpha}(\lambda) r_\alpha(\lambda), \tag{4.99}$$

cf. (4.85), and

$$\tau > \gamma := \sup_{\alpha,\lambda > 0} |s_\alpha(\lambda)|. \tag{4.100}$$

Consequently, in order for (4.99) to make sense, we need to impose a differentiability assumption on g_α and G_α. As a matter of fact, this parameter choice rule does not apply, e.g., to the truncated singular value expansion (cf. Example 4.8) since the corresponding functions $g_\alpha(\lambda)$ are not even continuous with respect to α.

The discrepancy principle is defined in a similar form, namely

$$\alpha(\delta, y^\delta) := \sup\{\alpha > 0 \mid \eta_\alpha \leq \tau \delta^2\}, \tag{4.101}$$

where η_α is given by (4.98) with

$$s_\alpha(\lambda) := r_\alpha^2(\lambda), \tag{4.102}$$

and τ has to fulfill (4.100). The somewhat more difficult formulation (4.101) is due to the fact that we do not want to impose more regularity on $r_\alpha(\lambda)$ than continuity from the left with respect to α. Note that for the method in Example 4.32 above (4.99) and (4.102) are equivalent.

We may now ask which property of these families $\{s_\alpha\}$ is responsible for the fact that we obtain order-optimal accuracy for the best possible range only with s_α defined as in (4.99), and for a limited range with s_α from (4.102). This is the topic of the next theorem. It will provide a deeper understanding of what makes the method (4.85) work better, and this understanding will later on (in Section 6.3) allow the construction of order-optimal parameter choice rules when the assumptions of Corollary 4.29 are not fulfilled.

Theorem 4.33. *Let the assumptions of Theorem 4.3 hold with*

$$\omega_\mu(\alpha) = c\alpha^\mu, \quad 0 < \mu \leq \mu_0 < \infty, \tag{4.103}$$

and some $c > 0$, and let G_α as defined in (4.14) fulfill

$$G_\alpha = O\left(\frac{1}{\alpha}\right). \tag{4.104}$$

Let $0 < \nu \leq \mu_0$, and $\{s_\alpha\}$ be a family of positive and piecewise continuous functions with

$$s_\alpha(\lambda) \sim \left(\frac{\alpha}{\lambda + \alpha}\right)^{2\nu+1}, \tag{4.105}$$

uniformly for $\lambda \in [0, \|T\|^2]$ and $\alpha \in \mathbb{R}^+$. Furthermore, assume that $s_\alpha(\lambda)$ is continuous from the left with respect to α. Then the parameter choice (4.101) with η_α

98   4. Continuous Regularization Methods

defined by (4.98) and τ subject to (4.100) is order-optimal in $\mathcal{X}_{\mu,\rho}$ for all $\rho > 0$ and $0 < \mu \leq \nu$.

Proof. The proof follows the line of argument of the proof of Theorem 4.17. First, we consider the case where for some δ and y^δ we have $\alpha(\delta, y^\delta) = +\infty$, i.e., $\eta_\alpha \leq \tau \delta^2$ for all $\alpha > 0$. Since (4.105) implies that $s_\alpha(\lambda)$ is bounded from below by some number $\varepsilon > 0$ uniformly for $\lambda \in [0, \|T\|^2]$ and $\alpha \geq 1$, it follows that

$$\eta_\alpha = \langle y^\delta, s_\alpha(TT^*)y^\delta \rangle \geq \varepsilon \|y^\delta\|^2, \qquad \alpha \geq 1,$$

and hence we have

$$\tau \delta^2 \geq \limsup_{\alpha \to \infty} \eta_\alpha \geq \varepsilon \|y^\delta\|^2.$$

Thus, it follows from $\|y^\delta - y\| \leq \delta$ that $y = 0$ if there is a sequence $\{(\delta_n, y^{\delta_n})\}$ with $\delta_n \to 0$, and $\alpha(\delta_n, y^{\delta_n}) = +\infty$ for all $n \in \mathbb{N}$. Since we have $x^{\delta_n}_{\alpha(\delta_n, y^{\delta_n})} = x^{\delta_n}_\infty = 0 = x^\dagger$ in this case, nothing remains to be shown.

In the sequel, we can therefore restrict our attention to the case $\alpha(\delta, y^\delta) < +\infty$ for all δ sufficiently small. Let $T^\dagger y = (T^*T)^\mu w$ with $\|w\| \leq \rho$ and $0 < \mu \leq \nu$. By (4.24) and assumption (4.103) we have – uniformly in $\alpha > 0$ –

$$|r_\alpha(\lambda)| \leq c \left(\frac{\lambda}{\alpha}\right)^{-\nu}, \qquad \lambda \in [0, \|T\|^2],$$

and because of the uniform boundedness of $\lambda g_\alpha(\lambda) = 1 - r_\alpha(\lambda)$, we also have

$$|r_\alpha(\lambda)| \leq c \qquad \text{for} \quad \lambda \in [0, \|T\|^2], \ \alpha > 0.$$

Thus, we may conclude from (4.105) that there is some other constant $c > 0$ with

$$|r_\alpha(\lambda)| \leq c\, s_\alpha^{\frac{\nu}{2\nu+1}}(\lambda), \qquad (4.106)$$

uniformly for $\lambda \in [0, \|T\|^2]$ and $\alpha > 0$. It follows that

$$\|x_\alpha - x^\dagger\| = \|r_\alpha(T^*T)(T^*T)^\mu w\| \leq c\|s_\alpha^{\frac{\nu}{2\nu+1}}(T^*T)(T^*T)^\mu w\|.$$

Denote by S the linear operator

$$S := s_\alpha^{\frac{1}{2\nu+1}}(T^*T)T^*T,$$

and consider the interpolation inequality (2.49) applied to

$$z := s_\alpha^{\frac{\nu-\mu}{2\nu+1}}(T^*T)w.$$

This yields

$$\begin{aligned}\|x_\alpha - x^\dagger\| &\leq c\|S^\mu z\| \\ &\leq c\|z\|^{\frac{1}{2\mu+1}} \|s_\alpha^{\frac{\mu+\frac{1}{2}+\nu-\mu}{2\nu+1}}(T^*T)(T^*T)^{\mu+\frac{1}{2}} w\|^{\frac{2\mu}{2\mu+1}} \quad (4.107) \\ &= c\|z\|^{\frac{1}{2\mu+1}} \|s_\alpha^{\frac{1}{2}}(TT^*)y\|^{\frac{2\mu}{2\mu+1}}.\end{aligned}$$

4.4. Improved A-posteriori Rules

Since $\{s_\alpha\}$ is uniformly bounded (cf. (4.105)) we have $\|z\| \leq c\rho$. Furthermore, with γ as in (4.100),
$$\|s_\alpha^{\frac{1}{2}}(TT^*)(y^\delta - y)\| \leq \sqrt{\gamma}\delta, \tag{4.108}$$
and hence, for $\alpha = \alpha(\delta, y^\delta)$,
$$\|s_\alpha^{\frac{1}{2}}(TT^*)y\|^2 \leq 2\eta_\alpha + 2\|s_\alpha^{\frac{1}{2}}(TT^*)(y^\delta - y)\|^2 \leq 2(\tau + \gamma)\delta.$$

Inserting these estimates into (4.107) we finally obtain
$$\|x_\alpha - x^\dagger\| \leq c\delta^{\frac{2\mu}{2\mu+1}}\rho^{\frac{1}{2\mu+1}} \quad \text{for} \quad \alpha = \alpha(\delta, y^\delta). \tag{4.109}$$

We now proceed by estimating $\|x_\alpha^\delta - x_\alpha\|$. By virtue of (4.16) and (4.104), this requires estimating $\alpha = \alpha(\delta, y^\delta)$ from below. From its definition (4.101), it follows that $\eta_{2\alpha} > \tau\delta^2$, and hence (4.108) and the triangle inequality yield
$$\|s_{2\alpha}^{\frac{1}{2}}(TT^*)y\| \geq \eta_{2\alpha}^{\frac{1}{2}} - \sqrt{\gamma}\delta \geq (\sqrt{\tau} - \sqrt{\gamma})\delta. \tag{4.110}$$

On the other hand, (4.105) implies that
$$\lambda^{2\mu+1}s_{2\alpha}(\lambda) \leq c\lambda^{2\mu+1}\left(\frac{2\alpha}{\lambda + 2\alpha}\right)^{2\nu+1} \leq c\lambda^{2\mu+1}\left(\frac{2\alpha}{\lambda + 2\alpha}\right)^{2\mu+1},$$
since $\mu \leq \nu$ by assumption. Therefore, we obtain
$$\lambda^{2\mu+1}s_{2\alpha}(\lambda) \leq c\left(\frac{2\alpha\lambda}{\lambda + 2\alpha}\right)^{2\mu+1} \leq c(2\alpha)^{2\mu+1},$$
uniformly for all $\lambda \in [0, \|T\|^2]$. Consequently, we have
$$\|s_{2\alpha}^{\frac{1}{2}}(TT^*)y\| = \|s_{2\alpha}^{\frac{1}{2}}(T^*T)(T^*T)^{\mu+\frac{1}{2}}w\| \leq c\alpha^{\mu+\frac{1}{2}}\rho. \tag{4.111}$$

Combining (4.110) and (4.111) yields
$$\delta = O(\alpha^{\mu+\frac{1}{2}}\rho),$$
and hence it follows from (4.16) that
$$\|x_\alpha - x_\alpha^\delta\| = O(\rho^{\frac{1}{2\mu+1}}\delta^{\frac{2\mu}{2\mu+1}}).$$

This, together with (4.109), implies the order-optimality of the parameter choice rule for $x^\dagger \in \mathcal{X}_{\mu,\rho}$. ∎

The crucial assumption of Theorem 4.33 is the asymptotic behaviour (4.105) of the functions $s_\alpha(\lambda)$. It guarantees, for example (cf. (4.106)), that
$$|r_\alpha(\lambda)| \leq c\, s_\alpha^{\frac{\nu}{2\nu+1}}(\lambda).$$

For $\{s_\alpha\}$ of (4.102) – the discrepancy principle – this only holds for $\nu \to +\infty$, which in turn requires $\mu_0 = +\infty$ because of the restriction $\nu \le \mu_0$. Concerning a regularization method with finite qualification $\mu_0 < \infty$, the assumptions of Theorem 4.33 are fulfilled for the discrepancy principle with equality in (4.105) if $\nu = \mu_0 - 1/2$. Thus, Theorem 4.33 contains Theorem 4.17 as a special case. From this, one can understand why the discrepancy principle fails to be order-optimal in $\mathcal{X}_{\mu_0}$: the reason is that the functions $s_\alpha = r_\alpha^2$ converge too slowly to zero as $\alpha \to 0$. The correct behaviour would be matched by the functions

$$s_\alpha(\lambda) = r_\alpha^{2+\frac{1}{\mu_0}}(\lambda). \tag{4.112}$$

The choice (4.112) for $\{s_\alpha\}$ has been proposed by Raus [233, 234] as a general rule for defining the regularization parameter from (4.101). As we will see in Section 5.1, for Tikhonov regularization the choice (4.112) essentially coincides with (4.99).

4.5. Heuristic Parameter Choice Rules

In the previous sections we have studied a number of a-posteriori parameter choice rules which all depend in one way or the other on the computed approximation – and on the given data error level δ. A perfect example to illustrate this general reasoning is the discrepancy principle (4.57) where reconstructions are discarded unless their data fit has the order of the noise level δ.

In real world examples such noise level information is not always available (or reliable). For instance, a given discrete data vector may consist of a finite number of measurements, for each of which we may or may not know the standard deviation and/or a worst-case error bound. Typically, the worst-case bound will be a severe overestimation, while the standard deviation might underestimate the true error. Both estimates may therefore lead to a significant loss of accuracy when used in these parameter choice rules. Another uncertainty problem arises if we are going to embed the discrete data into a continuous model by some interpolation or approximation process. Then we have to estimate the $\mathcal{L}^2$-norm of the difference between the constructed function and the true data function from the discrete noise information, and from a-priori assumed smoothness properties of the data.

Often it is therefore necessary to consider alternative (a-posteriori) parameter choice rules that avoid knowledge of the noise level, and to determine some realistic regularization parameter on the basis of the actual performance of the regularization method under consideration. Such heuristic parameter choice rules will be called *error free* further on. It must be emphasized, however, that in view of Theorem 3.3 error free parameter choice rules cannot provide a convergent regularization method in the strict sense of Definition 3.1. Still, there are examples where an error free rule leads to better reconstructions than some sophisticated order-optimal rule, cf., e.g., [117] for some numerical comparisons.

4.5. Heuristic Parameter Choice Rules

Most heuristic arguments for error free parameter choice rules are based on some kind of error estimation. In order to explain this general approach (for simplicity, under the assumption that $y = Qy$) we consider once again the discrepancy principle. There one monitors the norm of the residual

$$y^\delta - Tx_\alpha^\delta = r_\alpha(TT^*)y^\delta = r_\alpha(TT^*)Tx^\dagger + r_\alpha(TT^*)(y^\delta - y).$$

How does this relate to the actual error

$$x^\dagger - x_\alpha^\delta = r_\alpha(T^*T)x^\dagger + g_\alpha(T^*T)T^*(y - y^\delta)?$$

The above representations split into two components, corresponding to the exact solution and the data error, respectively. Concerning the former, the results from Section 4.1 readily lead to upper bounds provided $x^\dagger \in \mathcal{X}_{\mu,\rho}$:

$$\|r_\alpha(TT^*)Tx^\dagger\| \leq \rho\omega_{\mu+\frac{1}{2}}(\alpha), \qquad \|r_\alpha(T^*T)x^\dagger\| \leq \rho\omega_\mu(\alpha). \tag{4.113}$$

Assuming that the qualification of the regularization method is $\mu_0 = \infty$, and

$$\omega_\mu(\alpha) = O(\alpha^\mu), \qquad 0 \leq \mu < \infty, \tag{4.114}$$

we conclude that the upper bound for $\|r_\alpha(TT^*)Tx^\dagger\|$ decays faster by a factor of $\sqrt{\alpha}$ as $\alpha \to 0$; this extra factor is *independent* of the actual value μ in (4.113). Although (4.113) only provides upper bounds for the actual quantities, these bounds nevertheless illustrate pretty much what is really going on, because we know that up to a very small gap there is almost a one-to-one relation between the smoothness parameter μ and the decay rate of $\|r_\alpha(T^*T)x^\dagger\|$ (cf. Section 4.2).

Thus, even if we do not know the true value of μ, we can still expect that

$$\frac{1}{\sqrt{\alpha}} \|r_\alpha(TT^*)Tx^\dagger\| \sim \|r_\alpha(T^*T)x^\dagger\|.$$

It is therefore reasonable to rescale the residual by $1/\sqrt{\alpha}$, and to further investigate whether $\|y^\delta - Tx_\alpha^\delta\|/\sqrt{\alpha}$ can be used for error estimation.

To this end we also have to compare the propagated noise components of the rescaled residual and the approximation error, i.e.,

$$\frac{1}{\sqrt{\alpha}} r_\alpha(TT^*)(y^\delta - y) \quad \text{vs.} \quad g_\alpha(T^*T)T^*(y^\delta - y).$$

In view of the spectral theory we start with a qualitative comparison of the functions $r_\alpha(\lambda)/\sqrt{\alpha}$ and $\sqrt{\lambda}g_\alpha(\lambda)$. Figure 4.1 shows the two functions for the Showalter method (Example 4.7) and $\alpha = 0.05$.

It is obvious that the two functions look different: since $g_\alpha(\lambda) \approx 1/\lambda$ for larger values of λ and since there exists $c > 0$ such that $g_\alpha(\lambda) \geq c/\lambda$ for $\lambda \geq c\sqrt{\alpha}$, by virtue of (4.114), we have

$$\sqrt{\lambda}g_\alpha(\lambda) \approx \begin{cases} 0, & \lambda \ll \alpha, \\ \dfrac{1}{\sqrt{\alpha}}, & \lambda \sim \alpha, \\ \dfrac{1}{\sqrt{\lambda}}, & \lambda \gg \alpha, \end{cases}$$

102   4. Continuous Regularization Methods

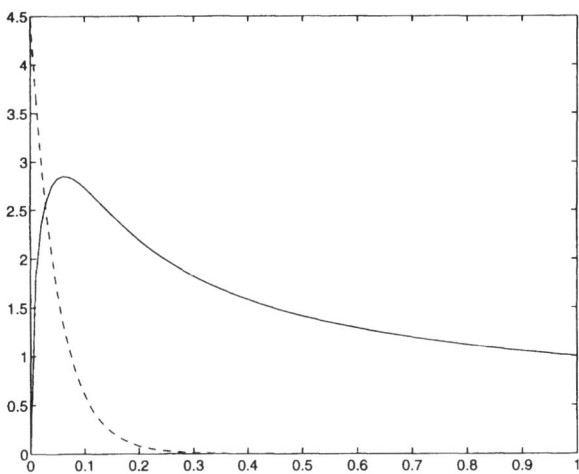

Figure 4.1: The functions $\lambda^{\frac{1}{2}} g_\alpha(\lambda)$ (solid) and $\alpha^{-\frac{1}{2}} r_\alpha(\lambda)$ (dashed).

while

$$\frac{1}{\sqrt{\alpha}} r_\alpha(\lambda) \approx \begin{cases} \dfrac{1}{\sqrt{\alpha}}, & \lambda \leq \alpha, \\ 0, & \lambda \gg \alpha. \end{cases}$$

(The latter follows from (4.114), since $|r_\alpha(\lambda)| < c_n (\alpha/\lambda)^n$ for all $n \in \mathbb{N}$).

We conclude that, although the two functions are quite different for larger values of λ, they are pretty similar for $\lambda \sim \alpha$ corresponding to high frequency noise components. Moreover, if $G_\alpha = O(\alpha^{-1})$ then the overall maxima of the two functions are both of the order of $1/\sqrt{\alpha}$, and are both attained for $\lambda \sim \alpha$.

To some extent it therefore seems to make sense to take $\|y^\delta - Tx_\alpha^\delta\|/\sqrt{\alpha}$ as error estimator, and to pick the regularization parameter as the one for which

$$\frac{1}{\sqrt{\alpha}} \|y^\delta - Tx_\alpha^\delta\| \to \min . \qquad (4.115)$$

The following result provides justification for this approach:

Theorem 4.34. *Assume that the functions $\{r_\alpha\}$ and $\{g_\alpha\}$ satisfy the assumptions of Theorem 4.1, and that*

$$\omega_\mu(\alpha) = O(\alpha^\mu), \quad 0 < \mu \leq \mu_0, \qquad G_\alpha = O\left(\frac{1}{\alpha}\right).$$

Without loss of generality, let $\delta = \|y^\delta - y\| < \|y\|$, and assume that $x^\dagger \in \mathcal{X}_{\mu,\rho}$ for some $0 < \mu \leq \mu_0 - 1/2$. If the global minimizer $\alpha(y^\delta)$ of (4.115) over $\alpha \in (0, \|T\|^2]$ exists, and if

$$\delta_* := \|y^\delta - Tx_{\alpha(y^\delta)}^\delta\| \neq 0,$$

4.5. Heuristic Parameter Choice Rules

then
$$\|x^\dagger - x^\delta_{\alpha(y^\delta)}\| \leq c(1 + \frac{\delta}{\delta_*})\max\{\delta, \delta_*\}^{\frac{2\mu}{2\mu+1}} \rho^{\frac{1}{2\mu+1}}. \tag{4.116}$$

Proof. As before we write
$$x^\dagger - x^\delta_\alpha = r_\alpha(T^*T)x^\dagger + g_\alpha(T^*T)T^*(y - y^\delta). \tag{4.117}$$

Introducing w with $x^\dagger = (T^*T)^\mu w$ the interpolation inequality yields

$$\begin{aligned}\|r_\alpha(T^*T)x^\dagger\| &\leq \|r_\alpha(TT^*)Tx^\dagger\|^{\frac{2\mu}{2\mu+1}} \|r_\alpha(T^*T)w\|^{\frac{1}{2\mu+1}} \\ &\leq c\bigl(\|r_\alpha(TT^*)y^\delta\| + \|r_\alpha(TT^*)(y^\delta - y)\|\bigr)^{\frac{2\mu}{2\mu+1}} \|w\|^{\frac{1}{2\mu+1}},\end{aligned}$$

where c depends only on the maximum of r_α over $[0, \|T\|^2]$. Since $r_\alpha(TT^*)y^\delta = y^\delta - Tx^\delta_\alpha$ we obtain with $\alpha = \alpha(y^\delta)$:

$$\|r_{\alpha(y^\delta)}(T^*T)x^\dagger\| \leq c(\delta_* + c\delta)^{\frac{2\mu}{2\mu+1}} \rho^{\frac{1}{2\mu+1}} \leq c_1 \max\{\delta_*, \delta\}^{\frac{2\mu}{2\mu+1}} \rho^{\frac{1}{2\mu+1}}. \tag{4.118}$$

It remains to estimate $\|g_{\alpha(y^\delta)}(T^*T)T^*(y^\delta - y)\|$. The standard estimate (4.16) of Theorem 4.2 yields

$$\|g_{\alpha(y^\delta)}(T^*T)T^*(y^\delta - y)\| \leq c\frac{\delta}{\sqrt{\alpha(y^\delta)}} = c\frac{\delta}{\delta_*}\frac{\|y^\delta - Tx^\delta_{\alpha(y^\delta)}\|}{\sqrt{\alpha(y^\delta)}}. \tag{4.119}$$

Now $\alpha(y^\delta)$ is defined as the minimizer of the fraction on the right-hand side over all $\alpha \in (0, \|T\|^2]$, and hence,

$$\frac{\|y^\delta - Tx^\delta_{\alpha(y^\delta)}\|}{\sqrt{\alpha(y^\delta)}} \leq \frac{\|y^\delta - Tx^\delta_{\alpha(\delta,y^\delta)}\|}{\sqrt{\alpha(\delta, y^\delta)}},$$

where $\alpha(\delta, y^\delta)$ is the parameter chosen by the discrepancy principle (4.57) (with some $\tau > 1$), provided $\alpha(\delta, y^\delta) \leq \|T\|^2$. In this case, a lower bound for $\alpha(\delta, y^\delta)$ has been derived in (4.71) which yields

$$\frac{\|y^\delta - Tx^\delta_{\alpha(y^\delta)}\|}{\sqrt{\alpha(y^\delta)}} \leq \frac{\tau\delta}{c\delta^{\frac{1}{2\mu+1}}\rho^{\frac{-1}{2\mu+1}}} = \frac{\tau}{c}\delta^{\frac{2\mu}{2\mu+1}}\rho^{\frac{1}{2\mu+1}}. \tag{4.120}$$

In the second case, where the parameter $\alpha(\delta, y^\delta)$ of the discrepancy principle is greater than $\|T\|^2$, we have $\|y^\delta - Tx^\delta_{\|T\|^2}\| \leq \tau\delta$ by (4.57). Since, by assumption,

$$\delta \leq \|y\| \leq \|T(T^*T)^\mu w\| \leq \|T\|^{2\mu+1}\rho,$$

we may use $\alpha = \|T\|^2$ to obtain the following inequality from the minimization property (4.115):

$$\frac{\|y^\delta - Tx^\delta_{\alpha(y^\delta)}\|}{\sqrt{\alpha(y^\delta)}} \leq \|T\|^{-1}\tau\delta \leq \|T\|^{-1}\tau\delta^{\frac{2\mu}{2\mu+1}}(\|T\|^{2\mu+1}\rho)^{\frac{1}{2\mu+1}} = \tau\delta^{\frac{2\mu}{2\mu+1}}\rho^{\frac{1}{2\mu+1}}.$$

4. Continuous Regularization Methods

Thus, (4.120) holds in either case, and inserting (4.120) into (4.119) we obtain

$$\|g_{\alpha(y^\delta)}(T^*T)T^*(y^\delta - y)\| \leq c\frac{\delta}{\delta_*}\delta^{\frac{2\mu}{2\mu+1}}\rho^{\frac{1}{2\mu+1}}.$$

Together with (4.117) and (4.118) the assertion follows. ∎

Several comments are in order. First, ignoring all the technical assumptions, the error bound (4.116) states that the approximation obtained from this heuristic parameter choice rule is order-optimal provided δ_* has about the order of δ. Note that $\alpha(y^\delta)$ is essentially the regularization parameter $\alpha(\delta_*/\tau, y^\delta)$ that would be obtained from the discrepancy principle with noise level information δ_*/τ (it need not fulfill the *sup* condition in (4.57), though). On the other hand, the knowledge of δ_* gives rise to the following a-posteriori check of this parameter choice rule: if $\delta_* \ll \delta$ (or what is believed to be δ), then one should be very cautious about the chosen parameter, since the factor δ/δ_* is large; if $\delta_* \gg \delta$, on the other hand, the situation is not critical and the magnitude of δ_* essentially determines the error.

As indicated by the assumptions of the theorem, the parameter choice rule fails if $\delta_* = 0$ or $\alpha(y^\delta) = 0$. When these degenerate cases occur we have $y^\delta \in \mathcal{D}(T^\dagger)$, and the error is given by $T^\dagger y^\delta - T^\dagger y$ which may be arbitrarily large (or small). Similarly, the assumption $\|y\| > \|y^\delta - y\|$ is the least we need for approximation purposes since otherwise the signal y is completely hidden by noise.

Finally, we emphasize the (already familiar) difficulties that originate from looking at the residual norm – as is done, for instance, in the discrepancy principle – for a regularization method with finite qualification μ_0. As in Theorem 4.17 error estimates with order-optimal exponent of δ can only be obtained if $x^\dagger \in \mathcal{X}_{\mu_0 - \frac{1}{2}}$. If $\mu_0 = \infty$, then there are no such problems. This drawback can be fixed in the same way as in Section 4.4 if instead of $\|y^\delta - Tx_\alpha^\delta\|$ some other quantity is employed in the parameter choice rule, e.g., if

$$\eta_\alpha = \langle y^\delta, s_\alpha(TT^*)y^\delta \rangle \tag{4.121}$$

of (4.98) is considered with s_α either as in (4.99), (4.112), or any other function s_α satisfying (4.105). In fact, it is no problem to combine the techniques in the proofs of Theorem 4.33 and Theorem 4.34 above to prove the following result from [121]:

Theorem 4.35. *Under the assumptions of Theorem 4.33, let η_α be defined by (4.121), and let $\alpha(y^\delta)$ be a minimizer of η_α/α in $(0, \|T\|^2)$. Furthermore, let $\delta = \|y^\delta - y\| < \|y\|$ and $x^\dagger \in \mathcal{X}_{\mu,\rho}$ for some $0 < \mu \leq \mu_0$. If*

$$\delta_* := \eta_{\alpha(y^\delta)}^{\frac{1}{2}} > 0,$$

then

$$\|x^\dagger - x_{\alpha(y^\delta)}^\delta\| \leq c(1 + \frac{\delta}{\delta_*})\max\{\delta, \delta_*\}^{\frac{2\mu}{2\mu+1}}\rho^{\frac{1}{2\mu+1}}.$$

4.5. Heuristic Parameter Choice Rules

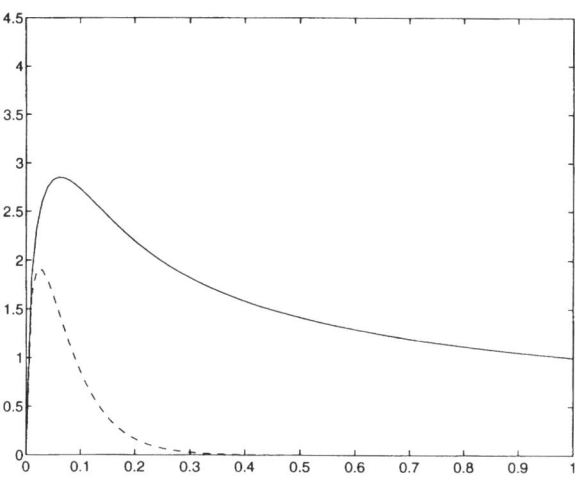

Figure 4.2: The functions $\lambda^{\frac{1}{2}} g_\alpha(\lambda)$ (solid) and $\frac{\lambda^{\frac{1}{2}}}{\alpha} r_\alpha(\lambda)$ (dashed).

Remark 4.36. Instead of looking at the usual residual norm as in Theorem 4.34 one could also look at the norm of the residual with respect to the normal equations, i.e.,

$$T^*(y^\delta - Tx_\alpha^\delta) = r_\alpha(T^*T)T^*Tx^\dagger + r_\alpha(T^*T)T^*(y^\delta - y).$$

Similar considerations as above suggest that under the same assumptions as before

$$\frac{1}{\alpha} \|T^*(y^\delta - Tx_\alpha^\delta)\|$$

may be a useful error estimator if $\mu_0 = \infty$. In fact, because of its root at the origin the function $\sqrt{\lambda} r_\alpha(\lambda)/\alpha$ resembles $\sqrt{\lambda} g_\alpha(\lambda)$ better than $r_\alpha(\lambda)/\alpha$ did before, cf. Figure 4.2. For this variant, however, no error bounds like (4.116) are known so far.

Another heuristic parameter choice rule which can be interpreted via some kind of error estimation is the method of *generalized cross-validation* introduced by Wahba (cf. [286] for the history of this method and a more detailed exposition). It applies to problems where $T = K$ is an operator into a finite-dimensional data space, e.g., a generalized moment problem (cf. Example 3.25)

$$(Kx)(s_i) = \int_\Omega k(s_i, t) x(t) \, dt = y(s_i) =: y_i, \qquad i = 1, \ldots, m.$$

To emphasize this point we shall further on write

$$\mathbf{y} = [y_1, \ldots, y_m]^T$$

106   4. Continuous Regularization Methods

for the entire data vector, and similarly, $\mathbf{y}^\delta$ for its perturbation.

The finite-dimensionality of $\mathcal{R}(K)$ is indeed an essential restriction, since the foundation of generalized cross-validation originates from statistical considerations, and strongly depends on the assumption that the data perturbation $\mathbf{y}^\delta - \mathbf{y}$ is discrete white noise, i.e.,

$$E[\mathbf{y}^\delta - \mathbf{y}] = 0 \quad \text{and} \quad E[(\mathbf{y}^\delta - \mathbf{y})(\mathbf{y}^\delta - \mathbf{y})^T] = \sigma^2 I, \tag{4.122}$$

where $E[\cdot]$ denotes the expectation. Note that this implies that

$$E[\,\|\mathbf{y}^\delta - \mathbf{y}\|^2\,] = m\sigma^2,$$

and hence, we make the identification

$$\delta = \sqrt{m}\,\sigma.$$

The advantage of the white noise assumption is that it allows for a more sophisticated analysis of the propagated data error. With the "deterministic approach" to the error that we have used so far in this book we can only estimate

$$\|x_\alpha^\delta - x_\alpha\|^2 = \|g_\alpha(K^*K)K^*(\mathbf{y}^\delta - \mathbf{y})\|^2 \leq \|\sqrt{\lambda}g_\alpha(\lambda)\|^2_{C[0,\|T\|^2]}\delta^2. \tag{4.123}$$

This (worst case) upper bound is only sharp if the data error is concentrated near the location of the maximum of the function $|\sqrt{\lambda}g_\alpha(\lambda)|$; typically, this maximum is attained near $\lambda \approx \alpha$ (cf. Figure 4.2), which varies with α. Hence, the estimate (4.123) will certainly not *always* be sharp. Using the white noise assumptions (4.122) we obtain (cf., e.g., [110])

$$\begin{aligned}
E[\,\|x_\alpha^\delta - x_\alpha\|^2\,] &= E[\,\langle \mathbf{y}^\delta - \mathbf{y}, g_\alpha^2(KK^*)KK^*(\mathbf{y}^\delta - \mathbf{y})\rangle\,] \\
&= \sigma^2 \,\text{trace}\{g_\alpha^2(KK^*)KK^*\} \tag{4.124} \\
&= \delta^2 \,\text{trace}\{\frac{1}{m} g_\alpha^2(KK^*)KK^*\}.
\end{aligned}$$

The disadvantage with (4.124) is that an analytical treatment of the trace term expression requires detailed information about the location of the singular values of K.

In generalized cross-validation the regularization parameter α is taken as the minimizer of the functional

$$V(\alpha) = \Big(\frac{\|\mathbf{y}^\delta - Kx_\alpha^\delta\|}{\text{trace}\{\frac{1}{m} r_\alpha(KK^*)\}}\Big)^2. \tag{4.125}$$

In contrast to the method of the previous section this cross-validation functional is not considered an error estimate, but rather an estimator of the so-called *predictive mean-square error*

$$T(\alpha) = \|\mathbf{y} - Kx_\alpha^\delta\|^2,$$

4.5. Heuristic Parameter Choice Rules

i.e., the residual with respect to the (unknown) *true data* **y**. Note that

$$\mathbf{y} - Kx_\alpha^\delta = \mathbf{y} - Kx_\alpha - g_\alpha(KK^*)KK^*(\mathbf{y}^\delta - \mathbf{y}), \qquad (4.126)$$

whereas

$$\mathbf{y}^\delta - Kx_\alpha^\delta = \mathbf{y} - Kx_\alpha + r_\alpha(KK^*)(\mathbf{y}^\delta - \mathbf{y}). \qquad (4.127)$$

The subtle influence of the trace term in the denominator of (4.125) is now most easily illustrated by looking at a particular regularization method, namely the truncated singular value expansion (cf. Example 4.8). Replacing the regularization parameter α by the number k of maintained singular values, it follows immediately that

$$V(k) = (1 - \frac{k}{m})^{-2} \|\mathbf{y}^\delta - Kx_k^\delta\|^2,$$

and – at least as far as expectation values are concerned – (4.127) yields

$$V(k) \approx (1 - \frac{k}{m})^{-2}\left(\|\mathbf{y} - Kx_k\|^2 + \delta^2 \operatorname{trace}\left\{\frac{r_\alpha^2(KK^*)}{m}\right\}\right)$$

$$= (1 - \frac{k}{m})^{-2}\left(\|\mathbf{y} - Kx_k\|^2 + \delta^2(1 - \frac{k}{m})\right).$$

From this it can be seen that the trace term guarantees that $V(k)$ attains its minimum for $k \ll m$. Moreover, assuming that $\|\mathbf{y} - Kx_k\|$ has a reasonable decay as a function of k we can estimate

$$V(k) \approx \|\mathbf{y} - Kx_k\|^2 + \delta^2(1 - \frac{k}{m})^{-1} \approx \|\mathbf{y} - Kx_k\|^2 + \delta^2(1 + \frac{k}{m}). \qquad (4.128)$$

On the other hand, for the expected value of $T(k)$ we obtain from (4.126)

$$\begin{aligned}T(k) &\approx \|\mathbf{y} - Kx_k\|^2 + \delta^2 \operatorname{trace}\{\frac{1}{m} g_\alpha^2(KK^*)(KK^*)^2\} \\ &= \|\mathbf{y} - Kx_k\|^2 + \delta^2 \frac{k}{m}.\end{aligned} \qquad (4.129)$$

Comparing (4.128) and (4.129) we conclude that, at least in the range of reasonable regularization parameters $k \ll m$, $V(k)$ and $T(k)$ behave likewise up to a shift by δ^2. Consequently, local minima of $T(k)$ in this range are likely to be global minima of $V(k)$.

Assuming specific decay rates for the singular values of K and the Fourier coefficients of **y**, this can be made more precise, and the asymptotic behaviour of k (as function of δ) can be deduced. We do not want to elaborate on this, since the corresponding convergence rates are necessarily different from what has been called order-optimal in the context of this book. This is due to the fact that the noise propagation rate (4.124) in the statistical context is not as severe as the deterministic worst-case estimate (4.123). We mention in passing that for *fixed δ* and increasing number m of data samples less regularization is required according to this statistical setting, i.e., the parameter k which minimizes the expected error goes to infinity.

108   4. Continuous Regularization Methods

Also, the expected accuracy improves as $m \to \infty$. While these results are comparatively easy to obtain for the truncated singular value decomposition (cf. [284]), the proofs are already extremely technical for Tikhonov regularization (cf. Wahba [286, Chapter 4], and Lukas [187] for further results).

As far as the implementation is concerned we like to point out that the approximation
$$V(\alpha) \approx T(\alpha) + \delta^2 = O(\delta^2)$$
holds for a comparatively wide range of parameters α near the "optimal" one. Consequently, $V(\alpha)$ often has a "flat" graph with a number of additional local minima in this range of critical regularization parameters. This leads to difficulties when the global minimum of $V(\alpha)$ is searched numerically.

While the trace term can be evaluated efficiently for Tikhonov regularization (see Section 9.4 for details) this remains a difficult problem for most other regularization methods because the matrix $r_\alpha(KK^*)$ is hardly ever formed explicitly. The computation would of course be easy if it could be based on the singular value expansion of K, but this is usually no appropriate approach. In particular, for iterative regularization methods, alternative implementations have to be considered.

One such alternative is the so-called *Monte Carlo cross-validation* technique of Girard [94]. The idea of Monte Carlo cross-validation is to replace the trace term by a Monte Carlo type estimate for it, which is based on the fact that
$$\text{trace}\{r_\alpha(KK^*)\} = E\left[\langle \mathbf{e}, r_\alpha(KK^*)\mathbf{e} \rangle\right],$$
where $\mathbf{e} \in \mathbb{R}^m$ is a normally distributed random vector with zero mean and covariance matrix I. Thus, in order to approximate the trace of $r_\alpha(KK^*)$ one can choose a pseudo white noise vector $\mathbf{e}$ with covariance matrix I, evaluate $\mathbf{r}_\alpha = r_\alpha(KK^*)\mathbf{e}$, and take $\mathbf{e}^T \mathbf{r}_\alpha$ as final estimate. Note that $\mathbf{r}_\alpha$ is the residual that is obtained from the respective regularization method if the right-hand side $\mathbf{y}^\delta$ is replaced by $\mathbf{e}$. It can be computed by running the regularization algorithm with the two right-hand side vectors $\mathbf{y}^\delta$ and $\mathbf{e}$ at the same time. This method for choosing the regularization parameter will therefore need the same amount of work in addition to what is already required for computing the approximation.

Another very popular error-free parameter choice rule has been advocated by Hansen [122]. Once again this method is based on an inspection of the residual norms of the computed approximations, this time by relating them to the norms of the approximations themselves. This is done by plotting $\|x_\alpha^\delta\|$ versus $\|y^\delta - Tx_\alpha^\delta\|$ in a log-log scale for a large range of α values. The motivation is the following. When α is moderate to large, $\|x_\alpha^\delta\|$ is typically of the same order of magnitude as $\|x^\dagger\|$, and it varies comparatively little as α decreases; at the same time the residual norm $\|y^\delta - Tx_\alpha^\delta\|$ is decreasing with a typical rate of $O(\alpha^\nu)$, where $\nu > 0$ depends on the smoothness of the solution and on the regularization method employed, cf. (4.25). This decay may be quite dramatic in this range of larger regularization parameters. As a consequence, the corresponding part of the graph of $(\|y^\delta - Tx_\alpha^\delta\|, \|x_\alpha^\delta\|)$ is very

4.5. Heuristic Parameter Choice Rules

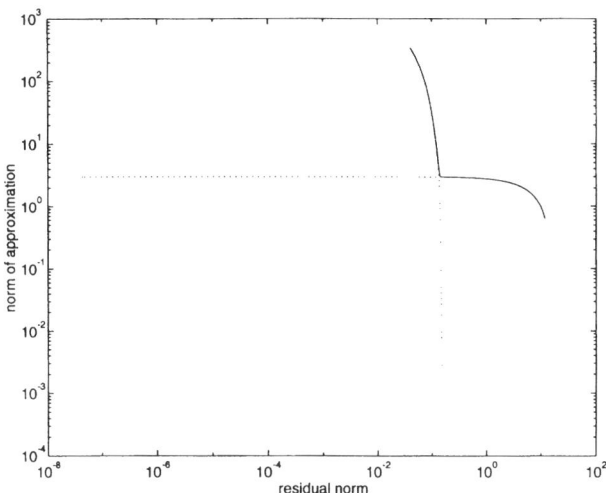

Figure 4.3: L-curve (solid line) made up by "superposition" of two components.

flat in general. Similarly, when α is small (as compared to its "optimal" value) then the residual norm is of the order of δ and changes very little, while the norms of the approximations blow up, e.g., like $O(\delta/\sqrt{\alpha})$. Thus, this results in a steep slope of the above graph, and therefore, the overall shape of the graph often looks like the letter "L", cf. Figure 4.3, hence its name *L-curve*.

For problems with artificially generated white noise it is possible to also plot the curves corresponding to the right-hand sides y and $y^\delta - y$, respectively, compare the dotted lines in Figure 4.3. This figure suggests that the horizontal part of the L-curve solely depends on y while the vertical part is mainly due to noise. The crossover point of the two curves is close to the "corner" of the "L", and it is natural to expect that the corresponding regularization parameter is a good compromise between data fitting and penalizing the norm of the reconstruction. This interpretation is particularly important for Tikhonov regularization (cf. Section 5.1) due to its variational foundation as given in Theorem 5.1.

Figure 4.3 corresponds to a certain model problem with Tikhonov regularization as regularization method. However, similar figures have been observed for many problems, also using other regularization methods instead of Tikhonov regularization, cf., e.g., [122, 123, 124]. Concerning general regularization methods, it seems like a necessary requirement of the L-curve method that the functions $\{r_\alpha\}$ and $\{g_\alpha\}$ are monotonic (increasing and decreasing, respectively) with respect to α for every fixed argument λ, since this implies that the graph of the L-curve is monotonically decreasing. Fortunately, except for the ν-methods (cf. Section 6.3) this is the case for all regularization methods considered in this book. Still though, strictly speaking this does not guarantee that the curve has the shape of an "L" (cf. [236] for a discus-

110   4. Continuous Regularization Methods

sion of this point). We also note that for the truncated singular value decomposition or for iterative regularization methods the L-curve plot is no continuous graph but a disconnected set of (countably many) points. However, after connecting the points with a linear or cubic spline, the main features of the graph are the same as for continuous regularization methods.

A practical difficulty with the L-curve method is the fact that there is typically quite a wide range of regularization parameters corresponding to points on the curve near its corner. Therefore, the location of the corner should not be chosen visually (like by trial and error), but rather by some numerical optimization routines. For example, for the important case where the functions $\{r_\alpha\}$ and $\{g_\alpha\}$ are twice differentiable with respect to α, Hansen and O'Leary [124] suggest an algorithm which seeks the point on the graph with maximal curvature

$$\kappa(\alpha) = \frac{\xi''(\alpha)\eta'(\alpha) - \xi'(\alpha)\eta''(\alpha)}{\left(\xi'(\alpha)^2 + \eta'(\alpha)^2\right)^{3/2}}, \qquad (4.130)$$

$$\xi(\alpha) = \log \|y^\delta - Tx_\alpha^\delta\|, \quad \eta(\alpha) = \log \|x_\alpha^\delta\|. \qquad (4.131)$$

Note, however, that the notion of a "corner" originates from a purely visual impression, and it is not at all obvious how to translate this impression into mathematics. In fact, maximizing the curvature (4.130) is just one option. An alternative definition has been mentioned by Regińska [236], again for the case where the functions $\{r_\alpha\}$ and $\{g_\alpha\}$ are differentiable with respect to α. She considers a point $C = (\xi(\alpha_*), \eta(\alpha_*))$ to be the corner of the L-curve $\mathcal{L}$ if

$$\begin{array}{c}\mathcal{L} \text{ is concave in a neighbourhood of } C, \text{ and} \\ \text{the tangent of } \mathcal{L} \text{ at } C \text{ has slope } -1.\end{array} \qquad (4.132)$$

(This does not imply, however, that this corner is "sharp"). As can be seen in Figure 4.3, the extra condition that $\mathcal{L}$ be concave at C is important, because the slope of $\mathcal{L}$ may also be -1 near its endpoints.

It turns out that the notion (4.132) of a corner leads to a new interpretation of the method, not only suitable for differentiable functions $\{r_\alpha\}$ and $\{g_\alpha\}$:

Proposition 4.37. *The point $C = (\xi(\alpha_*), \eta(\alpha_*))$ is a corner of the L-curve in the aforementioned sense (4.132) if and only if the function*

$$\psi(\alpha) := \|x_\alpha^\delta\| \, \|y^\delta - Tx_\alpha^\delta\| \qquad (4.133)$$

has a local minimum at $\alpha = \alpha_*$.

Proof. By (4.131) and (4.133) we have $\psi(\alpha) = \exp\{\xi(\alpha) + \eta(\alpha)\}$, and hence, if $\psi(\alpha)$ has a local extremum at α_*, then

$$\psi'(\alpha_*) = \bigl(\xi'(\alpha_*) + \eta'(\alpha_*)\bigr) \exp\{\xi(\alpha_*) + \eta(\alpha_*)\} = 0.$$

4.5. Heuristic Parameter Choice Rules

This implies that $\xi'(\alpha_*) + \eta'(\alpha_*) = 0$, which means that the tangent of $\mathcal{L}$ in C is orthogonal to the vector $[1,1]^T$. Consequently, the tangent has the form

$$\{(\xi,\eta) \mid \xi + \eta = \xi(\alpha_*) + \eta(\alpha_*)\}. \tag{4.134}$$

If α_* is furthermore a minimum of ψ, then $\xi(\alpha) + \eta(\alpha) = \log \psi(\alpha)$ also has a local minimum at $\alpha = \alpha_*$, i.e.,

$$\xi(\alpha) + \eta(\alpha) \geq \xi(\alpha_*) + \eta(\alpha_*) \tag{4.135}$$

for α near α_*. In other words, these points $(\xi(\alpha), \eta(\alpha))$ lie above the tangent (4.134).

Vice versa, if C is a corner of $\mathcal{L}$, then the tangent at C is given by (4.134), and the concavity condition (4.135) implies that ψ has a local minium at $\alpha = \alpha_*$. ∎

With this second definition of a corner the corresponding regularization parameter can therefore be found numerically in an easier way. Moreover, Proposition 4.37 reveals a connection between the L-curve method and the parameter choice method considered earlier in this section in Theorem 4.34. There we have multiplied the residual norm $\|y^\delta - Tx_\alpha^\delta\|$ by $\alpha^{-\frac{1}{2}}$ in order to estimate the total iteration error. In (4.133), instead of, $\alpha^{-\frac{1}{2}}$ the corresponding factor is $\|x_\alpha^\delta\|$. Note that one assumption in Theorem 4.34 was that $G_\alpha = O(\alpha^{-1})$, and hence $\alpha^{-\frac{1}{2}}$ is an upper bound for $\|x_\alpha^\delta\|$ by Theorem 4.2. On the other hand, one particular purpose of the factor $\alpha^{-\frac{1}{2}}$ was to simulate the propagation of noise. It seems that $\|x_\alpha^\delta\|$ serves as good as $\alpha^{-\frac{1}{2}}$ for this role, since for smaller parameters α the magnitude of $\|x_\alpha^\delta\|$ is essentially determined by the propagated noise. It follows that these two conceptually very different methods, namely the L-curve method and the error estimation technique (4.115) are more closely related than one would expect at first sight.

In spite of its use in several applications there still lacks a sound mathematical foundation of the L-curve method. In fact, recently, criticisms of the L-curve method have been published (cf. [71, 115, 285]) which go beyond the general observation of Theorem 3.3, which applies to all methods in this section. Whereas the results in [71] concern a non-logarithmic scaling of the axes, the other two papers discuss the doubly logarithmic plot as considered above. From [71] one may conclude that the use of doubly logarithmic scales is essential for a good performance of the L-curve method, since otherwise one can derive a completely unrealistic lower bound for the corresponding regularization parameter. In fact, only this type of scaling leads to a graph which is invariant with respect to rescalings of the equation.

Although the two papers [285] and [115] both consider the L-curve plot in the log-log scale, they differ in their concept of data perturbations and asymptotic analysis. In [285] the problem under consideration is an m-dimensional moment problem with discrete white noise as in (4.122), and the limiting process lets $m \to \infty$ with fixed noise variance σ. In [115] the problem is infinite-dimensional, and the limiting process lets $\|y^\delta - y\| \to 0$, as in Definition 3.1.

As a consequence the results are quite different. According to [285] the parameter determined by the L-curve criterion stagnates as $m \to \infty$, at least when the solution

112   4. Continuous Regularization Methods

$x^\dagger$ lacks smoothness, i.e., $x^\dagger \in \mathcal{X}_\mu$ with μ small. In [115], on the other hand, it is shown that the parameter converges too rapidly to zero (somewhat like δ^2). In view of Corollary 4.4 this suggests that the L-curve method does not really work well for smooth solutions, either (i.e., $x^\dagger \in \mathcal{X}_\mu$ with μ large). The reason for the discrepancy in these observations is the fact that in these two settings the corner of the L-curve originates from different phenomena:

In the model in [285] there is a comparatively large portion of noise in the singular subspaces corresponding to the very small singular values, so that the residual changes only very little in the logarithmic scale, once it has reached the noise level: this is the point where the L-curve bends to its vertical part. The regularization parameter where this level is attained depends only little on m; hence, for all m sufficiently large the corner of the L-curve is always at about the same point.

In [115] the noise is an element of the infinite-dimensional Hilbert space $\mathcal{X}$, and hence no such stagnation of the residual norm is observed in this model. (This might be different, though, if $\mathcal{R}(K)$ is not dense in $\mathcal{Y}$). Instead, in this second counterexample, the vertical part of the L-curve is due to the divergence of the approximations x_α. In the doubly logarithmic scale this divergence can first be detected when

$$\|x^\dagger\|^2 \approx \|x_\alpha^\delta - x_\alpha\| = \|g_\alpha(T^*T)T^*(y^\delta - y)\|^2 \leq cG_\alpha \delta^2 \,.$$

If $G_\alpha = O(\alpha^{-1})$ (as is assumed, e.g., in Corollary 4.4) and if the estimate on the right-hand side is reasonably sharp (as indicated by the results in Section 4.2), then this yields $\alpha \approx c\delta^2$ for the regularization parameter where divergence sets in, and hence, where the corner of the L-curve will appear. This is, in fact, the essence of what has been obtained rigorously in [115].

The same examples can be used to analyze the parameter choice criterion considered in Theorem 4.34. It turns out that the corresponding regularization parameter also stagnates in the setting of [285] when $m \to \infty$, whereas it converges to zero with the best possible rate in the continuous setting of [115] when $\delta \to 0$. On the other hand, while generalized cross-validation cannot be applied in this continuous setting, it yields asymptotically optimal regularization parameters in the setting of [285], cf. [187, 286].

4.6. Mollifier Methods

As indicated in Remark 4.6, one is frequently not really interested in the solution $x^\dagger$ of an ill-posed operator equation, but in some linear functional $\langle f, x^\dagger \rangle$ of the solution. As we have seen there, the problem of determining a linear functional can, depending on the functional, be much more well-behaved than the problem of determining $x^\dagger$ itself. This effect has been used, e.g., by Anderssen in what he calls the "linear functional strategy" [10].

4.6. Mollifier Methods

A slightly different point of view of the same general idea is the following one: one is interested in the solution $x^\dagger$ of $Tx = y$, but realizes that the problem is too ill-posed for being able to determine $x^\dagger$ accurately. Thus, one compromises by changing the problem into a more well-posed one, namely that of trying to determine a *mollified version* $E_\gamma x^\dagger$ of the solution, where E_γ is a suitable mollification operator depending on a parameter γ. The heuristic motivation is that the trouble usually comes from high frequency components of the data and of the solution, which are damped out by mollification (which, in fact, defines mollification). For treating inverse heat conduction problems, mollifier methods have been used for quite some time (see, e.g., [198, 199]).

In abstract terms, mollifier methods for linear ill-posed problems were introduced by Louis and Maaß [185] and further studied (also for some nonlinear problems) by Louis [184], whose approach we will describe shortly:

We consider (4.2) and introduce a smoothing operator $E_\gamma : \mathcal{X} \to \mathcal{X}$ such that

$$E_\gamma x \to x \quad \text{for all } x \in \mathcal{X} \text{ as } \gamma \to 0\,.$$

If $\mathcal{X}$ is a suitable function space, we represent E_γ by a *mollifier* e_γ via

$$(E_\gamma x)(s) =: \langle e_\gamma(s, \cdot), x \rangle. \qquad (4.136)$$

Instead of $x^\dagger$, we now look for $E_\gamma x^\dagger$ for some $\gamma > 0$. Now, assume that e_γ has a representation

$$T^* v_s^\gamma = e_\gamma(s, \cdot) \qquad (4.137)$$

with $v_s^\gamma \in \mathcal{Y}$. Then, if $Tx^\dagger = y$, we can compute $E_\gamma x^\dagger$ as follows: $(E_\gamma x^\dagger)(s) = \langle e_\gamma(s, \cdot), x^\dagger \rangle = \langle T^* v_s^\gamma, x^\dagger \rangle = \langle v_s^\gamma, Tx^\dagger \rangle = \langle v_s^\gamma, y \rangle$, i.e.,

$$(E_\gamma x^\dagger)(s) = \langle v_s^\gamma, y \rangle\,. \qquad (4.138)$$

Hence, the problem of solving (4.2) reduces to that of solving (4.137), which is also ill-posed as soon as (4.2) is. However, the right-hand side of (4.137) (which is actually a family of equations depending on the parameter s) is usually given analytically, since the mollifier e_γ is chosen. Hence, there is no (or much less) error in the data of (4.137), and this equation can be solved (by regularization) much better than (4.2) if it cannot be solved analytically. As soon as an approximation for v_s^γ has been computed, it can be used to solve (4.2) for any right-hand side y via (4.138).

Now, if we define the operator $S_\gamma : \mathcal{Y} \to \mathcal{X}$ via

$$(S_\gamma y)(s) := \langle v_s^\gamma, y \rangle, \qquad (4.139)$$

then, by (4.138), this operator maps right-hand sides of (4.2) to mollified solutions. This motivates the term *approximate inverse* (of T) used for S_γ in [184], also for the more general case that (4.137) is not solvable. In this case, (4.137) is replaced by

$$\|T^* v_s^\gamma - e_\gamma(s, \cdot)\| \to \min, \qquad (4.140)$$

114   4. Continuous Regularization Methods

which, by Theorem 2.6, is equivalent to

$$TT^*v_s^\gamma = Te_\gamma(s,\cdot). \qquad (4.141)$$

Note that one needs the requirement that (4.141) is solvable. The function v_s^γ is called *reconstruction kernel* in [184]; uniqueness can be enforced by solving (4.141) in the best-approximate sense, i.e., by selecting the solution of (4.141) with minimal norm: $v_s^\gamma = (T^*)^\dagger e_\gamma(s,\cdot)$.

Assume, for simplicity, that $\mathcal{R}(T)$ is dense in $\mathcal{Y}$, so that $(TT^*)^{-1}$ exists. Then we have with v_s^γ defined by (4.140):

$$\begin{aligned} S_\gamma y &= \langle (TT^*)^{-1}Te_\gamma(s,\cdot), y \rangle = \langle e_\gamma(s,\cdot), T^*(TT^*)^{-1}y \rangle \\ &= \langle e_\gamma(s,\cdot), (T^*T)^\dagger T^*y \rangle = (E_\gamma x^\dagger)(s), \end{aligned}$$

i.e., $S_\gamma y$ is the mollified version of the best-approximate solution of (4.2). This justifies (4.140).

Let T be compact with singular system $(\sigma_n; v_n, u_n)$ and let a regularization method be defined by (4.3), i.e.,

$$x_\alpha = \sum_{n=1}^\infty \sigma_n g_\alpha(\sigma_n^2) \langle y, u_n \rangle v_n. \qquad (4.142)$$

If we assume that $\mathcal{X}$ and $\mathcal{Y}$ are suitable function spaces, then (4.142) can be written as

$$x_\alpha(s) = \langle v_s^\gamma, y \rangle \qquad (4.143)$$

with

$$v_s^\gamma(t) = \sum_{n=1}^\infty \sigma_n g_\alpha(\sigma_n^2) u_n(t) v_n(s).$$

Now, v_s^γ can be written in the form (4.141) with

$$e_\gamma(s,t) = \sum_{n=1}^\infty \sigma_n^2 g_\alpha(\sigma_n^2) v_n(s) v_n(t). \qquad (4.144)$$

Hence, x_α can be considered as mollified solution $E_\gamma x^\dagger$ with E_γ given (in the sense of (4.136)) by the mollifier (4.144), so that linear regularization methods of the form (4.3) can also be viewed as mollifier methods.

In the arguments above, the "suitable function spaces" have to be such that point evaluation is continuous wherever used. The choice of the mollifier of course depends on what one wants to achieve; frequently used choices are (with suitable normalization constants c)

$$e_\gamma(x,y) = \begin{cases} c, & \|x - y\| \leq \gamma, \\ 0, & \text{otherwise}, \end{cases}$$

where averages of $x^\dagger$ over balls of radius γ are computed, the *sinc function* (cf. Section 1.4)

$$e_\gamma(x,y) = c \operatorname{sinc} \gamma(x-y),$$

4.6. Mollifier Methods

which cuts off high frequency components of $x^\dagger$, or the rapidly decaying *heat kernel*

$$e_\gamma(x,y) = c\exp\left(-\frac{\|x-y\|^2}{2\gamma^2}\right).$$

Note again that each of these mollifiers can be applied only to equations where (4.141) is solvable.

We close by mentioning a related method, the *Backus-Gilbert method* [17] for treating *moment problems* of the type

$$\langle x, k_i \rangle = y_i, \qquad i \in \{1,\ldots,n\}, \tag{4.145}$$

with given elements $k_i \in \mathcal{X}$, cf. Section 3.3. To be concrete, let $\mathcal{X} = \mathcal{L}^2(\Omega)$, so that (4.145) takes the form

$$\int_\Omega k_i(t)x(t)dt = y_i, \tag{4.146}$$

which can be thought of as resulting from discretizing an integral equation of the first kind

$$\int_\Omega k(s,t)x(t)dt = y(s)$$

by collocating at points $s_1,\ldots,s_n$, so that $k_i(t) := k(s_i,t)$, $y_i := y(s_i)$, cf. Example 3.25.

With $Tx := (\langle x,k_1\rangle,\ldots,\langle x,k_n\rangle)^T$, $y = (y_1,\ldots,y_n)$, (4.145) can be written in the form (4.2) with $\mathcal{X} = \mathcal{L}^2(\Omega)$, $\mathcal{Y} = \mathbb{R}^n$. In the Backus-Gilbert method, one looks for an approximate inverse $S : \mathbb{R}^n \to \mathcal{L}^2(\Omega)$ for T by defining

$$Sy = \sum_{i=1}^n y_i v_i \tag{4.147}$$

with functions $v_i \in \mathcal{L}^2(\Omega)$ to be determined as follows: since

$$(STx) = \sum_{i=1}^n \langle x, k_i\rangle v_i = \langle x, \sum_{i=1}^n k_i v_i \rangle,$$

i.e., for the concrete case (4.146)

$$(STx)(s) = \int_\Omega x(t)\left[\sum_{i=1}^n k_i(t)v_i(s)\right]dt,$$

one should aim at determining the functions v_i such that

$$\sum_{i=1}^n k_i(t)v_i(s) \sim \delta(|s-t|). \tag{4.148}$$

The question is how to formalize the requirement (4.148). In the Backus-Gilbert method, this is done by minimizing, for any *fixed* $s \in G$ and some chosen τ, the functional

$$(v_1,\ldots,v_n) \mapsto \int_\Omega |s-t|^{2\tau}\left|\sum_{i=1}^n k_i(t)v_i\right|^2 dt$$

116   4. *Continuous Regularization Methods*

under the normalization constraint

$$\int_\Omega \sum_{i=1}^n k_i(t) v_i \, dt = 1 ; \qquad (4.149)$$

we then take $v_i(s) := v_i$. The constraint (4.149) just says that for $x \equiv 1$, $(STx)(s) = x(s)$ holds. The parameter τ (in [17] $\tau = 1$) determines the concrete method.

The common feature between mollification and the Backus-Gilbert method is the following: in both cases, an approximate inverse (determined by v_s^γ or by the $v_i(s)$) is determined independently of the right-hand side of the equation, which can then be used to explicitly represent an approximate solution via (4.143) or via (4.147). Via Lagrange multipliers, the Backus-Gilbert basis functions $v_1, \ldots, v_n$ can be determined pointwise from the linear systems

$$\begin{pmatrix} G(s) & w \\ w^T & 0 \end{pmatrix} \begin{pmatrix} v \\ \lambda \end{pmatrix} = \begin{pmatrix} 0 \\ 1 \end{pmatrix}, \quad s \in \Omega,$$

with

$$G(s)_{ij} = \int_\Omega |s-t|^{2\tau} k_i(t) k_j(t) \, dt, \qquad i,j \in \{1,\ldots,n\},$$

$$w_i = \int_\Omega k_i(t) \, dt, \qquad i \in \{1,\ldots,n\}.$$

Note that the matrix of this system depends on s, while in the corresponding system (4.141) for mollifier methods, s enters only in the right-hand side.

A convergence analysis for the Backus-Gilbert method (which gives also information about suitable choices of τ) is given in [154].

5. Tikhonov Regularization

5.1. The Classical Theory

A special choice for g_α which fulfills the assumptions of Theorem 4.1 (with $C = 1$) is

$$g_\alpha(\lambda) := \frac{1}{\lambda + \alpha}. \tag{5.1}$$

With this choice, x_α^δ as defined by (4.4) has the form

$$x_\alpha^\delta = (T^*T + \alpha I)^{-1} T^* y^\delta, \tag{5.2}$$

i.e., is defined via the linear equation

$$T^*T x_\alpha^\delta + \alpha x_\alpha^\delta = T^* y^\delta, \tag{5.3}$$

which can be thought of a regularized form of the normal equation (2.18). We omit the index δ if the noisy data y^δ are replaced by the exact data y, i.e., (cf. (4.3))

$$x_\alpha = (T^*T + \alpha I)^{-1} T^* y.$$

This method is called *Tikhonov regularization* after the late A.N. Tikhonov ([269, 270]), sometimes it is also called *Tikhonov-Phillips regularization* [226]. Since $(T^*T + \alpha I)^{-1} T^* = T^* (TT^* + \alpha I)^{-1}$, it can also be computed as follows:

$$\begin{aligned} TT^* z_\alpha^\delta + \alpha z_\alpha^\delta &= y^\delta, \\ x_\alpha^\delta &= T^* z_\alpha^\delta. \end{aligned} \tag{5.4}$$

For a compact operator K with singular system $(\sigma_n; v_n, u_n)$, (5.2) has the form

$$x_\alpha^\delta = \sum_{n=1}^\infty \frac{\sigma_n}{\sigma_n^2 + \alpha} \langle y^\delta, u_n \rangle v_n. \tag{5.5}$$

A comparison with (2.24) clearly shows the stabilization: errors in $\langle y, u_n \rangle$ are not propagated with the factors $1/\sigma_n$, but only with the factors $\sigma_n/(\sigma_n^2 + \alpha)$ into the result; these factors stay bounded as $n \to \infty$.

Tikhonov regularization has the following *variational characterization*:

Theorem 5.1. *Let x_α^δ be as in (5.2). Then x_α^δ is the unique minimizer of the Tikhonov functional*

$$x \mapsto \|Tx - y^\delta\|^2 + \alpha \|x\|^2. \tag{5.6}$$

Proof. Let $f_\alpha(x)$ denote the functional in (5.6). For $\alpha > 0$, f_α is strictly convex, $\lim_{\|x\| \to +\infty} f_\alpha(x) = +\infty$. Hence, f_α has a unique minimizer, which can be characterized by the (in this case necessary and sufficient) first-order conditions

$$f_\alpha'(x) h = 0 \quad \text{for all} \quad h \in \mathcal{X}. \tag{5.7}$$

118   5. Tikhonov Regularization

Now,

$$f'_\alpha(x)h = 2(\langle Tx - y^\delta, Th \rangle + \alpha \langle x, h \rangle) = 2\langle T^*Tx - T^*y^\delta + \alpha x, h \rangle,$$

so that (5.7) is equivalent to (5.2). ∎

The Tikhonov functional in (5.6) sheds more light on the role of the regularization parameter. Minimization of (5.6) is a compromise between minimizing the residual norm, and keeping the "penalty term" $\|x\|$ small, i.e., enforcing stability.

While (5.2) makes sense only for a linear operator T, (5.6) can be formulated also for nonlinear operators. This is the key for defining Tikhonov regularization for nonlinear ill-posed problems (see Chapter 10).

The variational characterization can be used directly for proving results about Tikhonov regularization; we exemplify this for the following convergence result, although the result itself can of course be derived easily from the general framework of Chapter 4:

Theorem 5.2. *Let x^δ_α be defined by (5.2), $y \in \mathcal{R}(T)$, $\|y - y^\delta\| \leq \delta$. If $\alpha = \alpha(\delta)$ is such that*

$$\lim_{\delta \to 0} \alpha(\delta) = 0 \quad and \quad \lim_{\delta \to 0} \frac{\delta^2}{\alpha(\delta)} = 0, \tag{5.8}$$

then

$$\lim_{\delta \to 0} x^\delta_{\alpha(\delta)} = T^\dagger y. \tag{5.9}$$

Proof. Let $\delta_n \to 0$, $\alpha_n := \alpha(\delta_n)$, $y_n := y^{\delta_n}$, $x_n := x^{\delta_n}_{\alpha_n}$. By f_n, we denote the Tikhonov functional (5.6) for α_n, i.e., $f_n(x) = \|Tx - y_n\|^2 + \alpha_n \|x\|^2$. By Theorem 5.1, x_n is the unique minimizer of f_n. Hence, with $x^\dagger := T^\dagger y$,

$$\begin{aligned} \alpha_n \|x_n\|^2 &\leq f_n(x_n) \leq f_n(x^\dagger) \\ &= \|Tx^\dagger - y_n\|^2 + \alpha_n \|x^\dagger\|^2 \leq \delta_n^2 + \alpha_n \|x^\dagger\|^2, \end{aligned}$$

so that

$$\|x_n\|^2 \leq \frac{\delta_n^2}{\alpha_n} + \|x^\dagger\|^2. \tag{5.10}$$

Hence, x_n is bounded and thus has a weakly convergent subsequence

$$x_{n_k} \rightharpoonup z \in \mathcal{X}.$$

Since the bounded linear operator T is weakly sequentially continuous,

$$Tx_{n_k} \rightharpoonup Tz. \tag{5.11}$$

Again by Theorem 5.1, we obtain as above that

$$\|Tx_{n_k} - y_{n_k}\|^2 \leq f_{n_k}(x_{n_k}) \leq \delta_{n_k}^2 + \alpha_{n_k} \|x^\dagger\|^2 \to 0 \quad \text{as} \quad k \to \infty.$$

5.1. The Classical Theory   119

Together with (5.11), this implies

$$Tz = y. \tag{5.12}$$

Since any minimizer of f_n is in $\mathcal{N}(T)^\perp$ (a non-vanishing component in $\mathcal{N}(T)$ would increase the second term in f_n while leaving the first one unchanged), the x_n are in $\mathcal{N}(T)^\perp$, and hence $z \in \mathcal{N}(T)^\perp$. By Theorem 2.5 and its proof, this implies with (5.12) that $z = x^\dagger$, so that $x_{n_k} \rightharpoonup x^\dagger$. Applying the same argument to all subsequences, we obtain that

$$x_n \rightharpoonup x^\dagger. \tag{5.13}$$

Now, assume that there are an $\varepsilon > 0$ and a subsequence $\{x_{n_k}\}$ such that for all $k \in \mathbb{N}$, $\|x_{n_k}\| \leq \|x^\dagger\| - \varepsilon$. Then, this subsequence would have a further subsequence converging weakly to a z with $\|z\| \leq \|x^\dagger\| - \varepsilon$, contradicting (5.13). Hence,

$$\liminf_{n\to\infty} \|x_n\| \geq \|x^\dagger\|. \tag{5.14}$$

From (5.10) and (5.8), we obtain

$$\limsup_{n\to\infty} \|x_n\| \leq \|x^\dagger\|. \tag{5.15}$$

Now, (5.13), (5.14), and (5.15) imply that $x_n \to x^\dagger$. Since $\delta_n \to 0$ was arbitrary, (5.9) follows. ∎

Note that to prove (5.13), we needed only the conditions that $\alpha(\delta) \to 0$ and $\delta^2/\alpha(\delta)$ remains bounded as $\delta \to 0$.

We now apply the general results from Chapter 4 to Tikhonov regularization. G_α as defined by (4.14) has the form

$$G_\alpha = \frac{1}{\alpha}, \tag{5.16}$$

so that the stability estimate (4.16) becomes

$$\|x_\alpha - x_\alpha^\delta\| \leq \frac{\delta}{\sqrt{\alpha}}.$$

Moreover,

$$r_\alpha(\lambda) = \frac{\alpha}{\lambda + \alpha},$$

and as computed in Example 4.15, we can take

$$\omega_\mu(\alpha) = \begin{cases} \alpha^\mu, & \mu \leq 1, \\ c\alpha, & \mu > 1, \end{cases} \tag{5.17}$$

in (4.24). The qualification of Tikhonov regularization is $\mu_0 = 1$.
From (4.25), we obtain

$$\|x_\alpha - x^\dagger\| = O(\alpha^\mu) \tag{5.18}$$

120   5. Tikhonov Regularization

for $x^\dagger \in \mathcal{X}_\mu$, $\mu \leq 1$. From Corollary 4.4, we obtain that as long as $\mu \leq 1$, Tikhonov regularization with the a-priori parameter choice rule (4.29) is of optimal order in $\mathcal{X}_{\mu,\rho}$. The best possible convergence rate obtainable with this choice is that for $\mu = 1$, where

$$\alpha \sim \left(\frac{\delta}{\rho}\right)^{\frac{2}{3}}$$

and

$$\|x_\alpha^\delta - x^\dagger\| = O(\delta^{\frac{2}{3}}) \tag{5.19}$$

as soon as $x^\dagger \in \mathcal{X}_{1,\rho}$.

As already mentioned in Example 4.15, the results of Section 4.2 also apply to Tikhonov regularization, since (4.50) holds with $\mu = \mu_0 = 1$. Hence, Theorem 4.11 yields that the convergence rate $\|x_\alpha - x^\dagger\| = O(\alpha)$ can only be achieved if $x^\dagger \in \mathcal{X}_1 = \mathcal{R}(T^*T)$. Furthermore, Proposition 4.13 (note that (4.54) holds) and Lemma 4.12 imply that for $\mu < 1$, the rate $\|x_\alpha - x^\dagger\| = O(\alpha^\mu)$ can only hold if $x^\dagger \in \bigcap_{\rho < \mu} \mathcal{X}_\rho$.

Concerning the case of perturbed data, it follows from the converse results in [218] mentioned in Section 4.2, that the rate in (5.19) can only be achieved if $x^\dagger \in \mathcal{X}_1$. Moreover, since (4.50) holds, the corresponding saturation result implies that, for any parameter choice rule,

$$\sup\{\|x_\alpha^\delta - x^\dagger\| \mid \|Q(y - y^\delta)\| \leq \delta\} = o(\delta^{\frac{2}{3}}) \tag{5.20}$$

can only hold if $x^\dagger = 0$, as long as $\mathcal{R}(T)$ is non-closed. Since we did not prove this general saturation result in Section 4.2, we prove the special case for Tikhonov regularization with a compact operator here. The proof is taken from the book [103], which is exclusively dedicated to Tikhonov regularization.

Proposition 5.3. *Let K be compact with infinite-dimensional range, x_α^δ be defined by (5.2) with K instead of T. Let $\alpha = \alpha(\delta, y^\delta)$ be any parameter choice rule. Then (5.20) implies that $x^\dagger = 0$.*

Proof. Let $(\sigma_n; v_n, u_n)$ be a singular system for K. Since $\dim \mathcal{R}(K) = +\infty$, $\lim_{n \to \infty} \sigma_n = 0$. Let $\delta_n := \sigma_n^3$, $y_n := y + \delta_n u_n$, so that $\|y_n - y\| \leq \delta_n$. Moreover, we define $\alpha_n := \alpha(\delta_n, y_n)$, $x_n := x_{\alpha_n}$, $x_n^\delta := x_{\alpha_n}^{\delta_n}$. Then $x_n^\delta - x^\dagger = x_n - x^\dagger + \delta_n(K^*K + \alpha_n I)^{-1} K^* u_n$. Using (2.20) and (2.21), we obtain

$$\|x_n^\delta - x^\dagger\|^2 = \|x_n - x^\dagger\|^2 + \frac{2\delta_n \sigma_n}{\alpha_n + \sigma_n^2} \langle x_n - x^\dagger, v_n \rangle + \left(\frac{\sigma_n \delta_n}{\alpha_n + \sigma_n^2}\right)^2,$$

and hence, by the choice of δ_n,

$$(\delta_n^{-\frac{2}{3}} \|x_n^\delta - x^\dagger\|)^2 \geq \frac{2}{\alpha_n + \delta_n^{\frac{2}{3}}} \langle x_n - x^\dagger, v_n \rangle + \left(\frac{\delta_n^{\frac{2}{3}}}{\alpha_n + \delta_n^{\frac{2}{3}}}\right)^2. \tag{5.21}$$

5.1. The Classical Theory

Now, by (5.2) and (2.18),

$$(K^*K + \alpha_n I)(x^\dagger - x_n^\delta) = K^*y + \alpha_n x^\dagger - K^*y_n,$$

so that

$$\alpha_n \|x^\dagger\| = O(\delta_n + \|x^\dagger - x_n^\delta\|). \tag{5.22}$$

Since, by assumption, $\|x^\dagger - x_n^\delta\| = o(\delta_n^{\frac{2}{3}})$, it follows from (5.22) that, if $x^\dagger \neq 0$,

$$\lim_{n \to \infty} \alpha_n \delta_n^{-\frac{2}{3}} = 0. \tag{5.23}$$

Hence, the second term in the right-hand side of (5.21) tends to 1. Since, by assumption, the left-hand side of (5.21) tends to 0, we obtain

$$0 \geq 2 \limsup_{n \to \infty} \frac{\delta_n^{-\frac{2}{3}}}{1 + \alpha_n \delta_n^{-\frac{2}{3}}} \langle x_n - x^\dagger, v_n \rangle + 1. \tag{5.24}$$

Now, by assumption (5.20), also $\|x_n - x^\dagger\| = o(\delta_n^{\frac{2}{3}})$, so that, if $x^\dagger \neq 0$, we obtain from (5.24) with (5.23) the contradiction $0 \geq 1$. Hence, $x^\dagger = 0$. ∎

Thus, Tikhonov regularization for an ill-posed linear problem (with a compact operator) never yields a convergence rate which is faster than $O(\delta^{\frac{2}{3}})$, it *saturates* at this rate. This is not the case for other methods like truncated singular value expansion or the Showalter method.

We now turn to a-posteriori parameter choice methods, where we first give another motivation for the discrepancy principle of Section 4.3:

If we want to solve equation $Tx = y$ under the information that $\|y^\delta - y\| \leq \delta$, we have to accept any $x \in \mathcal{X}$ with

$$\|Tx - y^\delta\| \leq \delta \tag{5.25}$$

as an approximate solution, since it is compatible with the only knowledge we have on the data. However, if $\mathcal{R}(T)$ is non-closed, the set of all x fulfilling (5.25) is unbounded even if $\mathcal{N}(T) = \{0\}$, reflecting the ill-posedness of $Tx = y$. Since we are looking for the solution of $Tx = y$ with minimal norm, it makes sense to use this requirement also as a selection criterion, which solution of (5.25) we want to choose, i.e., we consider the problem

$$\|x\| \to \min, \quad \text{subject to } \|Tx - y^\delta\| \leq \delta. \tag{5.26}$$

Unless 0 is in the feasible set of this constrained optimization problem, i.e., $\|y^\delta\| \leq \delta$ (which means that there is less signal than noise), the minimum in (5.26) is achieved on the relative boundary of the feasible set, i.e., (5.26) is equivalent to

$$\|x\|^2 \to \min, \quad \text{subject to } \|Tx - y^\delta\|^2 = \delta^2. \tag{5.27}$$

5. Tikhonov Regularization

Using a Lagrange multiplier, this in turn is equivalent to

$$\|x\|^2 + \lambda \|Tx - y^\delta\|^2 \to \min,$$

i.e., to the minimization of (5.6) with $\alpha = 1/\lambda$. In addition to a different motivation for Tikhonov regularization itself, these considerations also provide a rule for choosing the regularization parameter: it should be chosen such that the constraint in (5.27) is fulfilled, which is just the discrepancy principle (4.57) (with $\tau = 1$) in a slightly different form. This provides an additional heuristic motivation (in addition to the one from Remark 4.16) for the discrepancy principle.

It follows from Theorem 4.17 that, if, for Tikhonov regularization, the regularization parameter is chosen according to the discrepancy principle (4.57), then the resulting method is convergent for all $y \in \mathcal{R}(T)$ and of optimal order in $\mathcal{X}_{\mu,\rho}$ for $\mu \in (0, 1/2]$. For $\mu < 1/2$, the convergence rate is in fact given by (4.72). The best convergence rate following from these arguments is the one for $\mu = 1/2$, i.e., $O(\sqrt{\delta})$. It follows from Proposition 4.20 that for a compact operator, this rate cannot be improved under any assumptions unless $\mathcal{R}(K)$ is finite-dimensional. Thus, Tikhonov regularization with the discrepancy principle is not of optimal order in $\mathcal{X}_{\mu,\rho}$ for $\mu \geq 1/2$, although for $\mu \in (1/2, 1]$, other choices of the regularization parameter (e.g., the a-priori choice (4.29)) exist which make Tikhonov regularization a method of optimal order.

We now turn to the a-posteriori parameter choice rule (4.85) for the special case of Tikhonov regularization. In fact, this rule has first been considered by Gfrerer [92] for Tikhonov regularization. For g_α as in (5.1), Assumption 4.23 is fulfilled with $K = 1$. Inserting the special form of g_α into (4.77) and (4.83), respectively, we obtain

$$f(\alpha, w) = 2\alpha^3 \|(TT^* + \alpha I)^{-\frac{3}{2}} Qw\|^2$$

and

$$h(w) = 2\|Qw\|^2.$$

The quantity L as defined by (4.87) turns out to be $L = 2$. Thus, the parameter choice strategy (4.85) has the form

$$\alpha^3 \|(TT^* + \alpha I)^{-\frac{3}{2}} Qy^\delta\|^2 = \tau \delta^2, \tag{5.28}$$

where τ has to fulfill

$$\tau \in (1, \|Qy^\delta\|^2 \delta^{-2}) \tag{5.29}$$

according to (4.86) and Theorem 4.27. Such a choice for τ is possible as long as the signal-to-noise-ratio condition

$$\|Qy^\delta\| > \delta \tag{5.30}$$

is fulfilled.

If $\alpha = \alpha(\delta, y^\delta)$ is chosen according to (5.28) and if (5.29) holds, we obtain the following convergence properties from (4.93) and Corollary 4.29:

$$\|x^\delta_{\alpha(\delta,y^\delta)} - T^\dagger y\| \leq \eta \inf\{\|x_\alpha - T^\dagger y\| + \frac{\delta}{\sqrt{\alpha}} \mid \alpha > 0\}, \tag{5.31}$$

and Tikhonov regularization with this parameter choice is of optimal order in $\mathcal{X}_{\mu,\rho}$ for $\mu \in (0,1]$; especially, if $x^\dagger \in \mathcal{X}_{1,\rho}$,

$$\|x^\delta_{\alpha(\delta,y^\delta)} - x^\dagger\| = O(\delta^{\frac{2}{3}}).$$

The assumptions of Theorem 4.31 are also fulfilled (with $\overline{\alpha}(\lambda) = \lambda$, $c_1 = 1/4$), so that the stronger optimality results from Theorem 4.31 (i) and (ii) hold, respectively.

Note that in (5.28), $TT^* + \alpha I$ appears instead of $T^*T + \alpha I$. Thus, it is advantageous to compute the regularized solution via (5.4) when using this parameter choice rule. Since

$$\|(TT^* + \alpha I)^{-\frac{3}{2}} Qy^\delta\|^2 = \langle (TT^* + \alpha I)^{-2} Qy^\delta, (TT^* + \alpha I)^{-1} Qy^\delta \rangle,$$

the computation of (5.28) can (in the case that $\mathcal{R}(T)$ is dense, i.e, $Qy^\delta = y^\delta$) be performed as follows, see also Section 9.4:

Let z^δ_α be defined by (5.4), and let $z^\delta_{\alpha,2}$ be defined by

$$TT^* z^\delta_{\alpha,2} + \alpha z^\delta_{\alpha,2} = y^\delta + \alpha z^\delta_\alpha, \tag{5.32}$$

then

$$\begin{aligned} \alpha^3 \|(TT^* + \alpha I)^{-\frac{3}{2}} y^\delta\|^2 &= \alpha^2 \langle (TT^* + \alpha I)^{-1} \alpha z^\delta_\alpha, z^\delta_\alpha \rangle \\ &= \alpha^2 \langle z^\delta_{\alpha,2} - z^\delta_\alpha, z^\delta_\alpha \rangle, \end{aligned} \tag{5.33}$$

so that the parameter choice rule (4.85) is equivalent to

$$\alpha^2 \langle z^\delta_{\alpha,2} - z^\delta_\alpha, z^\delta_\alpha \rangle = \tau \delta^2. \tag{5.34}$$

For computing $z^\delta_{\alpha,2}$ from (5.32), the same operator has to be inverted as the one for computing z^δ_α via (5.4). Thus, the computation of $z^\delta_{\alpha,2}$ does not require much additional work. It is actually the second step of what is called *iterated Tikhonov regularization*, which is defined as follows:

$$\begin{aligned} x^\delta_{\alpha,0} &:= 0, \\ T^*T x^\delta_{\alpha,i} + \alpha x^\delta_{\alpha,i} &= T^* y^\delta + \alpha x^\delta_{\alpha,i-1}, \quad i \in \mathbb{N}, \end{aligned} \tag{5.35}$$

or, equivalently,

$$\begin{aligned} z^\delta_{\alpha,0} &:= 0, \\ TT^* z^\delta_{\alpha,i} + \alpha z^\delta_{\alpha,i} &= y^\delta + \alpha z^\delta_{\alpha,i-1}, \quad i \in \mathbb{N}, \\ x^\delta_{\alpha,i} &:= T^* z^\delta_{\alpha,i}. \end{aligned} \tag{5.36}$$

Introducing $p^\delta_{\alpha,i} = x^\delta_{\alpha,i} - x^\delta_{\alpha,i-1}$, (5.35) yields

$$T^*T p^\delta_{\alpha,i} + \alpha p^\delta_{\alpha,i} = T^*(y^\delta - T x^\delta_{\alpha,i-1}),$$

and comparing this with (5.3), it follows from Theorem 5.1 that $p^\delta_{\alpha,i}$ is a minimizer of the functional

$$p \mapsto \|Tp - (y^\delta - T x^\delta_{\alpha,i-1})\|^2 + \alpha \|p\|^2,$$

124   5. Tikhonov Regularization

i.e., $x_{\alpha,i}^\delta$ minimizes the functional

$$x \mapsto \|Tx - y^\delta\|^2 + \alpha \|x - x_{\alpha,i-1}^\delta\|^2. \tag{5.37}$$

If one iterates n times, i.e., computes $x_{\alpha,n}^\delta$, this is called iterated Tikhonov regularization of order n. Note that each step of the iteration involves the same operator to be inverted. For $n = 1$ we obtain ordinary Tikhonov regularization. We consider iterated Tikhonov regularization here for a fixed order n and investigate convergence as $\alpha \to 0$, but it should be mentioned that (5.35) could also be considered as an iterative method for fixed $\alpha > 0$ with $n \to \infty$.

$x_{\alpha,n}^\delta$ can be written in the form (4.4) with

$$g_{\alpha,n}(\lambda) = \frac{(\lambda + \alpha)^n - \alpha^n}{\lambda(\lambda + \alpha)^n}. \tag{5.38}$$

Again, we can apply the results from Chapter 4. Note that now, (4.63) holds with $\mu_0 = n$, which is the qualification of this method. Thus, with the a-priori parameter choice (4.29), iterated Tikhonov regularization is of optimal order in $\mathcal{X}_{\mu,\rho}$ for $\mu \leq n$. The discrepancy principle (4.57) gives rise to an order-optimal regularization method in $\mathcal{X}_{\mu,\rho}$ for $\mu \leq n - 1/2$, while the parameter choice (4.85) is of optimal order in $\mathcal{X}_{\mu,\rho}$ for $\mu \leq n$. Also the stronger optimality results from Theorem 4.31 apply, since the assumptions of that theorem hold with $\overline{\alpha}(\lambda) = n\lambda$, $c_1 = [1 - (n/n+1)^n]^2$. Since the functional f from (4.77) becomes, for $g_{\alpha,n}$ as in (5.38),

$$f(\alpha, w) = 2\alpha^{2n+1} \langle (TT^* + \alpha I)^{-2n-1} Qw, Qw \rangle, \tag{5.39}$$

one sees, using (5.33) and induction, that, for the case that $\mathcal{R}(T)$ is dense in $\mathcal{Y}$, the parameter choice rule (4.85) has the form

$$\alpha^2 \langle z_{\alpha,n+1}^\delta - z_{\alpha,n}^\delta, z_{\alpha,n}^\delta - z_{\alpha,n-1}^\delta \rangle = \tau \delta^2$$

with

$$\tau \in (1, \|Qy^\delta\|^2 \delta^{-2}).$$

For $g_\alpha = g_{\alpha,n}$ as in (5.38), i.e., for iterated Tikhonov regularization of order n, (4.50) holds for the qualification $\mu = n$ (with $c = 1, \tau = 2^{-n}$). Thus, Theorem 4.11 implies that the convergence rate $\|x_{\alpha,n} - x^\dagger\| = O(\alpha^n)$ can only hold if $x^\dagger \in \mathcal{X}_n$. The other converse and saturation results mentioned at the end of Section 4.2 imply that with noisy data, the convergence rate $\|x_{\alpha,n}^\delta - x^\dagger\| = O(\delta^{\frac{2n}{1+2n}})$ implies that $x^\dagger \in \mathcal{X}_n$, and that, if $\mathcal{R}(T)$ is non-closed, the convergence rate $\|x_{\alpha,n}^\delta - x^\dagger\| = o(\delta^{\frac{2n}{1+2n}})$ holds only in the trivial case that $x^\dagger = 0$.

Thus, the best-possible convergence rate of Tikhonov regularization of order n is $O(\delta^{\frac{2n}{1+2n}})$, and the (small) additional effort of iterating Tikhonov regularization results in better convergence as long as the order n does not become larger than the next integer following the largest "smoothness index" μ for which $x^\dagger \in \mathcal{X}_\mu$ holds.

Recall that the convergence results concerning the above parameter choice for ordinary and iterated Tikhonov regularization required the knowledge of Qy^δ in (5.28)

5.1. The Classical Theory   125

and (5.39). If Qy^δ is not known and difficult to compute, one might prefer to use y^δ instead of Qy^δ. In this case, as is easily seen, the parameter choice coincides with the strategy (4.101), (4.98), with the corresponding function s_α given by (4.112). Consequently, Theorem 4.33 applies, showing that the strategies (5.28) and (5.39) with Qy^δ replaced by y^δ are at least order-optimal.

We close this section by mentioning a special error-free parameter choice rule for (iterated) Tikhonov regularization, namely the *quasioptimality criterion*, which has been used with considerable success in practice.

Because of (5.38), the functions $r_{\alpha,n}(\lambda) = 1 - \lambda g_{\alpha,n}(\lambda)$ for iterated Tikhonov regularization take the convenient form

$$r_{\alpha,n}(\lambda) = \left(\frac{\alpha}{\lambda + \alpha}\right)^n, \quad n \in \mathbb{N}.$$

It follows that for fixed α and $\lambda > 0$, $r_{\alpha,n}(\lambda)$ is decreasing in n, which implies that the approximation error is also decreasing:

$$\|x_{\alpha,n+1} - x^\dagger\| \leq \|x_{\alpha,n} - x^\dagger\|, \quad n \in \mathbb{N}. \tag{5.40}$$

Moreover, the larger α is, the more pronounced is the difference between the two sides of this inequality.

On the other hand, the functions $\{g_{\alpha,n}\}$ of (5.38) are monotonically increasing with respect to n: more precisely, we have

$$\begin{aligned} g_{\alpha,n+1}(\lambda) &= g_{\alpha,n}(\lambda) \frac{(\lambda+\alpha)^{n+1} - \alpha^{n+1}}{((\lambda+\alpha)^n - \alpha^n)(\lambda+\alpha)} \\ &= g_{\alpha,n}(\lambda) \left[1 + \frac{\alpha^n \lambda}{((\lambda+\alpha)^n - \alpha^n)(\lambda+\alpha)}\right] \\ &\leq g_{\alpha,n}(\lambda)\left(1 + \frac{1}{n}\right), \quad n \in \mathbb{N}. \end{aligned}$$

This implies the following relation between the propagated data errors:

$$\|x^\delta_{\alpha,n} - x_{\alpha,n}\| \leq \|x^\delta_{\alpha,n+1} - x_{\alpha,n+1}\| \leq (1 + \frac{1}{n})\|x^\delta_{\alpha,n} - x_{\alpha,n}\|. \tag{5.41}$$

Therefore, (5.40) and (5.41) show that as long as α is so large that the approximation error dominates $x^\delta_{\alpha,n} - x^\dagger$, $x^\delta_{\alpha,n+1}$ will be a better approximation of $x^\dagger$, and the absolute difference $\|x^\delta_{\alpha,n+1} - x^\delta_{\alpha,n}\|$ will decrease as α becomes smaller. For small values of the regularization parameter, however, $x^\delta_{\alpha,n+1}$ will be worse than $x^\delta_{\alpha,n}$ in view of (5.41), because the propagated data error will dominate both approximations. The smaller α gets, the larger will the absolute difference $\|x^\delta_{\alpha,n+1} - x^\delta_{\alpha,n}\|$ be. In summary, $\|x^\delta_{\alpha,n+1} - x^\delta_{\alpha,n}\|$ considered as a function of α will in general decrease as long as α is small, and increase for larger values of the parameter α. The minimum will be attained near the "cross-over point" where approximation error and propagated data error in $x^\delta_{\alpha,n}$ have about the same order of magnitude.

126   5. Tikhonov Regularization

The quasioptimality criterion [178, 179, 194, 272] is based on this observation: it chooses the regularization parameter for iterated Tikhonov regularization of order n as the location of the minimum of

$$\psi_{q,n}(\alpha) = \|x^\delta_{\alpha,n+1} - x^\delta_{\alpha,n}\|.$$

In particular, for ordinary Tikhonov regularization this amounts to minimizing

$$\begin{aligned}\psi_q(\alpha) &= \|x^\delta_{\alpha,2} - x^\delta_\alpha\| = \|\alpha(T^*T + \alpha I)^{-1} x^\delta_\alpha\| \\ &= \alpha \|(T^*T + \alpha I)^{-2} T^* y^\delta\|.\end{aligned} \quad (5.42)$$

Like the parameter choice rule (5.34), this can be implemented with only little extra work.

5.2. Regularization with Projection

For the numerical realization of Tikhonov regularization we have to approximate the space $\mathcal{X}$ by a finite-dimensional subspace $\mathcal{X}_n$. We produce then a finite-dimensional approximation of $x^\dagger$ by minimizing the Tikhonov functional

$$x \mapsto \|Tx - y^\delta\|^2 + \alpha \|x\|^2$$

over $\mathcal{X}_n$. It is obvious that this problem is equivalent to minimizing the functional

$$x \mapsto \|T_n x - y^\delta\|^2 + \alpha \|x\|^2$$

over $\mathcal{X}$, where $T_n := TP_n$ and P_n is the orthogonal projector onto the subspace $\mathcal{X}_n$. Therefore, the solution $x^\delta_{\alpha,n}$ to this minimization problem is given by

$$x^\delta_{\alpha,n} = (T_n^* T_n + \alpha I)^{-1} T_n^* y^\delta.$$

In the noise free case ($\delta = 0$) we use the notation $x_{\alpha,n}$. Assuming that we have an expanding sequence of finite-dimensional subspaces of $\mathcal{X}$ whose union is dense in $\mathcal{X}$, we want to get convergence

$$x_{\alpha,n} \to x^\dagger \quad \text{as} \quad \alpha \to 0 \text{ and } n \to \infty.$$

Since we know from Section 5.1 that $x_{\alpha,n} \to T_n^\dagger y$ as $\alpha \to 0$ and since $T_n^\dagger y$ does not converge towards $x^\dagger$ in general (see Example 3.19), convergence of the approximate solutions $x_{\alpha,n}$ towards $x^\dagger$ as $\alpha \to 0$ and $n \to \infty$ simultaneously is to be expected only if α and n are related in a proper way.

In Section 3.3 we have seen that for projection without regularization the choice $\mathcal{X}_n = T^* \mathcal{Y}_n$ and $T_n = Q_n T$ has many advantages, for instance, we have that $T_n^\dagger y = P_n x^\dagger$ (cf. Theorem 3.24). Thus, in the noise free case $T_n^\dagger y$ is the best possible approximation in $\mathcal{X}_n$ and there is no need for a further regularization. However, for

5.2. Regularization with Projection

noisy data the error estimate (3.53) shows that for severely ill-posed problems the dimension of the subspace has to be rather low to keep the total error small, since for these problems the smallest singular value of T_n decreases rapidly as n increases. To be able to use larger dimensions we have to combine the projection method with an additional regularization method, e.g., Tikhonov regularization. This approach is also possible for other general regularization methods as treated in Chapter 4, cf. [109].

For the convergence analysis for the approach $T_n = TP_n$ see [103]. In this section we will only present the results for the approach $T_n = Q_n T$, following [152] (see also [277]). Whenever appropriate, we will mention the advantages of this approach compared to the one where $T_n = TP_n$. For a combination of projections in $\mathcal{X}$ and $\mathcal{Y}$, i.e., with $T_n = Q_n T P_n$, see [229].

As in Section 3.3 we assume that $\{\mathcal{Y}_n\}$ is a sequence of finite-dimensional subspaces of $\mathcal{N}(T^*)^\perp$ with $\mathcal{Y}_n \subset \mathcal{Y}_{n+1}$ and that $\bigcup_{n\in\mathbb{N}} \mathcal{Y}_n$ is dense in $\mathcal{N}(T^*)^\perp$. Let again $Q_n : \mathcal{Y} \to \mathcal{Y}_n$ denote the orthogonal projector, then Q_n converges pointwise towards I on $\mathcal{N}(T^*)^\perp$ as $n \to \infty$. Actually, for our analysis we only need this pointwise convergence and not the inclusion $\mathcal{Y}_n \subset \mathcal{Y}_{n+1}$.

We consider the equation (3.48), i.e.,

$$T_n x = y_n, \quad T_n := Q_n T, \quad y_n := Q_n y.$$

We approximate $x^\dagger$ by

$$x_{\alpha,n}^\delta := (T_n^* T_n + \alpha I)^{-1} T_n^* y^\delta, \tag{5.43}$$

with $\alpha > 0$. Note that $x_{\alpha,n}^\delta$ is like x_α^δ an element of $\mathcal{R}(T^*)$, due to the special choice of T_n. As mentioned above, the definition of $x_{\alpha,n}^\delta$ may be extended to the case $\alpha = 0$ by $x_{0,n}^\delta := T_n^\dagger y^\delta$. Moreover, for the noise free case, $\alpha = 0$ is the best possible choice, since then $x_{\alpha,n} = x_{0,n} = P_n x^\dagger$. For noisy data y^δ, we will choose the regularization parameter α similar to the infinite-dimensional case (cf. (5.28)), i.e., as the solution of

$$\alpha^3 \langle (T_n T_n^* + \alpha I)^{-3} Q_n y^\delta, Q_n y^\delta \rangle = \tau \delta^2, \tag{5.44}$$

where $\tau \geq 1$. One can show as in Propositions 4.25 and 4.26 that (5.44) has a unique solution, $\alpha_n^\delta > 0$, if

$$\|Q_n(y - y^\delta)\| \leq \delta < \tfrac{1}{\tau}\|Q_n y^\delta\|, \tag{5.45}$$

and that then

$$(\sqrt{\tau} - 1)^2 \delta^2 \leq (\alpha_n^\delta)^3 \langle (T_n T_n^* + \alpha_n^\delta I)^{-3} Q_n y, Q_n y \rangle \leq (\sqrt{\tau} + 1)^2 \delta^2. \tag{5.46}$$

For the convergence analysis we need the following lemmata.

Lemma 5.4. *Let $\tau > 1$, let (5.45) hold and let α_n^δ be the solution of (5.44). Then for any $\alpha > 0$*

$$\|x_{\alpha_n^\delta,n}^\delta - x^\dagger\| \leq C[\alpha\|(T_n^* T_n + \alpha I)^{-1} x^\dagger\| + \frac{\delta}{\sqrt{\alpha}}],$$

128   5. Tikhonov Regularization

where
$$C = \max\{2(\sqrt{\tau}+1), \tfrac{1}{\sqrt{\tau}-1}\}.$$

Proof. Due to (5.46), there is a $\tilde\tau \in [(\sqrt{\tau}-1)^2, (\sqrt{\tau}+1)^2]$ such that
$$(\alpha_n^\delta)^3 \langle (T_n T_n^* + \alpha_n^\delta I)^{-3} Q_n y, Q_n y \rangle = \tilde\tau \delta^2. \tag{5.47}$$

Since the derivative of the functional
$$\alpha^2 \|(T_n^* T_n + \alpha I)^{-1} x^\dagger\|^2 + \frac{2\tilde\tau \delta^2}{\alpha}$$

with respect to α is given by
$$2\alpha \langle (T_n T_n^* + \alpha I)^{-3} Q_n y, Q_n y \rangle - \frac{2\tilde\tau \delta^2}{\alpha^2},$$

where we used the fact that $T_n x^\dagger = Q_n y$, (5.47) implies that α_n^δ is the global minimizer of this functional. This together with the estimate
$$\begin{aligned}
\|x_{\alpha,n}^\delta - x^\dagger\|^2 &\le 2\|x_{\alpha,n} - x^\dagger\|^2 + 2\|x_{\alpha,n}^\delta - x_{\alpha,n}\|^2 \\
&\le 2\alpha^2 \|(T_n^* T_n + \alpha I)^{-1} x^\dagger\|^2 + 2\frac{\delta^2}{\alpha} \\
&\le \max\{2, \tfrac{1}{\tilde\tau}\} (\alpha^2 \|(T_n^* T_n + \alpha I)^{-1} x^\dagger\|^2 + \frac{2\tilde\tau \delta^2}{\alpha})
\end{aligned}$$

implies the assertion. ∎

Lemma 5.5. *Let $\mu \in [0,1]$ and $v \in \mathcal{N}(T)^\perp$. Then*
$$\lim_{\substack{n \to \infty \\ \alpha \to 0}} \alpha^{1-\mu} \|(T_n^* T_n + \alpha I)^{-1} (T_n^* T_n)^\mu v\| = \begin{cases} 0, & \text{if } \mu < 1, \\ \|v\|, & \text{if } \mu = 1. \end{cases}$$

Moreover,
$$\lim_{\substack{n \to \infty \\ \alpha \to 0}} \alpha^{\frac{1}{2}} \|(T_n^* T_n + \alpha I)^{-1} T_n^* w\| = 0$$

for all $w \in \mathcal{N}(T^*)^\perp$.

Proof. Let $\{E_\lambda\}$ be the spectral family of $T_n^* T_n$. By the Hölder inequality we obtain that
$$\begin{aligned}
\alpha^{2-2\mu} \|(T_n^* T_n + \alpha I)^{-1} (T_n^* T_n)^\mu v\|^2 &= \int_0^\infty \frac{\alpha^{2(1-\mu)} \lambda^{2\mu}}{(\alpha + \lambda)^2} d\|E_\lambda v\|^2 \\
&\le \left[\int_0^\infty \left(\frac{\alpha}{\lambda + \alpha}\right)^2 d\|E_\lambda v\|^2 \right]^{1-\mu} \left[\int_0^\infty \left(\frac{\lambda}{\lambda + \alpha}\right)^2 d\|E_\lambda v\|^2 \right]^\mu \\
&\le \left[\alpha^2 \|(T_n^* T_n + \alpha I)^{-1} v\|^2 \right]^{1-\mu} \|v\|^{2\mu}.
\end{aligned}$$

5.2. Regularization with Projection

This together with

$$\|(T_n^*T_n + \alpha I)^{-1}(T_n^*T_n)v - v\| = \alpha\|(T_n^*T_n + \alpha I)^{-1}v\|$$

shows that the first assertion holds if we can prove that

$$\lim_{\substack{n\to\infty\\ \alpha\to 0}} \alpha\|(T_n^*T_n + \alpha I)^{-1}v\| = 0. \qquad (5.48)$$

Let us assume that $v = T^*z$, $z \in \mathcal{N}(T^*)^\perp$. Then, by (5.17),

$$\alpha\|(T_n^*T_n + \alpha I)^{-1}v\| \le \alpha\|(T_n^*T_n + \alpha I)^{-1}T_n^*z\| + \alpha\|(T_n^*T_n + \alpha I)^{-1}T^*(I - Q_n)z\|$$
$$\le \sqrt{\alpha}\|z\| + \|T\|\|(I - Q_n)z\|.$$

Thus, $\alpha\|(T_n^*T_n + \alpha I)^{-1}v\| \to 0$ as $\alpha \to 0$ and $n \to \infty$. Since $\alpha\|(T_n^*T_n + \alpha I)^{-1}\| \le 1$ and $\mathcal{R}(T^*)$ is dense in $\mathcal{N}(T)^\perp$, (5.48) now follows by the Banach-Steinhaus Theorem.

The proof of the second assertion is similar; note that $\|(T_n^*T_n + \alpha I)^{-1}T_n^*w\| = \|(T_nT_n^* + \alpha I)^{-1}(T_nT_n^*)^{\frac{1}{2}}w\|$. ∎

For the proof of convergence rates for compact operators the numbers

$$\gamma_n := \|(I - Q_n)T\| \qquad (5.49)$$

will play an important role. As the following result shows, the compactness of T guarantees that $\gamma_n \to 0$ as $n \to \infty$.

Lemma 5.6. *Let γ_n be defined as in (5.49). Then $\gamma_n \to 0$ as $n \to \infty$ if and only if T is compact.*

Proof. Since $\mathcal{Y}_n$ is finite-dimensional, the operator T_n is compact. Due to the fact that the limit of compact operators in the operator norm is compact again, T is compact if $\gamma_n \to 0$ as $n \to \infty$.

Let us now assume that T is compact. Due to the choice of the spaces $\mathcal{Y}_n$, the real valued continuous functions

$$f_n(y) := \|(I - Q_n)y\|$$

converge pointwise to zero on $\mathcal{N}(T^*)^\perp = \overline{\mathcal{R}(T)}$. Since $|f_n(y) - f_n(z)| \le \|y - z\|$, the convergence is uniform on compact subsets. Since T is compact, $f_n(y) \to 0$ uniformly for $y \in T\mathcal{B}$, where $\mathcal{B}$ is the unit ball in $\mathcal{X}$, hence $\gamma_n \to 0$. ∎

Lemma 5.7. *Let $\mu \in (0, 1)$ and $v \in \mathcal{X}$, $v \ne 0$. Then*

$$\|[(T_n^*T_n)^\mu - (T^*T)^\mu]v\| \le \frac{\sin\mu\pi}{\pi\mu(1-\mu)}\eta_n^\mu(v)\|v\| \le \frac{4}{\pi}\eta_n^\mu(v)\|v\| \le \frac{4}{\pi}\gamma_n^{2\mu}\|v\|,$$

where γ_n is defined by (5.49) and

$$\eta_n(v) := \sup_{t>0} t\|T^*(I - Q_n)T(T^*T + tI)^{-1}v\|\|v\|^{-1}. \qquad (5.50)$$

130   5. Tikhonov Regularization

Moreover,
$$\lim_{n\to\infty} \eta_n(v) = 0$$
(even for non-compact operators T).

Proof. We use the following identity (see [162])
$$(B^*B)^\mu - (A^*A)^\mu = -\frac{\sin\mu\pi}{\pi}\int_0^\infty t^\mu[(B^*B+tI)^{-1} - (A^*A+tI)^{-1}]\,dt \quad (5.51)$$

with $B = T_n$ and $A = T$. Together with
$$(T_n^*T_n + tI)^{-1} - (T^*T+tI)^{-1} = (T_n^*T_n+tI)^{-1}T^*(I-Q_n)T(T^*T+tI)^{-1}$$

we obtain that
$$\begin{aligned}
\|[(T_n^*T_n)^\mu - (T^*T)^\mu]v\| \\
\le \frac{\sin\mu\pi}{\pi}\Bigg[\int_0^{\eta_n(v)} t^\mu\|[(T_n^*T_n+tI)^{-1} - (T^*T+tI)^{-1}]v\|\,dt \\
+ \int_{\eta_n(v)}^\infty t^\mu\|(T_n^*T_n+tI)^{-1}T^*(I-Q_n)T(T^*T+tI)^{-1}v\|\,dt\Bigg].
\end{aligned} \quad (5.52)$$

Since for positive definite, selfadjoint operators A and B
$$\begin{aligned}
\|A - B\| &= \sup_{\|x\|=1} |\langle (A-B)x, x\rangle| \\
&= \sup_{\|x\|=1} |\|A^{\frac{1}{2}}x\|^2 - \|B^{\frac{1}{2}}x\|^2| \\
&\le \sup_{\|x\|=1} \max\{\|A^{\frac{1}{2}}x\|^2, \|B^{\frac{1}{2}}x\|^2\} \\
&= \max\{\|A\|, \|B\|\},
\end{aligned}$$

we obtain with $A = (T_n^*T_n + tI)^{-1}$ and $B = (T^*T+tI)^{-1}, t > 0$, that
$$\|(T_n^*T_n+tI)^{-1} - (T^*T+tI)^{-1}\| \le \frac{1}{t}.$$

Together with (5.50) and (5.52) we now get that
$$\begin{aligned}
\|[(T_n^*T_n)^\mu - (T^*T)^\mu]v\| &\le \frac{\sin\mu\pi}{\pi}\left[\int_0^{\eta_n(v)} t^{\mu-1}\,dt + \eta_n(v)\int_{\eta_n(v)}^\infty t^{\mu-2}\,dt\right]\|v\| \\
&= \frac{\sin\mu\pi}{\pi\mu(1-\mu)}\eta_n^\mu(v)\|v\| \le \frac{4}{\pi}\eta_n^\mu(v)\|v\|.
\end{aligned}$$

Obviously, $\eta_n(v) \le \gamma_n^2$. Since the mapping $h : \mathbb{R}^+ \to \mathcal{Y}$ with $h(t) := tT(T^*T+tI)^{-1}v$ is continuous and since
$$\lim_{t\to 0} h(t) = 0 \quad \text{and} \quad \lim_{t\to\infty} h(t) = Tv,$$

5.2. Regularization with Projection

the set $\{0\} \cup \{tT(T^*T + tI)^{-1}v \mid t > 0\} \cup \{Tv\}$ is compact. Thus, it follows as in the proof of Lemma 5.6 that $\eta_n(v) \to 0$ as $n \to \infty$. ∎

We are now in the position to prove the main result of this section:

Theorem 5.8. *Let $\tau > 1$, let (5.45) hold and let α_n^δ be the solution of (5.44). Moreover, let*
$$x^\dagger = (T^*T)^\mu v, \qquad v \in \mathcal{N}(T)^\perp,$$
for some $\mu \in [0,1]$. Then, as $\delta \to 0$ and $n \to \infty$,

$$\|x_{\alpha_n^\delta,n}^\delta - x^\dagger\| = \begin{cases} o(1), & \text{if } \mu = 0, \\ o(\delta^{\frac{2\mu}{2\mu+1}}) + O(\eta_n^\mu(v)), & \text{if } 0 < \mu < 1, \\ O(\delta^{\frac{2}{3}}) + O(\gamma_n\|(I - Q_n)Tv\|), & \text{if } \mu = 1, \end{cases}$$

where γ_n and $\eta_n(v)$ are defined by (5.49) and (5.50), respectively. For the special case that $\mu = 1/2$, we even obtain that
$$\|x_{\alpha_n^\delta,n}^\delta - x^\dagger\| = o(\delta^{\frac{1}{2}}) + O(\gamma_n\|(I - Q_n)z\|),$$
where $z \in \mathcal{N}(T^*)^\perp$ is such that $x^\dagger = (T^*T)^{\frac{1}{2}}v = T^*z$.

Proof. For $\mu = 0$ the assertion follows immediately with Lemma 5.4 and Lemma 5.5 (e.g., by choosing $\alpha = \delta$).

Let now $\mu > 0$. Then Lemma 5.4 implies that for any $\alpha > 0$

$$\|x_{\alpha_n^\delta,n}^\delta - x^\dagger\| = O(\alpha\|(T_n^*T_n + \alpha I)^{-1}(T_n^*T_n)^\mu v\| + \frac{\delta}{\sqrt{\alpha}} \qquad (5.53)$$
$$+ \|[(T_n^*T_n)^\mu - (T^*T)^\mu]v\|).$$

Obviously, we may assume that $v \neq 0$. This implies that $(T_n^*T_n)^\mu v \neq 0$ for n sufficiently large which we will assume in the following. Since then
$$\lim_{\alpha \to 0} \alpha^3 \|(T_n^*T_n + \alpha I)^{-1}(T_n^*T_n)^\mu v\|^2 = 0$$
and
$$\lim_{\alpha \to \infty} \alpha^3 \|(T_n^*T_n + \alpha I)^{-1}(T_n^*T_n)^\mu v\|^2 = \infty$$
and since $\alpha^3\|(T_n^*T_n + \alpha I)^{-1}(T_n^*T_n)^\mu v\|^2$ is continuous and strictly increasing with respect to α, by the Intermediate Value Theorem, there is a unique $\alpha_n \geq 0$ such that
$$\alpha_n^3\|(T_n^*T_n + \alpha_n I)^{-1}(T_n^*T_n)^\mu v\|^2 = \delta^2.$$
Together with (5.53) we now obtain that
$$\|x_{\alpha_n^\delta,n}^\delta - x^\dagger\| = O(\delta^{\frac{2\mu}{2\mu+1}}[\alpha_n^{1-\mu}\|(T_n^*T_n + \alpha_n I)^{-1}(T_n^*T_n)^\mu v\|]^{\frac{1}{2\mu+1}}$$
$$+ \|[(T_n^*T_n)^\mu - (T^*T)^\mu]v\|).$$

132   5. Tikhonov Regularization

The assertion now follows with Lemma 5.5, Lemma 5.7 and the fact that $\alpha_n \to 0$ as $n \to \infty$ and $\delta \to 0$. The proof for the special case that $\mu = 1/2$ again follows with Lemma 5.5 similar to the one above; note that $\|(T_n^* - T^*)z\| \leq \gamma_n \|(I - Q_n)z\|$. ∎

Remark 5.9. Note that the convergence rate in Theorem 5.8 is optimal with respect to δ (cf. Section 5.1).

One advantage of the approach presented here compared to the one, where a compact operator T is approximated by TP_n is that α_n^δ only depends on δ and not on γ_n. This means that α_n^δ can go to zero independently of γ_n. For the approach, where $T_n = TP_n$ and γ_n in (5.49) is replaced by $\|(I - P_n)T^*\|$, α_n^δ is not allowed to go faster to zero than $\|(I - P_n)T^*\|^2$ (cf. [103, Theorem 4.2.4]).

A further advantage is that for compact operators the fastest rate with respect to the finite-dimensional approximation in our case is $\|(I - Q_n)T\|^2$ (cf. Theorem 5.8), whereas for $T_n = TP_n$ the rate $\|(I - P_n)T^*\|$ is best possible. Since, for many operators T, $\|(I - P_n)T^*\|$ behaves as $\|(I - Q_n)T\|$ if $\mathcal{X}_n$ and $\mathcal{Y}_n$ are chosen as the same spaces, this means that for our approach the square of the best possible convergence rate for $T_n = TP_n$ may be obtained.

We now show that also $\|(I - P_n)x^\dagger\|$ may be estimated from above by the second error term (concerning the rate with respect to the finite-dimensional approximation) of Theorem 5.8.

Lemma 5.10. *If* $x^\dagger = (T^*T)^\mu v$, $v \neq 0$, *and* $\mu \in (0,1]$, *then*

$$\|(I - P_n)x^\dagger\| \leq \begin{cases} \dfrac{4}{\pi}\eta_n^\mu(v)\|v\|, & \text{if } \mu < 1, \\ \gamma_n \|(I - Q_n)z\|, & \text{if } \mu = \tfrac{1}{2}, \\ \gamma_n \|(I - Q_n)Tv\|, & \text{if } \mu = 1, \end{cases}$$

where $x^\dagger = (T^*T)^{\frac{1}{2}}v = T^*z$ *in the special case that* $\mu = 1/2$, *and* γ_n *and* $\eta_n(v)$ *are defined by* (5.49) *and* (5.50), *respectively.*

Proof. Using (5.51) with $B = T$ and $A = 0$, we obtain for $0 < \mu < 1$ that

$$(I - P_n)(T^*T)^\mu = -\frac{\sin \mu\pi}{\pi}\int_0^\infty t^\mu (I - P_n)[(T^*T + tI)^{-1} - (tI)^{-1}]\,dt$$

$$= \frac{\sin \mu\pi}{\pi}\int_0^\infty t^{\mu-1}(I - P_n)T^*T(T^*T + tI)^{-1}\,dt.$$

Due to (3.49), $(I - P_n)T^*T = (I - P_n)T^*(I - Q_n)T$. Thus, we obtain as in Lemma 5.7

$$\|(I - P_n)(T^*T)^\mu v\| \leq \frac{\sin \mu\pi}{\pi}\left[\int_0^{\eta_n(v)} t^{\mu-1}\,dt + \eta_n(v)\int_{\eta_n(v)}^\infty t^{\mu-2}\,dt\right]\|v\|$$

$$\leq \frac{4}{\pi}\eta_n^\mu(v)\|v\|.$$

For the special case that $\mu = 1/2$ and that $x^\dagger = T^*z$, we obtain that

$$\begin{aligned} \|(I - P_n)T^*z\| &= \|(I - P_n)T^*(I - Q_n)z\| \\ &\leq \gamma_n \|(I - Q_n)z\| \, . \end{aligned}$$

Finally, for $\mu = 1$ the result follows as above with z replaced by Tv. ∎

Remark 5.11. The estimates in Lemma 5.10 are reasonably sharp for compact operators as the following argument shows:

Let T be a compact operator with singular system $(\sigma_n; v_n, u_n)$ and assume that $\mathcal{Y}_n := \mathrm{span}\{u_1, \ldots, u_n\}$. Without loss of generality, we assume that $\sigma_{n+1} < \sigma_n$ for all $n \in \mathbb{N}$. Then for $x^\dagger = (T^*T)^\mu v$ we obtain with Lemma 5.10 and $\eta_n(v) \leq \gamma_n^2$ that

$$\|(I - P_n)x^\dagger\| = O(\gamma_n^{2\mu}) = O(\sigma_{n+1}^{2\mu})$$

(compare the proof of Proposition 3.29). Let us now assume that this estimate holds for an element $x^\dagger \in \mathcal{N}(T)^\perp$. Since for this special choice of spaces $\mathcal{Y}_n$ the projection method coincides with the truncated singular value expansion, we have that

$$\|E_t x^\dagger\| = \|(I - P_n)x^\dagger\| = O(\sigma_{n+1}^{2\mu}) = O(t^\mu)$$

for all $t \in (\sigma_{n+1}^2, \sigma_n^2]$. Hence, (4.52) is satisfied. Now by the converse result Lemma 4.12 we obtain that

$$x^\dagger \in \bigcap_{\nu < \mu} \mathcal{X}_\nu \, .$$

In practice, we want to know how $\tau \in [1, \infty)$ in the parameter choice strategy (5.44) should be chosen to make the error as small as possible for fixed $\delta > 0$ and fixed n. This question is solved by the next proposition. The result shows that the smallest error is obtained for the choice $\tau = 1$. Therefore, the rates of Theorem 5.8 are even valid if $\tau = 1$.

Proposition 5.12. *Let n, δ be fixed and let (5.45) (with $\tau = 1$) hold. Moreover, let α_n^δ be the unique solution of (5.44) (with $\tau = 1$). Then for all $\alpha \geq \alpha_n^\delta$*

$$\|x_{\alpha_n^\delta, n}^\delta - x^\dagger\| \leq \|x_{\alpha, n}^\delta - x^\dagger\|$$

holds.

Proof. Let $e(\alpha)$ be defined by

$$e(\alpha) := \frac{1}{2} \|x_{\alpha, n}^\delta - x^\dagger\|^2 \, .$$

Using $T_n x^\dagger = Q_n y$ we obtain that

134   5. Tikhonov Regularization

$$\begin{aligned}
e'(\alpha) &= \langle x^\delta_{\alpha,n} - x^\dagger, \frac{d}{d\alpha} x^\delta_{\alpha,n} \rangle \\
&= -\langle (T_n^* T_n + \alpha I)^{-1} T_n^* y^\delta - x^\dagger, (T_n^* T_n + \alpha I)^{-2} T_n^* y^\delta \rangle \\
&= -\langle (T_n T_n^* + \alpha I)^{-1} T_n T_n^* y^\delta - Q_n y, (T_n T_n^* + \alpha I)^{-2} Q_n y^\delta \rangle \\
&= \alpha \| (T_n T_n^* + \alpha I)^{-\frac{3}{2}} Q_n y^\delta \|^2 - \langle Q_n(y^\delta - y), (T_n T_n^* + \alpha I)^{-2} Q_n y^\delta \rangle \\
&\geq \| (T_n T_n^* + \alpha I)^{-\frac{3}{2}} Q_n y^\delta \| (\alpha \| (T_n T_n^* + \alpha I)^{-\frac{3}{2}} Q_n y^\delta \| \\
&\qquad\qquad - \| (T_n^* T_n + \alpha I)^{-\frac{1}{2}} Q_n (y^\delta - y) \|) \\
&\geq \| (T_n T_n^* + \alpha I)^{-\frac{3}{2}} Q_n y^\delta \| (\alpha \| (T_n T_n^* + \alpha I)^{-\frac{3}{2}} Q_n y^\delta \| - \frac{\delta}{\sqrt{\alpha}}).
\end{aligned}$$

By the definition of α_n^δ and the monotonicity of $\alpha^{\frac{3}{2}} \| (T_n T_n^* + \alpha I)^{-\frac{3}{2}} Q_n y^\delta \|$, which follows from spectral theory, this shows that

$$e'(\alpha) \geq 0 \quad \text{for} \quad \alpha \geq \alpha_n^\delta,$$

which proves the assertion. ∎

This proposition shows that with our parameter selection method we always obtain a solution, where α_n^δ is larger than the actual minimizer of $\| x^\delta_{\alpha,n} - x^\dagger \|$ but that the error is worse for any $\alpha \geq \alpha_n^\delta$. Note that for an $\alpha \geq \alpha_n^\delta$ the error is always bounded by $\| x^\dagger \|$. Only if $\alpha > 0$ is too close to 0, the error may explode, which can never happen for our parameter choice.

The results above may be slightly improved if the smallest eigenvalue of $T_n T_n^*$ (or equivalently of $T_n^* T_n$) is used in the estimates and the parameter choice; see [152] for details. There also numerical results can be found verifying the theoretical ones. For implementation of Tikhonov regularization in finite dimensions see Section 9.3.

It was shown in [109] that the parameter choice (4.85) applied to a general regularization method for solving (3.48) again yields optimal rates with respect to the data noise level δ and with respect to the finite-dimensional approximation. A convergence analysis for regularization methods where the data y and the operator T are only known approximately can be found, e.g., in [214, 277].

5.3. Maximum Entropy Regularization

As we will see also in Chapter 8, it makes sense to use other penalty terms instead of $\|x\|^2$ in ordinary Tikhonov regularization as treated in Section 5.1 under some circumstances. The choice of the regularization term depends on qualitative assumptions about the solution and influences the convergence properties.

All regularization terms treated in Chapter 8 are still quadratic in the unknown to be determined (as in ordinary Tikhonov regularization), so that the first order

5.3. Maximum Entropy Regularization

necessary conditions are still linear equations. These properties do not hold any more for another widely-used class of regularization terms, which involve what is called the (information-theoretic) *entropy*. We give a motivation for this kind of regularization here and make some preliminary remarks about the behaviour of such methods. For a detailed analysis, we will need results about Tikhonov regularization for nonlinear problems to be described in Section 10.2, so that we defer this analysis to Section 10.6.

Entropy is a probabilistic concept which appears in information theory and in statistical mechanics (see, e.g., [130, 140, 141, 149]). Hence, we have to motivate maximum entropy regularization in a probabilistic setting. We do this by considering the problem of determining a probability distribution from incomplete data. For simplicity, we first consider the case that the underlying random variable has only finitely many possible realizations. The case of continuous distributions follows then via a quite usual limiting argument.

Assume we have a random variable $\underline{x}$ with finitely many possible realizations $x_1, \ldots, x_n$, and assume that we make some prior assumptions about the probabilities of the outcomes in the form of a probability measure P^* such that

$$P^*(\{x_i\}) = p_i^* \qquad (5.54)$$

is the probability that $\underline{x}$ has the realization x_i. P^* is called the *prior distribution*. Now, assume we obtain (e.g., by measurements) additional data about $\underline{x}$, which let us change our prior distribution into a new probability assignment P with

$$P(\{x_i\}) = p_i. \qquad (5.55)$$

Of course, we want to do this in such a way that as only additional information we use the new data. This simple statement contains the key to the maximum entropy method. Before explaining this, we have to formalize what we mean by "information". Originally due to Shannon, this is done in the following axiomatic way:

Let $I(P; P^*) = I(p_1, \ldots, p_n; p_1^*, \ldots, p_n^*)$ denote the *information of P relative to P^**, i.e., the information content of the probability assigment (5.55) when the prior probabilities were given by (5.54). One postulates that this information satisfies the following axioms:

$$I \text{ is continuous}, \qquad (5.56)$$

$$I(P; P) = 0, \qquad (5.57)$$

$$\text{for all permutations } \pi, \quad I(\pi(P); \pi(P^*)) = I(P; P^*), \qquad (5.58)$$

$$I\left(\tfrac{1}{m}, \ldots, \tfrac{1}{m}, 0, \ldots, 0; \tfrac{1}{n}, \ldots, \tfrac{1}{n}\right) \text{ is increasing in } n \text{ and}$$
$$\text{decreasing in } m \text{ for all } m, n \in \mathbb{N}, m \leq n, \qquad (5.59)$$

5. Tikhonov Regularization

with $q_1^{(*)} := \sum_{i=1}^{r} p_i^{(*)}$, $q_2^{(*)} := \sum_{i=r+1}^{n} p_i^{(*)}$, the following holds:

$$I(p_1,\ldots,p_n;p_1^*,\ldots,p_n^*) = I(q_1,q_2;q_1^*,q_2^*) + \quad (5.60)$$
$$q_1 I\left(\frac{p_1}{q_1},\ldots,\frac{p_r}{q_1};\frac{p_1^*}{q_1^*},\ldots,\frac{p_r^*}{q_1^*}\right) + q_2 I\left(\frac{p_{r+1}}{q_2},\ldots,\frac{p_n}{q_2};\frac{p_{r+1}^*}{q_2^*},\ldots,\frac{p_n^*}{q_2^*}\right).$$

Here and below, we use the notation $p_i^{(*)}$ etc. to indicate that we mean both p_i and p_i^*. Continuity (5.56) and invariance under changes of the labeling of the possible realizations (5.58) are obvious requirements, while (5.57) states that, if the new data do not change the probability assignment, then they do not have any information content, which is also reasonable.

In (5.59), situations are compared where the new data lead to an exclusion of $n-m$ realizations, leaving all other realizations equally likely. The more realizations can be excluded, the higher the information content of the data.

Finally, (5.60) is a composition rule that says that, if one divides the possible realizations into two groups $\{x_1,\ldots,x_r\}$ and $\{x_{r+1},\ldots,x_n\}$, then the total information can also be obtained by calculating the information contained in the probabilities of the two subgroups and adding (weighted by the probabilities of the subgroups) the information contained in the conditional probabilities within these subgroups. Also this seems to be quite reasonable.

It is remarkable that, if for all n and all $p_1,\ldots,p_n,p_1^*,\ldots,p_n^* \geq 0$ with

$$\sum_{i=1}^{n} p_i^{(*)} = 1, \quad (5.61)$$

a function I fulfilling (5.56) – (5.60) is defined, then I has to have the form

$$I(p_1,\ldots,p_n;p_1^*,\ldots,p_n^*) = k \sum_{i=1}^{n} p_i \log\left(\frac{p_i}{p_i^*}\right) \quad (5.62)$$

with some positive constant k. For a proof see [130, Appendix A].

The negative of the expression on the right-hand side of (5.62) is called the *entropy of P* (with prior distribution P^*); $I(P;P^*)$ is always nonnegative, and vanishes if and only if $P = P^*$.

Now, we return to the statement made after (5.55) that we want to make the new probability assignment P in such a way that the only information we use is the new data. Using the concept of information just introduced, this means that we choose P such that $I(P;P^*)$ is minimized subject to the data, i.e., due to (5.62), we maximize the entropy

$$E(P;P^*) := -k \sum_{i=1}^{n} p_i \log\left(\frac{p_i}{p_i^*}\right)$$

with the data and (5.61) as constraints. This is called the *maximum entropy method* (for this simple case).

5.3. Maximum Entropy Regularization

For a discussion of the maximum entropy method in image reconstruction from noisy data see [67]. From the vast literature on the maximum entropy method applied to a variety of problems in physics we quote [256].

It is now clear how to carry over these arguments to continuous distributions via limiting arguments (see, e.g., [130, pp. 40]): let, e.g., $x : [a, b] \to \mathbb{R}$ be a nonnegative function with

$$\int_a^b x(t)\, dt = 1, \tag{5.63}$$

which can be interpreted as an unknown probability distribution. Our "prior knowledge" is assumed to be incorporated into a "prior distribution" x^* (cf. [142]). As data, we assume to have a (possibly nonlinear) operator equation

$$F(x) = y \tag{5.64}$$

on a suitable function space, e.g., on $\mathcal{L}^2([a,b])$. The information contained in x relative to x^* is then the negative of the entropy

$$E(x; x^*) := -k \int_a^b x(t) \log \frac{x(t)}{x^*(t)}\, dt. \tag{5.65}$$

Thus, if we want to find x such that *only the data* (5.64) *are used*, we have to minimize the information subject to the data, i.e., we have to maximize the entropy E subject to the constraint (5.64). Via Lagrange multipliers, this motivates the auxiliary problem

$$\|F(x) - y\|^2 + \alpha \int_a^b x(t) \log \frac{x(t)}{x^*(t)}\, dt \to \min. \tag{5.66}$$

One easily sees that the global maximizer of $E(\cdot, x^*)$ is x^*/e, and x^* if the restriction (5.63) is imposed, which reiterates the role of x^* as a-priori information without additional data. If we now replace (5.64) by a feasibility constraint

$$\|F(x) - y^\delta\| \leq \delta \tag{5.67}$$

and the global minimizer does not fulfill this constraint, then the convex functional $-E$ assumes its minimum on the boundary of the feasible set described by (5.67), i.e., suitable properties of F assumed, on the set described by

$$\|F(x) - y^\delta\| = \delta,$$

which leads to (5.66). (5.66) now resembles Tikhonov regularization, but with the quadratic regularization term replaced by a non-quadratic one.

So far, we still assumed (5.63), both for x and for x^*. Since $\log z \geq 1 - 1/z$ holds for all $z > 0$, we then obtain, by setting $z := x(t)/x^*(t)$, that $-E(x; x^*)$ is nonnegative. This need no longer be true if (5.63) is omitted. Then, one sometimes replaces the negative entropy in (5.66) by

$$\int_a^b |x(t)| \log \left|\frac{x(t)}{x^*(t)}\right| dt,$$

138   5. Tikhonov Regularization

but the information-theoretic motivation for this "maximum entropy method" is certainly lost.

In the literature, slightly different concepts of and names for entropy are used. E.g., in [140], the "extended maximum entropy method" is defined as maximizing, subject to the data, the entropy-like quantity

$$\int_a^b \left[\frac{x^*(t)}{x(t)} + \log\frac{x(t)}{x^*(t)} - 1\right] dt,$$

which is called *Burg's entropy* in [60] (cf. [37] for the discrete version and [144] for a comparison). The entropy we use in (5.65), then, is called *cross entropy* (also *relative entropy*, *Kullback-Leibler number*, *directed divergence*) in [145], while the term *cross entropy* is used for

$$\int_a^b [x(t) \log\frac{x(t)}{x^*(t)} + x^*(t) - x(t)] \, dt$$

in [60], which is of course the same as (5.65) as long as x^* and x are normalized by specifying $\int_a^b x^{(*)}(t)\,dt$.

In emission tomography, a class of iterative algorithms called *expectation minimization* (EM) algorithms for computing maximum likelihood reconstructions is frequently used (see, e.g., [160, 254]). It turns out that these algorithms can be analyzed using the entropy (5.65) (see [197] and [139]).

The first rigorous study of (5.66) as a regularization method seems to be [155]. There, for linear F, the authors prove stability of the minimization problem (5.66) for fixed $\alpha > 0$ (for the weak topology in $\mathcal{L}^2$, as far as the minimizer is concerned). The first convergence result as $\alpha \to 0$ (without rates) is contained in [9]. Convergence rates were provided first in [60] and [75]: in these papers, rather similar results are proven with completely different methods. We will present the methods and results of [75] in Section 10.6, since, as mentioned above, (5.66) is a nonlinear (non-quadratic) problem even if F is linear. Here, we give a short outline of the method of proof used in [60]:

Let, for bounded domains $\Omega_1, \Omega_2 \subseteq \mathbb{R}^n$,

$$(Kx)(s) := \int_{\Omega_1} k(s,t)x(t)\,dt \quad (s \in \Omega_2)$$

be an integral operator with continuous kernel k, considered as a compact operator from $\mathcal{L}^1(\Omega_1)$ into $\mathcal{L}^2(\Omega_2)$. Then, for $y \in \mathcal{L}^2(\Omega_2)$, the equation

$$Kx = y \quad (5.68)$$

is solved approximately by considering, for $\alpha > 0$, the minimization problem

$$l(x) := \|Kx - y^\delta\|^2_{\mathcal{L}^2(\Omega_2)} + \alpha D(x; x^*) \to \min, \quad x \in \mathcal{L}^1(\Omega_1), x \geq 0 \quad (5.69)$$

with

$$D(x, x^*) := \int_{\Omega_1} \left[x(t) \log\frac{x(t)}{x^*(t)} + x^*(t) - x(t)\right] dt\,.$$

5.3. Maximum Entropy Regularization

An essential point is that the cross-entropy D can be written as

$$D(x, x^*) = d(x) - d(x^*) - \langle d'(x^*), x - x^* \rangle$$

with

$$d(x) := \int_{\Omega_1} x(t) \log x(t)\, dt\,.$$

This and the convexity of d imply the following estimate (originally due to Bregman [35]): if x_α^δ is a solution of (5.69), then, for all $x \in \mathcal{L}^1(\Omega_1)$ with $x \geq 0$,

$$\|K(x - x_\alpha^\delta)\|^2 + \alpha D(x, x_\alpha^\delta) \leq l(x) - l(x_\alpha^\delta)\,. \tag{5.70}$$

From this, the following estimate follows: if x_α and z_α solve (5.69) with y^δ replaced by y and v, respectively, then

$$\|K(x_\alpha - z_\alpha)\|^2 + \alpha D(x_\alpha, z_\alpha) \leq 4\|y - v\|^2$$

holds. This already shows that Kx_α depends continuously on the right-hand side (uniformly in α) and that the same stability result holds for x_α for any fixed $\alpha > 0$ with respect to the $\mathcal{L}^1$-norm, since

$$\|x_\alpha - z_\alpha\|^2_{\mathcal{L}^1(\Omega_1)} \leq CD(x_\alpha, z_\alpha)$$

holds with (cf. [60, (1.7)], [33])

$$C = \frac{4}{3}\int_{\Omega_1} x_\alpha(t)\, dt + \frac{2}{3}\int_{\Omega_1} z_\alpha(t)\, dt\,.$$

The existence of solutions of (5.69) is based on the fact that level sets of the cross-entropy, i.e., sets of the form $\{x \in \mathcal{L}^1(\Omega_1) \mid x \geq 0, D(x, x^*) \leq M\}$ are weakly compact in $\mathcal{L}^1(\Omega_1)$; the uniqueness follows from the strict convexity of l.

Again based on (5.70), the following convergence result is shown in [60]: if (5.68) has a nonnegative least-squares solution in $\mathcal{L}^1(\Omega_1)$ that minimizes $x \to D(x, x^*)$, then the minimizers x_α^δ of (5.69) converge in $\mathcal{L}^1(\Omega_1)$ to this (unique) maximum entropy least-squares solution provided that $\delta/\sqrt{\alpha} \to 0$ and $\|y - y^\delta\| \leq \delta$. For $x^* \equiv 1$, this has already been shown in [9]. In addition, in [60] also the convergence rate $O(\sqrt{\delta})$ of x_α^δ towards the maximum entropy least-squares solution $x_E^\dagger$, again in the $\mathcal{L}^1$-norm, is proved provided that the source condition

$$\log\left(\frac{x_E^\dagger}{x^*}\right) \in R(K^*) \tag{5.71}$$

holds.

In Section 10.6, we will prove rather similar results (for the entropy used in (5.66)) by a different approach, which has the advantage that it carries over to the case that F in (5.66) is nonlinear.

In spite of the information-theoretic motivation outlined here, it is not easy to compare maximum entropy methods with linear regularization methods like Tikhonov

140   5. Tikhonov Regularization

regularization. In many practical problems, maximum entropy methods work well, but so does Tikhonov regularization. In our opinion, clear criteria when to prefer one over the other are not available.

Another variant of Tikhonov regularization is *total variation regularization*, where (5.66) is replaced by

$$\|F(x) - y\|^2 + \alpha \int_\Omega |\nabla x(t)|\, dt \to \min;$$

here, x is a function defined on a multi-dimensional domain. This variant has the advantage over Tikhonov regularization that it enhances sharp features in x, which is useful, e.g., in image reconstruction (cf. [55, 239]). A technical difficulty is the non-differentiability of the penalty term. For a preliminary convergence analysis see [1].

5.4. Convex Constraints

If one solves a linear problem $Tx = y$, one frequently knows that the solution satisfies some additional conditions, e.g., it is clear that density functions will never assume negative values. This means that one knows a-priori that the solution is an element of a certain subset of $\mathcal{X}$. On the other hand, one is sometimes not interested in the solution $T^\dagger y$, but in the best-approximate solution on a certain set $\mathcal{C}$, which we assume to be closed and convex in the following. Therefore, we consider the problem

$$Tx = y \wedge x \in \mathcal{C}, \tag{5.72}$$

where we assume that $T \in \mathcal{L}(\mathcal{X}, \mathcal{Y})$. Note that in spite of the linearity of T, problem (5.72) is now nonlinear because of the constraint.

The concept of solution for problem (5.72) is the $\mathcal{C}$-*best-approximate solution* $x_\mathcal{C}^\dagger$, i.e.,

$$\|Tx_\mathcal{C}^\dagger - y\| = \inf\{\|Tx - y\| \mid x \in \mathcal{C}\} \tag{5.73}$$

and

$$\|x_\mathcal{C}^\dagger\| = \min\{\|x\| \mid \|Tx - y\| = \|Tx_\mathcal{C}^\dagger - y\| \wedge x \in \mathcal{C}\}. \tag{5.74}$$

Thus, similar to the solution $x^\dagger = T^\dagger y$ of linear unconstrained problems, a $\mathcal{C}$-best-approximate solution minimizes the norm of the residuals on $\mathcal{C}$ and has minimal norm among all minimizers. For the proof of the existence and uniqueness of $x_\mathcal{C}^\dagger$ we need the following lemma.

Lemma 5.13. *Let $f : \mathcal{X} \to \mathbb{R}$ be a convex and Fréchet-differentiable functional with gradient ∇f and let $\mathcal{C}$ be a closed and convex subset of $\mathcal{X}$. Then $x \in \mathcal{C}$ is a solution of the minimization problem*

$$f(x) \to \min, \quad x \in \mathcal{C} \tag{5.75}$$

5.4. Convex Constraints

if and only if

$$\langle \nabla f(x), h - x \rangle \geq 0 \quad \text{for all } h \in \mathcal{C} \tag{5.76}$$

holds. Moreover, if f is strictly convex and

$$\lim_{\|u\| \to \infty} f(u) = \infty, \tag{5.77}$$

then problem (5.75) has a unique solution.

Proof. Let $x \in \mathcal{C}$ be a solution of (5.75) and let $h \in \mathcal{C}$ be arbitrary, but fixed. Since $\mathcal{C}$ is convex, $(th + (1-t)x) \in \mathcal{C}$ for all $t \in [0,1]$. Together with the minimizing property of x this implies that

$$0 \leq \lim_{t \to 0^+} \frac{f(x + t(h-x)) - f(x)}{t} = \langle \nabla f(x), h - x \rangle,$$

and thus (5.76). On the other hand, let now $x \in \mathcal{C}$ be such that (5.76) holds, then we obtain with the convexity of f that

$$0 \leq \langle \nabla f(x), h - x \rangle = \lim_{t \to 0^+} \frac{f(x + t(h-x)) - f(x)}{t} \leq f(h) - f(x)$$

for all $h \in \mathcal{C}$, meaning that x solves (5.75).

Let us now assume that f is strictly convex and that (5.77) holds. Since f is Fréchet-differentiable, it is also continuous. Together with the convexity of f this implies that f is weakly lower semicontinuous. Due to condition (5.77) we may restrict the search for a minimizer of (5.75) to a bounded, closed and convex subset of $\mathcal{C}$. Since this set is then weakly compact (cf. [58]), the weak lower semicontinuity and strict convexity of f imply the existence of a unique minimizer of (5.75). ∎

The existence and uniqueness of $x_\mathcal{C}^\dagger$ can be now characterized as follows:

Proposition 5.14. *Let $Q_\mathcal{C}$ be the metric projector of $\mathcal{Y}$ onto $\overline{T(\mathcal{C})}$. Then a $\mathcal{C}$-best-approximate solution exists if and only if $Q_\mathcal{C} y \in T(\mathcal{C})$; it is then unique.*

Proof. Since $\|Tx - y\|^2$ is convex and Fréchet-differentiable with gradient $2T^*(Tx - y)$, Lemma 5.13 implies that a minimizer of $\|Tx - y\|^2$ may be characterized via

$$\langle T^*(Tx - y), h - x \rangle \geq 0 \quad \text{for all } h \in \mathcal{C}$$

or equivalently, since $T(\mathcal{C})$ is dense in $\overline{T(\mathcal{C})}$, via

$$\langle Tx - y, u - Tx \rangle \geq 0 \quad \text{for all } u \in \overline{T(\mathcal{C})}. \tag{5.78}$$

Since $\overline{T(\mathcal{C})}$ is closed and convex and $\|u - y\|^2$ is a strictly convex and Fréchet-differentiable functional with gradient $2(u - y)$ satisfying condition (5.77), Lemma 5.13 again implies that $Q_\mathcal{C} y$ is the unique element in $\overline{T(\mathcal{C})}$ for which

$$\langle Q_\mathcal{C} y - y, u - Q_\mathcal{C} y \rangle \geq 0 \quad \text{for all } u \in \overline{T(\mathcal{C})} \tag{5.79}$$

142   5. Tikhonov Regularization

holds.

Let us now assume that a $\mathcal{C}$-best-approximate solution $x_\mathcal{C}^\dagger$ exists. Then (5.73), (5.78) and (5.79) imply that

$$Tx_\mathcal{C}^\dagger = Q_\mathcal{C} y \tag{5.80}$$

and hence $Q_\mathcal{C} y \in T(\mathcal{C})$. If, on the other hand, $Q_\mathcal{C} y \in T(\mathcal{C})$, then, due to (5.78) and (5.79), the set $\mathcal{M} := \{\bar{x} \in \mathcal{C} \mid \bar{x} \text{ minimizes } \|Tx - y\| \text{ over } \mathcal{C}\}$ is not empty. Since $\mathcal{C}$ is closed and convex and since $\|Tx - y\|$ is a convex functional, $\mathcal{M}$ is also closed and convex. Therefore, there exists a unique element of minimal norm in $\mathcal{M}$ (cf. Lemma 5.13). Due to (5.73) and (5.74) this is the $\mathcal{C}$-best-approximate solution $x_\mathcal{C}^\dagger$. ∎

Note that, if $\mathcal{X} = \mathcal{C}$, the condition that $Q_\mathcal{C} y \in T(\mathcal{C})$ is equivalent to $Qy \in \mathcal{R}(T)$, where Q is the orthogonal projector of $\mathcal{Y}$ onto $\overline{\mathcal{R}(T)}$.

We know from the unconstrained case that the solutions do not depend continuously on the data if $\mathcal{R}(T)$ is not closed (cf. Proposition 2.4). The question arises, whether the convex constraints stabilize the problem or if it is still necessary to regularize problem (5.72). There is a well known result by Tikhonov saying that the restriction of an injective (not necessarily linear) operator T to the set $\mathcal{C}$ has a continuous inverse if $\mathcal{C}$ is compact. On the other hand, one can show that $x_\mathcal{C}^\dagger$ does not depend continuously on the data y if T is compact and injective and if a sequence exists in $\mathcal{C}$ converging weakly but not strongly towards $x_\mathcal{C}^\dagger$, meaning that $\mathcal{C}$ is infinite-dimensional around $x_\mathcal{C}^\dagger$ (see Proposition 10.1). Therefore, in general, problem (5.72) is ill-posed. Thus, we have to regularize it. Since $x_\mathcal{C}^\dagger \in \mathcal{C}$, it is quite natural to require that the regularized solutions are in $\mathcal{C}$, too.

If $x_\mathcal{C}^\dagger = x^\dagger$, then there is a fast way to compute regularized solutions in $\mathcal{C}$. First one determines a regularized solution x_α^δ of the unconstrained problem, using any method of Section 4.1. Then one computes the metric projection of x_α^δ onto the set $\mathcal{C}$. Since $\mathcal{C}$ is closed and convex, it follows from Lemma 5.13 that the metric projector $P_\mathcal{C}$, is non-expansive. Therefore,

$$\|x_\mathcal{C}^\dagger - P_\mathcal{C} x_\alpha^\delta\| = \|P_\mathcal{C} x^\dagger - P_\mathcal{C} x_\alpha^\delta\| \leq \|x^\dagger - x_\alpha^\delta\| \,.$$

This means that, in this case, all results concerning stability, convergence and convergence rates of Section 4.1 also hold for the constrained case. For some sets $\mathcal{C}$, the metric projection can be computed without having to solve a nonlinear optimization problem, e.g., the metric projection of a function $f \in \mathcal{L}^2$ onto the set of nonnegative functions in $\mathcal{L}^2$, is given by

$$P_\mathcal{C} f = \max\{f, 0\}\,. \tag{5.81}$$

If $x_\mathcal{C}^\dagger \neq x^\dagger$, we have to choose a different regularization method. In this section we show that constrained Tikhonov regularization stabilizes problem (5.72). Moreover, we prove convergence and present conditions that guarantee convergence rates. Thereafter, we show that the discrepancy principle yields, as in the unconstrained case, the rate $O(\sqrt{\delta})$ if $x_\mathcal{C}^\dagger$ satisfies a smoothness condition.

5.4. Convex Constraints

Tikhonov regularization in the constrained case amounts to solve the minimization problem

$$\|Tx - y^\delta\|^2 + \alpha \|x\|^2 \to \min, \quad x \in \mathcal{C}. \tag{5.82}$$

As in the unconstrained case, y^δ are perturbations of y. Again the solutions of (5.82) are denoted by x_α^δ or x_α if y^δ is replaced by y.

It was shown in Morozov [194] that problem (5.82) has a stable solution and that it converges towards $x_\mathcal{C}^\dagger$ if the regularization parameter is chosen appropriately. In [215] conditions have been presented that guarantee the same convergence rates as for the unconstrained problem. Since, due to the nonlinearity of the problem, the analysis can no longer be based on spectral theory, which requires proof methods different from the unconstrained case, we include most of the proofs following the presentation in [211, 215]. First we show that problem (5.82) has a unique solution.

Theorem 5.15. *Let $y^\delta \in \mathcal{Y}, \alpha > 0$, and $\mathcal{C}$ be closed and convex. Then problem (5.82) has a unique solution x_α^δ. x_α^δ remains the solution if y^δ is replaced by Qy^δ.*

Proof. Since the functional in (5.82) is strictly convex, tending towards ∞ as $\|x\| \to \infty$, and since it is Fréchet-differentiable with gradient $2(T^*Tx + \alpha x - T^*y^\delta)$, it follows with Lemma 5.13 that problem (5.82) has a unique solution x_α^δ, which is characterized as the unique element in $\mathcal{C}$ satisfying the variational inequality

$$\langle T^*Tx_\alpha^\delta + \alpha x_\alpha^\delta - T^*y^\delta, h - x_\alpha^\delta \rangle \geq 0 \quad \text{for all } h \in \mathcal{C} \tag{5.83}$$

Since $T^* = T^*Q$, x_α^δ also minimizes (5.82) with y^δ replaced by Qy^δ. ∎

In the next theorem we show that the constrained Tikhonov regularized solution x_α^δ depends Lipschitz continuously on the data y^δ. Therefore, the problem of solving (5.82) is stable.

Theorem 5.16. *Let $\alpha > 0$, and let x_α^δ and $\overline{x}_\alpha^\delta$ be constrained Tikhonov regularized solutions for the right-hand sides $y^\delta, \overline{y}^\delta \in \mathcal{Y}$ of problem (5.72), respectively. Then*

$$\|x_\alpha^\delta - \overline{x}_\alpha^\delta\| \leq \frac{\|Q(y^\delta - \overline{y}^\delta)\|}{\sqrt{\alpha}} \quad \text{and} \quad \|T(x_\alpha^\delta - \overline{x}_\alpha^\delta)\| \leq \|Q(y^\delta - \overline{y}^\delta)\|$$

hold.

Proof. With (5.83) we get

$$\langle T^*Tx_\alpha^\delta + \alpha x_\alpha^\delta - T^*y^\delta, \overline{x}_\alpha^\delta - x_\alpha^\delta \rangle \geq 0$$

and

$$\langle T^*T\overline{x}_\alpha^\delta + \alpha \overline{x}_\alpha^\delta - T^*\overline{y}^\delta, x_\alpha^\delta - \overline{x}_\alpha^\delta \rangle \geq 0.$$

Adding both inequalities, we obtain

$$\langle (T^*T + \alpha I)(x_\alpha^\delta - \overline{x}_\alpha^\delta) + T^*(\overline{y}^\delta - y), \overline{x}_\alpha^\delta - x_\alpha^\delta \rangle \geq 0$$

144   5. Tikhonov Regularization

and hence with $T^* = T^*Q$

$$\|T(\bar{x}_\alpha^\delta - x_\alpha^\delta)\|^2 + \alpha\|\bar{x}_\alpha^\delta - x_\alpha^\delta\|^2 \leq \langle Q(\bar{y}^\delta - y^\delta), T(\bar{x}_\alpha^\delta - x_\alpha^\delta)\rangle$$
$$\leq \|Q(\bar{y}^\delta - y^\delta)\|\,\|T(\bar{x}_\alpha^\delta - x_\alpha^\delta)\|\,.$$

Our assertions now follow from the last inequality. ∎

Now we address the question of convergence. In view of Theorem 5.16 we split the error term $\|x_\alpha^\delta - x_C^\dagger\|$, as in the unconstrained case, into the approximation error $\|x_\alpha - x_C^\dagger\|$, where x_α is, as mentioned above, the regularized solution with exact right-hand side y, and the propagated data error $\|x_\alpha^\delta - x_\alpha\|$.

Theorem 5.17. *Let $y \in \mathcal{Y}$. Then the constrained Tikhonov regularized solutions x_α converge to an element in C for $\alpha \to 0$ if and only if $Q_C y \in \overline{T(C)}$. Moreover, $Q_C y \in T(C)$ implies that*

$$\lim_{\alpha \to 0} x_\alpha = x_C^\dagger \quad \text{and} \quad \|Tx_\alpha - Q_C y\| = o(\sqrt{\alpha})\,.$$

Proof. Let us first assume that $x_\alpha \to \bar{x} \in C$ as $\alpha \to 0$. Then it follows with (5.83) (with $y = y^\delta$) that

$$\langle T^*T\bar{x} - T^*y, h - \bar{x}\rangle = \langle T\bar{x} - y, Th - T\bar{x}\rangle \quad \text{for all } h \in C$$

and hence, by the density of $T(C)$ in $\overline{T(C)}$, that

$$\langle T\bar{x} - y, u - T\bar{x}\rangle \geq 0 \quad \text{for all } u \in \overline{T(C)}\,.$$

This together with (5.79) implies that $T\bar{x} = Q_C y$ and hence that $Q_C y \in T(C)$.

Now we assume that $Q_C y \in T(C)$. We shall show that x_α converges towards $x_C^\dagger$ as $\alpha \to 0$. By Proposition 5.14 we know that $x_C^\dagger$ exists. Now the definition of $Q_C y$ and (5.80) imply that

$$\|Tx - y\| \geq \|Tx_C^\dagger - y\| \quad \text{for all } x \in C\,.$$

Together with the definition of x_α, we now obtain that

$$\alpha\|x_\alpha\|^2 \leq \|Tx_\alpha - y\|^2 - \|Tx_C^\dagger - y\|^2 + \alpha\|x_C^\dagger\|^2 \leq \alpha\|x_C^\dagger\|^2$$

and hence that

$$\|x_\alpha\| \leq \|x_C^\dagger\| \quad \text{for all } \alpha > 0\,. \tag{5.84}$$

The variational inequality (5.79) (with $u = Tx_\alpha$) and (5.80) imply that

$$\langle T^*y - T^*Tx_C^\dagger, x_C^\dagger - x_\alpha\rangle \geq 0\,.$$

This together with (5.83) (with $h = x_C^\dagger, y = y^\delta$) implies that

$$\langle T^*Tx_\alpha + \alpha x_\alpha - T^*Tx_C^\dagger, x_C^\dagger - x_\alpha\rangle \geq 0$$

and hence with (5.80) and (5.84) that

$$\|Tx_\alpha - Q_C y\|^2 \leq \alpha \langle x_\alpha, x_C^\dagger - x_\alpha \rangle \leq 2\alpha \|x_C^\dagger\|^2. \quad (5.85)$$

Thus,

$$\lim_{\alpha \to 0} Tx_\alpha = Q_C y. \quad (5.86)$$

Let now $\{\alpha_n\}$ be an arbitrary, but fixed, sequence with $\alpha_n \searrow 0$ for $n \to \infty$. Due to (5.84), there exist a subsequence (again denoted by $\{\alpha_n\}$) and an element u with $\|u\| \leq \|x_C^\dagger\|$ such that $x_{\alpha_n} \rightharpoonup u$. Since $\mathcal{C}$ is closed and convex, it is also weakly closed, and therefore, $u \in \mathcal{C}$. By the weak continuity of T, (5.86) implies that $Tu = Q_C y$. Since $x_C^\dagger$ is the unique element of minimal norm among all elements $x \in \mathcal{C}$ with $Tx = Q_C y$, we now obtain that $u = x_C^\dagger$. Therefore, we have shown that

$$x_\alpha \rightharpoonup x_C^\dagger \quad \text{as} \quad \alpha \to 0.$$

Together with (5.84) we now obtain that

$$\begin{aligned}\|x_\alpha - x_C^\dagger\|^2 &= \|x_\alpha\|^2 + \|x_C^\dagger\|^2 - 2\langle x_\alpha, x_C^\dagger \rangle \\ &\leq 2\langle x_C^\dagger - x_\alpha, x_C^\dagger \rangle \to 0 \quad \text{as} \quad \alpha \to 0.\end{aligned}$$

Thus, we have shown that

$$\lim_{\alpha \to 0} x_\alpha = x_C^\dagger.$$

This together with (5.85) implies that

$$\|Tx_\alpha - Q_C y\| = o(\sqrt{\alpha}). \quad \blacksquare$$

The next corollary, an immediate consequence of the last two theorems, shows that the constrained regularized solutions x_α^δ have the same convergence behaviour as the unconstrained ones (cf. Theorem 5.2).

Corollary 5.18. *Let $Q_C y \in T(\mathcal{C})$ and $y^\delta \in \mathcal{Y}$ be such that $\|Q(y - y^\delta)\| \leq \delta$. If $\alpha = \alpha(\delta)$ is such that $\alpha \to 0$ and $\delta^2/\alpha \to 0$ as $\delta \to 0$, then*

$$\lim_{\delta \to 0} x_\alpha^\delta = x_C^\dagger$$

holds.

Proof. Using the triangle inequality, the proof follows immediately from Theorem 5.16 and Theorem 5.17. ∎

We know from the unconstrained case that regularized solutions can converge arbitrarily slowly. However, as we have seen in Section 4.1 if the exact solution satisfies certain source conditions, one can guarantee convergence rates. We show in the next theorem that the condition that $x^\dagger \in \mathcal{R}(T^*)$, which is sufficient for the convergence rate $o(\sqrt{\alpha})$ in the noise free unconstrained case, can be replaced by

5. Tikhonov Regularization

$x_C^\dagger \in \mathcal{R}(P_C T^*)$ in the constrained case, where P_C is the metric projector onto C. The theorem also contains a converse result.

Theorem 5.19. *Let $Q_C y \in T(C)$. If $x_C^\dagger \in \mathcal{R}(P_C T^*)$, then*

$$\|x_\alpha - x_C^\dagger\| = O(\sqrt{\alpha}) \quad \text{and} \quad \|T x_\alpha - Q_C y\| = O(\alpha).$$

If, in addition, $Qy = Q_C y$, we even obtain that

$$\|x_\alpha - x_C^\dagger\| = o(\sqrt{\alpha}).$$

Moreover, $\|T x_\alpha - Q_C y\| = O(\alpha)$ implies that $x_C^\dagger \in \mathcal{R}(P_C T^*)$.

Proof. Let us assume that $x_C^\dagger \in \mathcal{R}(P_C T^*)$. Since P_C is a metric projector, by Lemma 5.13, we obtain that the non-empty set $\{u \in \overline{\mathcal{R}(T)} \mid P_C T^* u = x_C^\dagger\}$ is closed and convex. Hence, it has an element w of minimal norm. Moreover,

$$\langle T^* w - x_C^\dagger, x_C^\dagger - x_\alpha \rangle \geq 0.$$

Together with (5.85) and (5.80) we now obtain that

$$\begin{aligned}
\|T x_\alpha - Q_C y\|^2 + \alpha \|x_\alpha - x_C^\dagger\|^2 &\leq \alpha \langle x_C^\dagger, x_C^\dagger - x_\alpha \rangle \\
&\leq \alpha \langle T^* w, x_C^\dagger - x_\alpha \rangle \\
&= \alpha \langle w, Q_C y - T x_\alpha \rangle. \quad (5.87)
\end{aligned}$$

This immediately yields that

$$\|T x_\alpha - Q_C y\| \leq \alpha \|w\| \quad (5.88)$$

and

$$\|x_\alpha - x_C^\dagger\| \leq \sqrt{\alpha} \|w\| \quad (5.89)$$

hold.

Let now $Qy = Q_C y$. Note that this implies that $T^* y = T^* Q y = T^* Q_C y$. We will show that

$$w_\alpha := \frac{Q_C y - T x_\alpha}{\alpha} \to w \quad \text{as} \quad \alpha \to 0. \quad (5.90)$$

Let $\{\alpha_n\}$ be an arbitrary, but fixed, sequence with $\alpha_n \searrow 0$ as $n \to \infty$. Then, by estimate (5.88), there exist a subsequence (again denoted by $\{\alpha_n\}$) and an element $g \in \overline{\mathcal{R}(T)}$ such that $\|g\| \leq \|w\|$ and $w_{\alpha_n} \rightharpoonup g$ as $n \to \infty$. Together with (5.89), (5.83) (with $y^\delta = y$) and $T^* y = T^* Q_C y$, we obtain that

$$0 \leq \langle x_{\alpha_n} - T^* \frac{Q_C y - T x_{\alpha_n}}{\alpha_n}, h - x_{\alpha_n} \rangle \to \langle x_C^\dagger - T^* g, h - x_C^\dagger \rangle$$

for all $h \in C$ as $n \to \infty$. Hence, $x_C^\dagger = P_C T^* g$. Since $\|g\| \leq \|w\|$, by the definition of w, this implies that $g = w$. Thus, (5.90) holds.

5.4. Convex Constraints

With (5.88) and (5.90) we now get that

$$\|w_\alpha - w\|^2 = \|w_\alpha\|^2 + \|w\|^2 - 2\langle w_\alpha, w\rangle$$
$$\leq 2\langle w - w_\alpha, w\rangle \to 0 \quad \text{as} \quad \alpha \to 0.$$

Thus, we have shown that

$$\frac{Q_C y - T x_\alpha}{\alpha} \to w \quad \text{as} \quad \alpha \to 0. \tag{5.91}$$

Rewriting (5.87) we get that

$$\alpha \|x_\alpha - x_C^\dagger\|^2 \leq \alpha \langle w, Q_C y - T x_\alpha \rangle - \|T x_\alpha - Q_C y\|^2$$

and hence that

$$\|x_\alpha - x_C^\dagger\|^2 \leq \left\langle w - \frac{Q_C y - T x_\alpha}{\alpha}, Q_C y - T x_\alpha \right\rangle$$
$$\leq \left\| w - \frac{Q_C y - T x_\alpha}{\alpha} \right\| \|Q_C y - T x_\alpha\|.$$

With (5.88) and (5.91) this implies that $\|x_\alpha - x_C^\dagger\| = o(\sqrt{\alpha})$.

Let us now assume that $Qy = Q_C y$ and that $\|T x_\alpha - Q_C y\| = O(\alpha)$. Then it follows analogously to the proof of (5.90) that $x_C^\dagger \in \mathcal{R}(P_C T^*)$. ∎

It is also possible to show that the condition $x^\dagger \in \mathcal{R}((T^*T)^\nu)$, $\nu < 1/2$, for the convergence rate $o(\alpha^\nu)$ in the unconstrained case can be replaced by an analogous condition for $x_C^\dagger$ in the constrained case (see [39, 211]).

It is much more difficult to find an analogous condition to $x^\dagger \in \mathcal{R}(T^*T)$, which implies the convergence rate $O(\alpha)$ in the constrained case, too. The condition that $x_C^\dagger \in \mathcal{R}(P_C T^*T)$ is only necessary but not sufficient for the rate $O(\alpha)$ (cf. [211, Theorem 4.5]). Only if we require further conditions on the set C and the exact solution $x_C^\dagger$, we can guarantee the rate $O(\alpha)$ in the constrained case. Note that obviously, if $x_C^\dagger \in \overset{\circ}{C}$, then $x_C^\dagger = x^\dagger = T^\dagger y$ and x_α is also the unconstrained regularized solution for $\alpha > 0$ sufficiently small, and therefore, all results from the unconstrained case hold for the constrained regularized solutions, too. Thus, different results are only to be expected if $x_C^\dagger \in \partial C$. Since the proofs and conditions of some general results, which have been developed in [211], are rather involved, they will be omitted here. We will only state two special cases.

The first result deals with sets C that can be described by a finite number of inequality constraints.

Theorem 5.20. *Let $N \in \mathbb{N}$, $f_n(\neq 0) \in \mathcal{X}, c_n \in \mathbb{R}$ for all $n = 1, \ldots, N$, and let C be defined as*

$$C := \bigcap_{n=1}^{N} \{x \in \mathcal{X} \mid \langle f_n, x \rangle \leq c_n\}.$$

5. Tikhonov Regularization

Moreover, let $Qy = Q_C y \in T(C)$, $x_C^\dagger \in \partial C \cap \mathcal{R}(P_C T^*)$ and $w \in \mathcal{R}(TP)$, where w is the unique element of minimal norm among all elements u satisfying $x_C^\dagger = P_C T^* u$, P is the orthogonal projector onto

$$\mathcal{V} := \bigcap_{n \in \mathcal{I}} \{h \in \mathcal{X} \mid \langle f_n, h \rangle = 0\}$$

and $\mathcal{I} \subset \{1, \ldots, N\}$ is the set of active constraints of $x_C^\dagger$. Then

$$\|x_\alpha - x_C^\dagger\| = O(\alpha).$$

Proof. Let $\mathcal{I}_\alpha$ denote the set of active constraints of x_α and let

$$\mathcal{V}_\alpha := \bigcap_{n \in \mathcal{I}_\alpha} \{h \in \mathcal{X} \mid \langle f_n, h \rangle = 0\}.$$

If $\mathcal{I}_\alpha = \emptyset$, then $\mathcal{V}_\alpha = \mathcal{X}$. Since $x_\alpha \to x_C^\dagger$ as $\alpha \to 0$ and since $\mathcal{I}_\alpha \subset \{1, \ldots, N\}$ is finite, $\mathcal{I}_\alpha \subset \mathcal{I}$ or equivalently $\mathcal{V}_\alpha \supset \mathcal{V}$ for $\alpha > 0$ sufficiently small, which we assume to hold in the following. If P_α denotes the orthogonal projector onto $\mathcal{V}_\alpha$, then it follows with (5.83) (with $y^\delta = y$) that

$$P_\alpha(T^*T x_\alpha + \alpha x_\alpha - T^* y) = 0. \tag{5.92}$$

On the other hand, since $\mathcal{V} \subset \mathcal{V}_\alpha$

$$P_\alpha(x_\alpha - x_C^\dagger) = x_\alpha - x_C^\dagger. \tag{5.93}$$

Therefore, $Qy = Q_C y$ and (5.80) imply that

$$\begin{aligned}(P_\alpha T^*T P_\alpha + \alpha I) P_\alpha x_\alpha &= P_\alpha T^* y - P_\alpha T^* T(I - P_\alpha) x_\alpha \\ &= P_\alpha T^* (Q_C y - T(I - P_\alpha) x_C^\dagger) \\ &= P_\alpha T^* T P_\alpha x_C^\dagger.\end{aligned} \tag{5.94}$$

Let us now assume that $\overline{\mathcal{I}} \subset \mathcal{I}$ is such that $\mathcal{I}_\alpha = \overline{\mathcal{I}}$ and $P_\alpha = \overline{P}$ for infinitely many $\alpha > 0$ accumulating in 0. Then, by (5.91) and (5.92),

$$0 = \overline{P}(x_\alpha - T^* \frac{Q_C y - T x_\alpha}{\alpha}) \to \overline{P}(x_C^\dagger - T^* w). \tag{5.95}$$

Since $w \in \mathcal{R}(TP)$, there is an element $v \in \mathcal{V}$ such that $w = Tv$. Now $\mathcal{V} \subset \mathcal{V}_\alpha$ and (5.95) imply that

$$\overline{P} x_C^\dagger = \overline{P} T^* T \overline{P} v.$$

This together with (5.93) and (5.94) implies that

$$\|x_\alpha - x_C^\dagger\| = \|\overline{P}(x_\alpha - x_C^\dagger)\| = \alpha \|(\overline{P} T^* T \overline{P} + \alpha I)^{-1} \overline{P} T^* T \overline{P} v\| \leq \alpha \|v\|.$$

Since $\overline{\mathcal{I}}$ was arbitrary, this proves the assertion. ∎

If C has a twice continuously Fréchet-differentiable boundary in a neighbourhood of $x_C^\dagger$, one can prove the following result. The lengthy proof is omitted here; see [211] for details.

Theorem 5.21. *Let ∂C be twice continuously Fréchet-differentiable in a neighbourhood of $x_C^\dagger \in \partial C$. Moreover, let $Qy = Q_C y \in T(C)$, $x_C^\dagger \in \mathcal{R}(P_C T^*)$ and $x_C^\dagger \neq T^* w$, where w is as in Theorem 5.20. If $P x_C^\dagger \in \mathcal{R}(PT^*TP)$, where P is the orthogonal projector onto $\{h \in \mathcal{X} \mid \langle x_C^\dagger - T^* w, h \rangle = 0\}$, then*

$$\|x_\alpha - x_C^\dagger\| = O(\alpha).$$

The condition for the rate $O(\alpha)$ is similar to the one in the unconstrained case with T restricted to a linear subspace parallel to the tangential plane in $x_C^\dagger$.

Remark 5.22. The question arises if there is any connection between the rate of convergence of constrained and unconstrained regularized solutions. If the operator T is injective and $Qy = Q_C y$, which implies that $x_C^\dagger = x^\dagger$, then the condition that $x_C^\dagger \in \mathcal{R}(P_C T^*)$ for the convergence rate $o(\sqrt{\alpha})$ in the constrained case is weaker than the condition that $x^\dagger \in \mathcal{R}(T^*)$ in the unconstrained case, since there exist examples (see, e.g., [211, Example 3.4]) where $x^\dagger \notin \mathcal{R}(T^*)$ and the unconstrained regularized solutions do not converge with the rate $O(\sqrt{\alpha})$, but $x_C^\dagger = x^\dagger \in \mathcal{R}(P_C T^*)$, which due to Theorem 5.19 implies that the constrained regularized solutions converge with the rate $o(\sqrt{\alpha})$. This means that in this case the constrained regularized solutions converge faster than the unconstrained ones. Obviously, the converse implication "$x^\dagger \in \mathcal{R}(T^*) \Rightarrow x^\dagger \in \mathcal{R}(P_C T^*)$" always holds if $x^\dagger \in C$. An analogous assertion does not hold for the rate $O(\alpha)$, since there are examples (cf. [211, Example 4.7]) where on the one hand only the constrained and on the other hand only the unconstrained regularized solutions converge with the rate $O(\alpha)$. This shows that the widely held belief that a known constraint should always be used is technically not justified.

We will show, in the next proposition that, as in the unconstrained case, $O(\alpha)$ is the best possible convergence rate.

Proposition 5.23. *Let $Qy = Q_C y \in T(C)$ and $\|x_\alpha - x_C^\dagger\| = o(\alpha)$. Then $x_C^\dagger = x_C$, where x_C is the unique element of minimal norm in C.*

Proof. It follows with (5.83) (with $y^\delta = y$) and $\|x_\alpha - x_C^\dagger\| = o(\alpha)$ that

$$\langle x_C^\dagger, h - x_C^\dagger \rangle = \lim_{\alpha \to 0} \langle x_\alpha, h - x_\alpha \rangle \geq \langle T^*T \frac{x_C^\dagger - x_\alpha}{\alpha}, h - x_\alpha \rangle = 0$$

for all $h \in C$. Now, by Lemma 5.13, this shows that $x_C^\dagger = x_C$. ∎

Combining Theorem 5.16 with Theorems 5.19 – 5.21, one can obtain the convergence rates $O(\sqrt{\delta})$ and $O(\delta^{\frac{2}{3}})$ for the perturbed regularized solutions, respectively,

150   5. Tikhonov Regularization

as in the unconstrained case. Of course, the question arises, how to choose α in dependence on δ without needing any information about the exact solution. We will show next that Morozov's discrepancy principle always yields convergence and that we obtain the rate $O(\sqrt{\delta})$ if $x_C^\dagger \in \mathcal{R}(P_C T^*)$. Note that even in the unconstrained case this is the best possible rate to be expected by this discrepancy principle (cf. Proposition 4.20).

As in the other sections on the discrepancy principle, we assume that $y = Qy$. We even assume in the following that $y = Qy = Q_C y$. We will choose the regularization parameter $\alpha = \alpha(\delta) \in \mathbb{R}^+ \cup \{\infty\}$ by

$$\alpha(\delta) := \sup\{\alpha > 0 \mid \|Tx_\alpha^\delta - y^\delta\| \leq \tau\delta\}, \qquad (5.96)$$

for some $\tau > 1$. For the proof that $\alpha(\delta)$ is well defined we need the following proposition.

Proposition 5.24. *Let $y^\delta \in \mathcal{Y}$ and let x_α^δ be the unique solution of (5.82). Then the following assertions hold.*

(i) *x_α^δ depends continuously on α.*

(ii) *$\|x_\alpha^\delta\|$ is monotonically decreasing and $\|Tx_\alpha^\delta - y^\delta\|$ is monotonically increasing with respect to α.*

(iii) *Let x_C be as in Proposition 5.23. Then*

$$\lim_{\alpha \to \infty} x_\alpha^\delta = x_C.$$

Proof. (i) Let x_α^δ and x_β^δ, $\alpha \neq \beta$, be two regularized solutions in C. Then the variational inequalities for both solutions (see (5.83)) yield

$$\langle T^*Tx_\alpha^\delta + \alpha x_\alpha^\delta - T^*y^\delta, x_\beta^\delta - x_\alpha^\delta \rangle \geq 0$$

and

$$\langle T^*Tx_\beta^\delta + \beta x_\beta^\delta - T^*y^\delta, x_\alpha^\delta - x_\beta^\delta \rangle \geq 0.$$

Adding both inequalities yields

$$\|Tx_\alpha^\delta - Tx_\beta^\delta\|^2 + \beta \|x_\alpha^\delta - x_\beta^\delta\|^2 \leq |\alpha - \beta| \|x_\alpha^\delta\| \|x_\alpha^\delta - x_\beta^\delta\|.$$

This implies that x_α^δ depends continuously on α.

(ii) By definition of x_α^δ and x_β^δ, we obtain that

$$\|Tx_\alpha^\delta - y^\delta\|^2 + \alpha \|x_\alpha^\delta\|^2 \leq \|Tx_\beta^\delta - y^\delta\|^2 + \alpha \|x_\beta^\delta\|^2 \qquad (5.97)$$

and

$$\|Tx_\beta^\delta - y^\delta\|^2 + \beta \|x_\beta^\delta\|^2 \leq \|Tx_\alpha^\delta - y^\delta\|^2 + \beta \|x_\alpha^\delta\|^2.$$

5.4. Convex Constraints   151

Adding both inequalities yields
$$\alpha\|x_\alpha^\delta\|^2 + \beta\|x_\beta^\delta\|^2 \leq \alpha\|x_\beta^\delta\|^2 + \beta\|x_\alpha^\delta\|^2$$
and hence
$$(\beta - \alpha)\|x_\beta^\delta\|^2 \leq (\beta - \alpha)\|x_\alpha^\delta\|^2.$$
This yields that $\|x_\alpha^\delta\|$ is monotonically decreasing with respect to α. Now (5.97) implies that $\|Tx_\alpha^\delta - y^\delta\|$ is monotonically increasing with respect to α.

(iii) The variational inequality for x_C (cf. the proof of Proposition 5.23) yields
$$\alpha \langle x_C, x_\alpha^\delta - x_C \rangle \geq 0.$$
By adding this inequality and (5.83) (with $h = x_C$), we obtain that
$$\alpha\|x_\alpha^\delta - x_C\|^2 \leq \langle T^*Tx_\alpha^\delta - T^*y^\delta, x_C - x_\alpha^\delta \rangle$$
and hence that
$$\|x_\alpha^\delta - x_C\| \leq \frac{1}{\alpha}(\|T\|^2\|x_\alpha^\delta\| + \|T^*y^\delta\|).$$
Together with (ii) this implies the assertion. ∎

Note that, in the unconstrained case, $\|x_\alpha^\delta\|$ is even strictly monotonically decreasing if $x_C^\dagger \neq x_C$. This is not true in the constrained case. Even if $x_C^\dagger \neq x_C$, it is possible that $x_\alpha^\delta = x_\beta^\delta$ for $\alpha \neq \beta$.

We are now in the position to prove that $\alpha(\delta)$ in (5.96) is well defined.

Lemma 5.25. *Let $\tau > 1$, $\|y - y^\delta\| \leq \delta$ and $y = Qy = Q_Cy \in T(C)$. Then $\alpha(\delta)$ in (5.96) is well defined. If $x_C^\dagger \neq x_C$ and if δ is sufficiently small, then $\alpha(\delta) < \infty$. Moreover, the set of all parameters α, satisfying*
$$\|Tx_\alpha^\delta - y^\delta\| = \tau\delta,$$
is a non-empty closed interval.

Proof. If $x_C^\dagger = x_C$, then, by the definition of x_α^δ and x_C, we obtain that
$$\begin{aligned}\|Tx_\alpha^\delta - y^\delta\|^2 + \alpha\|x_\alpha^\delta\|^2 &\leq \|Tx_C - y^\delta\|^2 + \alpha\|x_C\|^2 \\ &\leq \delta^2 + \alpha\|x_\alpha^\delta\|^2\end{aligned}$$
and hence $\alpha(\delta) = \infty$ for all $\delta > 0$.

Let us now assume that $x_C^\dagger \neq x_C$. Due to Proposition 5.24 (iii)
$$\lim_{\alpha \to \infty}\|Tx_\alpha^\delta - y^\delta\| = \|Tx_C - y^\delta\| \geq \|Tx_C - y\| - \delta.$$
Since $y = Q_Cy$, (5.80) implies that $\|Tx_C - y\| > 0$. Therefore,
$$\lim_{\alpha \to \infty}\|Tx_\alpha^\delta - y^\delta\| > \tau\delta, \qquad (5.98)$$

152 5. Tikhonov Regularization

if
$$\delta < \frac{\|Tx_C - y\|}{\tau + 1}. \tag{5.99}$$

On the other hand, by the definition of x_α^δ
$$\|Tx_\alpha^\delta - y^\delta\|^2 + \alpha \|x_\alpha^\delta\|^2 \leq \|Tx_C^\dagger - y^\delta\|^2 + \alpha \|x_C^\dagger\|^2$$
$$\leq \delta^2 + \alpha \|x_C^\dagger\|^2 \tag{5.100}$$

and hence
$$\limsup_{\alpha \to 0} \|Tx_\alpha^\delta - y^\delta\| \leq \delta.$$

This together with the continuity of x_α^δ and the monotonicity of $\|Tx_\alpha^\delta - y^\delta\|$ with respect to α (see Proposition 5.24 (i) and (ii)) implies that $\alpha(\delta)$ is well defined. Moreover, by (5.98), the set $\{\alpha > 0 \mid \|Tx_\alpha^\delta - y^\delta\| = \tau\delta\}$ is a non-empty closed bounded interval if δ satisfies (5.99). $\alpha(\delta)$ is then the supremum of this set, and hence, $\alpha(\delta) < \infty$. ∎

We will show in the next proposition that Morozov's discrepancy principle yields convergence and that we obtain the rate $O(\sqrt{\delta})$ if $x_C^\dagger \in \mathcal{R}(P_C T^*)$.

Proposition 5.26. Let $\tau > 1$, $\|y - y^\delta\| \leq \delta$ and $y = Qy = Q_C y \in T(\mathcal{C})$. If $\alpha(\delta)$ is defined as in (5.96), then
$$\lim_{\delta \to 0} x_{\alpha(\delta)}^\delta = x_C^\dagger.$$

If, in addition, $x_C^\dagger \in \mathcal{R}(P_C T^*)$, then
$$\|x_{\alpha(\delta)}^\delta - x_C^\dagger\| = O(\sqrt{\delta}).$$

Proof. If $x_C^\dagger = x_C$, then it was shown in the proof of Lemma 5.25 that $\alpha(\delta) = \infty$ for all $\delta > 0$. Due to Proposition 5.23 (iii), this implies that $x_{\alpha(\delta)}^\delta = x_C = x_C^\dagger$ for all $\delta > 0$. Thus, all results trivially hold.

Let us now assume that $x_C^\dagger \neq x_C$ and that $\delta > 0$ is so small that $\alpha(\delta) < \infty$. This is possible due to Lemma 5.25. Now (5.100) and the definition of $\alpha(\delta)$ imply that
$$\|x_{\alpha(\delta)}^\delta\| \leq \|x_C^\dagger\| \quad \text{and} \quad \lim_{\delta \to 0} Tx_{\alpha(\delta)}^\delta = y.$$

This implies, as in the proof of Theorem 5.17 that
$$\lim_{\delta \to 0} x_{\alpha(\delta)}^\delta = x_C^\dagger.$$

If $x_C^\dagger = P_C T^* w$, then, by the definition of x_α^δ, we obtain with (5.80) and the variational characterization of $P_C T^* w$ that
$$\|Tx_\alpha^\delta - y^\delta\|^2 + \alpha \|x_\alpha^\delta - x_C^\dagger\|^2 \leq \delta^2 + 2\alpha \langle x_C^\dagger, x_C^\dagger - x_\alpha^\delta \rangle$$
$$\leq \delta^2 + 2\alpha \langle T^* w, x_C^\dagger - x_\alpha^\delta \rangle$$
$$\leq \delta^2 + 2\alpha \|w\| \|Tx_\alpha^\delta - y\|$$

and hence for $\alpha = \alpha(\delta)$ that

$$\tau^2\delta^2 + \alpha(\delta)\|x^\delta_{\alpha(\delta)} - x^\dagger_\mathcal{C}\|^2 \leq \delta^2 + 2\alpha(\delta)\|w\|(\tau+1)\delta.$$

Thus,

$$\|x^\delta_{\alpha(\delta)} - x^\dagger_\mathcal{C}\| < \sqrt{2(\tau+1)\|w\|\delta}. \quad\blacksquare$$

A combination of Tikhonov regularization with finite-dimensional approximation of the set $\mathcal{C}$, where the convergence of the sets $\mathcal{C}_n$ towards $\mathcal{C}$ has to be understood in the sense of Mosco (cf. [196]), was treated in [211, 212].

6. Iterative Regularization Methods

We have mentioned in Chapter 1 that in most "real-world inverse problems" the direct problem is much better understood than the inverse problem. Typically, there even exists sophisticated software for solving the direct problem. In the operator framework of the foregoing chapters this would correspond to software for evaluating Tx for a given element $x \in \mathcal{X}$. Of course, one may be tempted to use such software for solving the inverse problem by successive iteration.

Note that there is a long history of iteration methods for well-posed problems. In the sequel, we mainly focus our study of iteration methods on their regularizing properties. We will see, similarly to our observations concerning projection methods in Section 3.3, that many iterative methods exhibit a "self-regularizing property" in the sense that early termination of the iterative process has a regularizing effect. In other words, the iteration index plays the role of the regularization parameter α, and the stopping rule plays the role of the parameter selection method.

6.1. Landweber Iteration

Most iterative methods for approximating $T^\dagger y$ are based on a transformation of the normal equation into equivalent fixed point equations like

$$x = x + T^*(y - Tx). \tag{6.1}$$

If $\|T\|^2 < 2$ then the corresponding fixed point operator $I - T^*T$ is nonexpansive and one may apply the method of successive approximations. It must be emphasized that $I - T^*T$ is no contraction if our basic problem is ill-posed, since the spectrum of T^*T clusters at the origin. Nevertheless, this fixed point iteration has long tradition in the context of inverse problems: in 1951, Landweber [176] settled its strong convergence provided T is compact and $y \in \mathcal{D}(T^\dagger)$; Fridman [88], in 1956, studied equations with compact, selfadjoint positive semidefinite operators T, in which case the adjoint operator T^* in (6.1) can be avoided; being obviously unaware of Landweber's work, Bialy [26] finally extended both results to not necessarily compact operators T in 1959. Still other names are associated with essentially equivalent procedures in the engineering literature; for example, there is Cimmino's method, SIRT, in computerized tomography (cf. [147]), the Gerchberg-Papoulis algorithm for extrapolation of bandlimited functions in signal processing (cf. Section 1.4), and the van Cittert iteration in image reconstruction (cf. [170]).

In this section we give a first, elementary treatment of what we further on shall call the Landweber iteration, to explain the intrinsic properties of iterative regularization methods. In the following sections we introduce a framework for analyzing more general iterative methods, and we will show that the number of iterations required by Landweber's method can be significantly reduced using more sophisticated algorithms without losing any accuracy. Exemplarily, we study the so-called ν-methods in greater detail.

Besides Landweber iteration, other iterative methods are used for inverse problems that do not fit into the framework of this chapter. Most prominent is the method of conjugate gradients, a study of which is devoted to a separate chapter. ART (cf., e.g., [207]) is another algorithm which is used mainly in image reconstruction. Although ART frequently shows regularizing effects, this has not yet been established rigorously.

The Landweber iteration can be supplied with an initial guess x^* which plays the same role as in Tikhonov regularization, that is, it selects the particular solution which will be approximated in case of ambiguity; using $x_0^\delta = x^*$ the iteration computes further approximations $\{x_k^\delta\}$ recursively,

$$x_k^\delta = x_{k-1}^\delta + T^*(y^\delta - T x_{k-1}^\delta), \quad k \in \mathbb{N}. \tag{6.2}$$

As in the former sections we always write x_k instead of x_k^δ when the iteration with precise data $y^\delta = y$ is concerned. Throughout the analysis it will turn out convenient to have $\|T\| \leq 1$. If this were not the case, then we would introduce a relaxation parameter $0 < \omega \leq \|T\|^{-2}$ in front of T^* in (6.2), i.e., we would iterate

$$x_k^\delta = x_{k-1}^\delta + \omega T^*(y^\delta - T x_{k-1}^\delta), \quad k \in \mathbb{N}.$$

Of course, this has the same effect as premultiplying the underlying equation $Tx = y^\delta$ by $\omega^{\frac{1}{2}}$ and iterating with (6.2). Hence, without loss of generality, we preassume $\|T\| \leq 1$ and drop the relaxation parameter ω. It turns out that we obtain the same sequence of iterates $\{x_k^\delta\}$ if we take $x_0^\delta = 0$ and right-hand side $y^\delta - Tx^*$ instead of y^δ. Therefore, we can assume, without loss of generality, in the sequel that $x_0^\delta = 0$.

Theorem 6.1. *If $y \in \mathcal{D}(T^\dagger)$, then $x_k \to T^\dagger y$ as $k \to \infty$. If $y \notin \mathcal{D}(T^\dagger)$, then $\|x_k\| \to \infty$ as $k \to \infty$.*

Proof. By induction, the iterates x_k may be expressed non-recursively through

$$x_k = \sum_{j=0}^{k-1} (I - T^*T)^j T^* y. \tag{6.3}$$

Now let $y \in \mathcal{D}(T^\dagger)$. Then we have $T^*y = T^*Tx^\dagger$ for $x^\dagger = T^\dagger y \in \mathcal{X}$, and hence,

$$x^\dagger - x_k = x^\dagger - T^*T \sum_{j=0}^{k-1} (I - T^*T)^j x^\dagger.$$

The identity

$$T^*T \sum_{j=0}^{k-1} (I - T^*T)^j = I - (I - T^*T)^k \tag{6.4}$$

thus yields
$$x^\dagger - x_k = (I - T^*T)^k x^\dagger.$$

Using the terminology of Chapter 4, we introduce functions

$$g_k(\lambda) = \sum_{j=0}^{k-1}(1-\lambda)^j \quad \text{and} \quad r_k(\lambda) = (1-\lambda)^k, \tag{6.5}$$

and write
$$x_k = g_k(T^*T)T^*y, \qquad x^\dagger - x_k = r_k(T^*T)x^\dagger.$$

Since $\|T\| \leq 1$ by assumption, we consider $\lambda \in (0,1]$: in this interval $\lambda g_k(\lambda) = 1 - r_k(\lambda)$ is uniformly bounded, and $g_k(\lambda)$ converges to $1/\lambda$ as $k \to \infty$ because $r_k(\lambda)$ converges to zero. We can therefore apply Theorem 4.1 to prove the assertion. Note that k^{-1} now takes the role of α. ∎

According to Theorem 6.1 the sequence $\{x_k\}$ converges to a least-squares solution of $Tx = y$ when $y \in \mathcal{D}(T^\dagger)$. We may ask next, in which way the data error deteriorates this sequence. On one hand, according to Theorem 6.1 the iterates must diverge when we have perturbed data y^δ with $y^\delta \notin \mathcal{D}(T^\dagger)$; on the other hand, the kth iterate x_k^δ (k fixed) obviously depends continuously on the data so that the data error cannot be arbitrarily large. In the following we derive a simple estimate for the error propagation in the Landweber iteration; this can alternatively be interpreted as worst-case rate of divergence when $y^\delta \notin \mathcal{D}(T^\dagger)$.

Lemma 6.2. *Let y, y^δ be a pair of right-hand side data with $\|y^\delta - y\| \leq \delta$, and let $\{x_k\}$ and $\{x_k^\delta\}$ be the corresponding two iteration sequences, cf. (6.2). Then we have*
$$\|x_k - x_k^\delta\| \leq \sqrt{k}\delta, \qquad k \geq 0.$$

Proof. By (6.3) we have
$$x_k - x_k^\delta = \sum_{j=0}^{k-1}(I - T^*T)^j T^*(y - y^\delta) =: R_k(y - y^\delta),$$

and it is the norm of R_k what is searched for. From (6.4) follows

$$\|R_k\|^2 = \|R_k R_k^*\| = \|\sum_{j=0}^{k-1}(I-T^*T)^j(I - (I - T^*T)^k)\| \leq \|\sum_{j=0}^{k-1}(I-T^*T)^j\|,$$

where we have used that $I - TT^*$ is positive semidefinite with $\|I - TT^*\| \leq 1$. Obviously, the right-hand side is bounded by k, and the assertion follows. ∎

This completes the picture of the behaviour of the Landweber iteration in the presence of perturbed data $y^\delta \notin \mathcal{D}(T^\dagger)$, $\|y^\delta - y\| \leq \delta$. Rewriting

$$T^\dagger y - x_k^\delta = T^\dagger y - x_k + x_k - x_k^\delta,$$

6.1. Landweber Iteration

we observe that the total error has two components, an *approximation error* converging (slowly) to zero (assuming $x^* = 0$) and a *data error* of the order of at most $\sqrt{k}\delta$. Consequently, for small values of k the data error is negligible and the iteration seems to converge to the exact solution $T^\dagger y$. When $\sqrt{k}\delta$ reaches the order of magnitude of the approximation error, the propagated data error is no longer hidden in x_k^δ, and the approximations change to the worse.

This general observation is denoted *semiconvergence*; it follows that the regularizing properties of iterative methods for ill-posed problems ultimately depend on reliable stopping rules for detecting the transient from convergence to divergence. We therefore conclude that the iteration index plays the role of the regularization parameter, and the stopping criterion is the counterpart of the parameter choice rule in continuous regularization methods. Consequently, appropriate stopping rules must be based on additional information such as the noise level δ, cf. Theorem 3.3.

For example, for the residual $y^\delta - Tx_k^\delta$ one has

$$y^\delta - Tx_k^\delta = y^\delta - Tx_{k-1}^\delta - TT^*(y^\delta - Tx_{k-1}^\delta) = (I - TT^*)(y^\delta - Tx_{k-1}^\delta), \quad (6.6)$$

and hence, from the nonexpansivity of $I - TT^*$ follows that the residual norm is monotonically decreasing during the iteration. However, we have seen in Theorem 6.1 that the iterates may nevertheless diverge to infinity if $y^\delta \notin \mathcal{D}(T^\dagger)$. Therefore, a small residual norm does not necessarily indicate a good approximation. On the other hand, if the noise level δ is known, then – as for continuous regularization methods, cf. Section 4.3 – the discrepancy principle is a useful parameter choice rule (i.e., a stopping rule), at least in the attainable case $y \in \mathcal{R}(T)$. According to the discrepancy principle the iteration is terminated with $k = k(\delta, y^\delta)$ when for the first time

$$\|y^\delta - Tx_{k(\delta,y^\delta)}^\delta\| \leq \tau\delta, \quad (6.7)$$

with $\tau > 1$ fixed. We shall briefly motivate its usefulness with a simple observation due to Defrise and de Mol [51].

Proposition 6.3. *Let $y \in \mathcal{R}(T)$ and consider any solution x of $Tx = y$. If $\|y^\delta - Tx_k^\delta\| > 2\delta$, then x_{k+1}^δ is a better approximation of $x^\dagger$ than x_k^δ.*

Proof. We estimate

$$\begin{aligned}
\|x^\dagger - x_{k+1}^\delta\|^2 &= \|x^\dagger - x_k^\delta - T^*(y^\delta - Tx_k^\delta)\|^2 \\
&= \|x^\dagger - x_k^\delta\|^2 - 2\langle x^\dagger - x_k^\delta, T^*(y^\delta - Tx_k^\delta)\rangle \\
&\quad + \langle y^\delta - Tx_k^\delta, TT^*(y^\delta - Tx_k^\delta)\rangle \\
&= \|x^\dagger - x_k^\delta\|^2 - 2\langle y - y^\delta, y^\delta - Tx_k^\delta\rangle - \|y^\delta - Tx_k^\delta\|^2 \\
&\quad + \langle y^\delta - Tx_k^\delta, (TT^* - I)(y^\delta - Tx_k^\delta)\rangle.
\end{aligned}$$

As $TT^* - I$ is negative semidefinite we obtain

$$\|x^\dagger - x_k^\delta\|^2 - \|x^\dagger - x_{k+1}^\delta\|^2 \geq \|y^\delta - Tx_k^\delta\| \left(\|y^\delta - Tx_k^\delta\| - 2\delta\right).$$

158 6. Iterative Regularization Methods

If $\|y^\delta - Tx_k^\delta\| > 2\delta$, then the right-hand side is positive which was to be shown. ∎

In other words, the iteration should not be terminated before (6.7) with $\tau = 2$ is violated; the discrepancy principle with $\tau = 2$ determines the beginning of the transition of the iteration from convergence to divergence. If the discrepancy principle is employed with some $\tau \in (1,2)$, then it depends on the particular noise sample whether the chosen stopping index is at the beginning or at the end of the transition phase. Note that τ should be bigger than 1 since otherwise the residual norms might never reach the given tolerance; for the case $\tau > 1$ one has the following simple bound for the stopping index defined by (6.7).

Proposition 6.4. *If $\tau > 1$ in (6.7) is fixed, then the discrepancy principle determines a finite stopping index $k(\delta, y^\delta)$ for the Landweber iteration, with $k(\delta, y^\delta) = O(\delta^{-2})$.*

Proof. We consider for the moment the sequence $\{x_k\}$ corresponding to the Landweber iteration with exact right-hand side y. As in the proof of Proposition 6.3 we obtain

$$\|x^\dagger - x_j\|^2 - \|x^\dagger - x_{j+1}\|^2 = \|y - Tx_j\|^2 + \langle y - Tx_j, (I - TT^*)(y - Tx_j)\rangle$$
$$\geq \|y - Tx_j\|^2.$$

Adding up these inequalities from $j = 1$ through k yields

$$\|x^\dagger - x_1\|^2 - \|x^\dagger - x_{k+1}\|^2 \geq \sum_{j=1}^k \|y - Tx_j\|^2 \geq k\|y - Tx_k\|^2,$$

where the final inequality follows from the monotonicity of the residual norms. From (6.6) we obtain by induction

$$y - Tx_k = (I - TT^*)^k(y - Tx_0),$$

and conclude

$$\|(I - TT^*)^k(y - Tx_0)\| = \|y - Tx_k\| \leq k^{-\frac{1}{2}}\|x^\dagger - x_1\|.$$

Turning now to the estimation of the real residual $y^\delta - Tx_k^\delta$ we find

$$\|y^\delta - Tx_k^\delta\| = \|(I - TT^*)^k(y^\delta - Tx_0)\|$$
$$\leq \|(I - TT^*)^k(y^\delta - y)\| + \|(I - TT^*)^k(y - Tx_0)\|$$
$$\leq \delta + k^{-\frac{1}{2}}\|x^\dagger - x_1\|.$$

Consequently, the right-hand side is below $\tau\delta$ as soon as $k > (\tau - 1)^{-2}\|x^\dagger - x_1\|^2 \delta^{-2}$, and hence $k(\delta, y^\delta) \leq c\delta^{-2}$, where c depends on τ only. ∎

We conclude this section with a proof of the order-optimality of the discrepancy principle (6.7) with fixed $\tau > 1$.

6.1. Landweber Iteration   159

Theorem 6.5. *If $y \in \mathcal{R}(T)$, then the Landweber iteration with the discrepancy principle (6.7) ($\tau > 1$ fixed) is an order optimal regularization method; if $T^\dagger y \in \mathcal{X}_\mu$ with $\mu > 0$, then $k(\delta, y^\delta) = O(\delta^{-\frac{2}{2\mu+1}})$.*

Proof. We will make use of the detailed analysis of the discrepancy principle from Section 4.3. To apply Theorem 4.17 we have to study the residual polynomials $r_k(\lambda) = (1 - \lambda)^k$ of the Landweber iteration, and check the assumptions (4.54), (4.62), and (4.63), where we make the identification $\alpha = k^{-1}$. From (6.5) and the general assumption $\|T\| \leq 1$ we see at once that $|g_k(\lambda)| \leq k$ in $[0, 1]$, and hence, G_α as defined in (4.14) is bounded by α^{-1}. Furthermore, the maximum of $\lambda^\mu r_k(\lambda)$ over $[0, 1]$ is attained at $\lambda = \mu/(\mu + k)$, and hence

$$0 \leq \lambda^\mu r_k(\lambda) \leq \left(\frac{\mu}{\mu+k}\right)^\mu \left(\frac{k}{\mu+k}\right)^k \leq \left(\frac{\mu}{\mu+k}\right)^\mu, \quad 0 \leq \lambda \leq 1.$$

This leads to numbers $\omega_\mu(k)$ as introduced in Theorem 4.3:

$$\lambda^\mu r_k(\lambda) \leq \omega_\mu(k) := \begin{cases} (k+1)^{-\mu}, & \mu \leq 1, \\ \mu^\mu(k+1)^{-\mu}, & \mu > 1. \end{cases} \quad (6.8)$$

Consequently, (4.63) holds for $0 < \mu < \infty$. In the iterative case, however, the regularization parameters form a countable set, and the stopping index does not precisely correspond to the parameter $\alpha(\delta, y^\delta)$ defined in (4.57). However, we can make use of Remark 4.18 since the stopping index from (6.7) fulfills precisely the assumptions of this remark at least if $k(\delta, y^\delta) \neq 1$. Thus, if $k(\delta, y^\delta) \neq 1$, the order-optimality follows from Theorem 4.17 and Remark 4.18; the bound on $k(\delta, y^\delta)$ corresponds to (4.71). It remains to consider the case $k(\delta, y^\delta) = 1$. In this case we have with $x^\dagger = (T^*T)^\mu w$

$$\tau \delta < \|y^\delta\| = \|T(T^*T)^\mu w + y^\delta - y\| \leq \|w\| + \delta,$$

and it follows that

$$\frac{\|w\|}{\delta} > \tau - 1.$$

In other words, (4.71) holds with $c = (\tau - 1)^{\frac{2}{2\mu+1}}$, which is enough for the proof of Theorem 4.17 to go through in this final case. ∎

Note that we have made no efforts in optimizing the upper bounds in (6.8). We emphasize that the discrepancy principle in Landweber's iteration does not saturate as opposed to Tikhonov's method, that is, the Landweber iteration is order-optimal for all $\mathcal{X}_\mu$ with $\mu > 0$. The reason is that the qualification of the method, cf. Section 4.1, is $\mu_0 = \infty$, as can be seen from the numbers $\omega_\mu(k)$ given in (6.8). However, as mentioned before, Landweber iteration is rarely used in practice, since it usually requires far too many iterations until the stopping criterion (6.7) is met. For instance, in Theorem 6.5 we have obtained the estimate $k(\delta) = O(\delta^{-\frac{2}{2\mu+1}})$

provided $T^\dagger y \in \mathcal{X}_\mu$, and it can be shown with essentially the same argument as in Theorem 6.9 below that the exponent $2/(2\mu+1)$ cannot be improved in general. We mention that another order-optimal stopping rule has been proposed in [70], but it requires a similar number of iterations. However, as we will see in the subsequent sections, the number of iterations can be reduced to about the square root, i.e., to $O(\delta^{-\frac{1}{2\mu+1}})$ if more sophisticated iterative methods are used.

We conclude this section with a remark concerning the implementation of Landweber iteration in the presence of convex constraints as considered in Section 5.4 for Tikhonov regularization.

Remark 6.6. Assume that $\mathcal{C}$ is a closed convex set and that the metric projector, $P_\mathcal{C}$, onto $\mathcal{C}$ is easy to calculate; for instance, for the set of nonnegative functions in $\mathcal{L}^2$ the metric projection is given by (5.81). Eicke [61] proposed a constrained Landweber iteration by projecting the iterates after each step onto $\mathcal{C}$, i.e.,

$$x_{k+1} = P_\mathcal{C}(x_k - T^*(Tx_k - y)).$$

Since T and $P_\mathcal{C}$ are continuous, the iterates obviously depend continuously on the data y. Eicke showed that x_k converges weakly towards a solution of $Tx = Q_\mathcal{C} y$ if $Q_\mathcal{C} y \in T(\mathcal{C})$, where $Q_\mathcal{C}$ is the metric projector of $\mathcal{Y}$ onto $\overline{T(\mathcal{C})}$. Only for some special sets $\mathcal{C}$ he proved that this convergence takes also place in the norm topology. But even then, it might happen that x_k converges to a solution of $Tx = Q_\mathcal{C} y$ in $\mathcal{C}$ being different from $x_\mathcal{C}^\dagger$. Therefore, Eicke investigated an alternative approach, namely

$$z_{k+1} = z_k - T^*(Tx_k - y), \quad x_k = P_\mathcal{C} z_k.$$

For this method he proved convergence only for smooth solutions $x_\mathcal{C}^\dagger$. He showed that, if $Q_\mathcal{C} y = Qy$ and if $x_\mathcal{C}^\dagger \in \mathcal{R}(P_\mathcal{C} T^*)$ (cf. Theorem 5.19), then

$$\sum_{k=0}^\infty \|x_k - x_\mathcal{C}^\dagger\|^2 < \infty.$$

This implies convergence of x_k towards $x_\mathcal{C}^\dagger$. Note that, if $\|x_k - x_\mathcal{C}^\dagger\|$ is monotonically decreasing, this even implies that

$$\|x_k - x_\mathcal{C}^\dagger\| = O(k^{-\frac{1}{2}})$$

as in the unconstrained case under the equivalent condition that $x^\dagger \in \mathcal{R}(T^*)$, compare Theorem 4.3 and (6.8).

6.2. Accelerated Landweber Methods

The major drawback of Landweber iteration is its comparatively slow rate of convergence. Far too many iterations are required to reduce the residual to the order of

6.2. Accelerated Landweber Methods

the noise level. In the past decade, more sophisticated iteration methods have been developed on the basis of so-called *semiiterative methods*, cf. [111, 249, 281]: a basic step of a semiiterative method consists of one step of iteration (6.2), followed by an averaging process over all or some of the previously obtained approximations. Semi-iterative methods fit into the framework of Chapter 4, since they can be analyzed with spectral theoretic tools. The functions g_k and r_k (the regularization parameter remains the iteration index, as in Landweber iteration) are now polynomials of degree $k-1$ and k, respectively.

We require a few preliminary definitions. Let $A : \mathcal{X} \to \mathcal{X}$ be a linear operator and x some element of $\mathcal{X}$. Then the kth *Krylov subspace* $\mathcal{K}_k(x, A)$ is defined as the linear space

$$\mathcal{K}_k(x, A) = \text{span}\{x, Ax, A^2 x, \ldots, A^{k-1} x\}.$$

Any element $z \in \mathcal{K}_k(x, A)$ can be expanded as

$$z = \sum_{j=0}^{k-1} \zeta_j A^j x = \Big(\sum_{j=0}^{k-1} \zeta_j A^j\Big) x = g(A) x,$$

where g is a polynomial of degree $k - 1$ or less, i.e.,

$$g(\lambda) = \zeta_0 + \zeta_1 \lambda + \ldots + \zeta_{k-1} \lambda^{k-1}.$$

If A is selfadjoint and if x *cannot* be written as the sum of $k-1$ eigenvectors of A, then the monomials $\{x, Ax, \ldots, A^{k-1} x\}$ form a basis of $\mathcal{K}_k(x, A)$; the same is true for any other sequence $\{z_j = g_j(A) x \mid j = 1, \ldots, k\}$ provided all polynomials g_j have precise degree $j - 1$, respectively for $j = 1, \ldots, k$.

The kth Landweber iterate x_k^δ belongs to the Krylov subspace $\mathcal{K}_k(T^* y^\delta, T^* T)$; in (6.5) we have seen that $x_k^\delta = g_k(T^* T) T^* y^\delta$ with

$$g_k(\lambda) = \sum_{j=0}^{k-1} (1-\lambda)^j = \frac{1 - (1-\lambda)^k}{\lambda}. \tag{6.9}$$

We may consider g_k as approximation of $1/\lambda$ just as we did with the function $g_\alpha(\lambda) = (\lambda + \alpha)^{-1}$ in Tikhonov regularization.

Any sequence of polynomials $\{g_k\}$ where each g_k has precise degree $k - 1$, respectively, defines a semiiterative method, i.e., a sequence $\{x_k^\delta\}$ with

$$x_k^\delta = g_k(T^* T) T^* y^\delta. \tag{6.10}$$

The idea behind (6.10) is the following: polynomials g_k (of degree $k - 1$) different from the one in (6.9) may be better approximations of $1/\lambda$ and may thus lead to more favorable approximations x_k^δ of $T^\dagger y$ in the corresponding Krylov space.

Because of its role in (6.10) we call g_k *iteration polynomial*. Associated with it is the polynomial

$$r_k(\lambda) := 1 - \lambda g_k(\lambda) \tag{6.11}$$

of degree k, which is called *residual polynomial* since it determines the residual $y^\delta - Tx_k^\delta$:
$$y^\delta - Tx_k^\delta = y^\delta - Tg_k(T^*T)T^*y^\delta = r_k(TT^*)y^\delta \,;$$
for the Landweber iteration we have had
$$r_k(\lambda) = (1-\lambda)^k \,.$$
The residual polynomials approximate zero where g_k approximates $1/\lambda$ but, cf. (6.11), the residual polynomials satisfy the constraint
$$r_k(0) = 1 \,.$$
Consequently, the residual polynomials cannot approximate zero uniformly on the spectrum of TT^*, since the spectrum of TT^* clusters at the origin $\lambda = 0$. Any polynomial r_k of degree k satisfying this constraint may serve as residual polynomial; it uniquely defines an iteration polynomial g_k via (6.11).

We stress that for unperturbed right-hand side $y \in \mathcal{D}(T^\dagger)$ with associated iterates $\{x_k\}$ and limit $x^\dagger = T^\dagger y$,
$$x^\dagger - x_k = x^\dagger - g_k(T^*T)T^*Tx^\dagger = r_k(T^*T)x^\dagger \,. \tag{6.12}$$
In other words, the residual polynomials also determine the *approximation error*. On the other hand, comparing x_k and x_k^δ, cf. (6.10),
$$x_k - x_k^\delta = g_k(T^*T)T^*(y - y^\delta) \,, \tag{6.13}$$
i.e., the iteration polynomials determine the *propagated data error*.

Particularly important is the case when the residual polynomials $\{r_k\}$ form a sequence of orthogonal polynomials with respect to some measure over $\mathbb{R}^+$. When this is the case, the residual polynomials satisfy a three-term recurrence relation (cf. Appendix A.2),
$$r_k(\lambda) = r_{k-1}(\lambda) + \mu_k(r_{k-1}(\lambda) - r_{k-2}(\lambda)) - \omega_k \lambda r_{k-1}(\lambda) \,, \qquad k \geq 2 \,,$$
and straightforward computation shows that this recursion carries over to the iterates $\{x_k^\delta\}$ of the associated semiiterative method:
$$x_k^\delta = x_{k-1}^\delta + \mu_k(x_{k-1}^\delta - x_{k-2}^\delta) + \omega_k T^*(y - Tx_{k-1}^\delta) \,, \qquad k \geq 2 \,. \tag{6.14}$$
Thus, the implementation of such methods is extremely efficient.

Obviously, semiiterative methods belong to the general class of approximation methods studied in Chapters 4. For instance, on the basis of the transformation $\alpha = 1/k$ we can apply Theorem 4.1 to prove the following fundamental analogue of Theorem 6.1.

Theorem 6.7. *Let $\{r_k\}$ be a sequence of residual polynomials, uniformly bounded on $[0, \|T\|^2]$ and converging pointwise to zero on $(0, \|T\|^2]$. Then the Krylov subspace approximations $\{x_k^\delta\}$ defined in (6.10) and (6.11) either converge to $T^\dagger y^\delta$ or $\|x_k^\delta\| \to \infty$ as $k \to \infty$, depending on whether $y^\delta \in \mathcal{D}(T^\dagger)$ or not.*

6.2. Accelerated Landweber Methods

We mention that in essentially all semiiterative methods that can be found in the literature (cf. [111] for an overview)

$$|r_k(\lambda)| \leq 1, \qquad 0 \leq \lambda \leq 1, \quad k \in \mathbb{N}_0; \tag{6.15}$$

recall that we preassume $\|T\| \leq 1$ in this chapter.

An essential difference between the continuous regularization methods of Chapter 4 and semiiterative methods is that polynomials – as encountered in the latter – have additional properties which lead to stronger results. One such property is Markov's inequality (cf. [54]) which states that the derivative of a polynomial r_k fulfilling (6.15) cannot be arbitrarily large, namely

$$|r_k'(\lambda)| \leq 2k^2, \qquad 0 \leq \lambda \leq 1. \tag{6.16}$$

This gives rise to the following simple a-priori stopping rule:

Theorem 6.8. *If the residual polynomials $\{r_k\}$ of a semiiterative method satisfy (6.15) and converge to zero pointwise on $(0, 1]$, then the semiiterative method is a regularization method if the iteration is stopped with $k = k(\delta)$, where*

$$k(\delta) \to \infty, \qquad \delta k(\delta) \to 0,$$

as $\delta \to 0$. Moreover, we have the following estimate for the propagated data error:

$$\|x_k - x_k^\delta\| \leq 2\, k\delta.$$

Proof. We first derive an upper bound for $|g_k(\lambda)|$, cf. (6.11), over $[0, 1]$. Since $r_k(0) = 1$ we can find some $\tilde{\lambda} \in [0, \lambda]$ from the Mean Value Theorem with

$$g_k(\lambda) = \frac{1 - r_k(\lambda)}{\lambda} = -r_k'(\tilde{\lambda}).$$

Combining this with Markov's inequality (6.16) yields

$$|g_k(\lambda)| \leq \sup_{0 \leq \tilde{\lambda} \leq 1} |r_k'(\tilde{\lambda})| \leq 2k^2, \qquad 0 \leq \lambda \leq 1.$$

The assertion now follows from Proposition 3.7 and Theorem 4.2; recall that we have $R_k = g_k(T^*T)T^*$ by (6.10). ∎

Let us now restrict our attention to the approximation error of a semiiterative method, i.e., to its performance in case of precise data $y \in \mathcal{D}(T^\dagger)$. Recall that the rate of convergence of any kind of approximation to the solution of ill-posed problems is arbitrarily slow in general. Consequently, a specific rate of convergence in turn implies certain properties of the solution relative to the initial guess, similar to what we have observed in Section 4.2. In particular, the convergence rate $O(k^{-\nu})$ of a semiiterative method implies smoothness of $T^\dagger y$ in terms of the sets $\mathcal{X}_\mu$.

Theorem 6.9. *Let the assumptions of Theorem 6.7 be satisfied. If, for some $x \in \mathcal{X}$,*
$$\|x - x_k\| = O(k^{-2\nu}) \quad \text{as} \quad k \to \infty,$$
then $y \in \mathcal{D}(T^\dagger)$, $x = T^\dagger y$, and $x \in \bigcap_{0 < \mu < \nu} \mathcal{X}_\mu$.

Proof. By Theorem 6.7, we have $y \in \mathcal{D}(T^\dagger)$ and $x = T^\dagger y$. As before, we denote by $\{r_k\}$ the residual polynomials of the iteration. Since the residual polynomials are uniformly bounded over $[0, 1]$, Markov's inequality (6.16) yields
$$|r_k'(\lambda)| = O(k^2), \quad 0 \le \lambda \le 1.$$
Because of the constraint $r_k(0) = 1$ we can therefore find $\varepsilon > 0$ such that
$$r_k(\lambda) \ge \frac{1}{2}, \quad 0 \le \lambda \le \varepsilon k^{-2}.$$
Using (6.12) this gives rise to the following lower bound for the kth iteration error:
$$\|x - x_k\|^2 = \int_0^{1+} r_k^2(\lambda)\, d\|E_\lambda x\|^2 \ge \frac{1}{4}\|E_{\varepsilon k^{-2}} x\|^2.$$
Now, given a fixed $\lambda \in (0, \varepsilon]$ we can find $k \in \mathbb{N}$ such that $(k+1)^{-2} \le \lambda/\varepsilon \le k^{-2}$, hence, by assumption,
$$\|E_\lambda x\|^2 \le \|E_{\varepsilon k^{-2}} x\|^2 \le 4\|x - x_k\|^2 \le ck^{-4\nu}$$
for some $c > 0$, independent of k. Since $k + 1 \le 2k$, this yields
$$\|E_\lambda x\|^2 \le c\left(\frac{k+1}{2}\right)^{-4\nu} \le c\left(\frac{4}{\varepsilon}\right)^{2\nu} \lambda^{2\nu},$$
uniformly for $0 < \lambda \le \varepsilon$. Thus, we have established (4.52), and the assertion follows from Lemma 4.12. ∎

One way of interpreting this result is as follows. Assume that we want to solve a problem $Tx = y$ whose exact solution $x^\dagger = T^\dagger y$ does not belong to $\mathcal{X}_{\mu'}$ for some $\mu > \mu' > 0$: then either we supply an initial guess x^* which compensates for the disturbing components of $x^\dagger$, or no semiiterative method satisfying (6.15) can converge to $x^\dagger$ with convergence rate $O(k^{-2\mu})$. Consequently, when a problem has the exact solution $x^\dagger \in \mathcal{X}_\mu$ and no further prior information is given (i.e., when we use $x^* = 0$), then we cannot expect to get a faster rate of convergence than $o(k^{-2\mu})$.

For $x \in \mathcal{X}_\mu$, Theorem 4.3 provides a sufficient condition for the convergence rate $O(k^{-2\mu})$, namely
$$\|\lambda^\mu r_k(\lambda)\|_{C[0,1]} = \omega_\mu(k) = O(k^{-2\mu}), \quad k \in \mathbb{N}. \tag{6.17}$$

If (6.17) is fulfilled for some largest $\mu_0 > 0$ (the qualification) then we say that the *approximation error* of the semiiterative method has *optimal rate of convergence*

6.2. Accelerated Landweber Methods

(with respect to k) for $x^\dagger \in \mathcal{X}_{\mu_0}$. The ν-methods which will be introduced in the following section constitute examples of semiiterative methods satisfying (6.17) for appropriate values of μ.

We now present some important consequences of (6.17).

Proposition 6.10. *Any sequence of residual polynomials $\{r_k\}$ fulfilling (6.17) for some $\mu > 0$ is uniformly bounded on $[0,1]$ and converges pointwise to zero on $(0,1]$.*

Proof. Rewriting (6.17) we can find $c > 0$, independent of $k \in \mathbb{N}$, with

$$|r_k(\lambda)| \leq c(\lambda k)^{-2\mu}, \qquad 0 < \lambda \leq 1.$$

If λ is fixed in the given interval, and k goes to infinity, then the right-hand side goes to zero. Thus, r_k converges to zero, pointwise on $(0,1]$. Moreover, the right-hand side is always less than c if we restrict λ to $[k^{-2}, 1]$. Consequently, Bernstein's inequality (cf. [54]) asserts that $|r_k|$ is bounded on the entire interval $[0,1]$ by $O((1+k^{-2})^k) = O(1)$. ∎

Proposition 6.10 implies that semiiterative methods with residual polynomials fulfilling (6.17) can always be made regularization methods, e.g., by choosing the simple stopping rule from Theorem 6.8. A more sophisticated stopping rule could again be based on the discrepancy principle:

Theorem 6.11. *Let $y \in \mathcal{R}(T)$, and let the residual polynomials satisfy (6.17) for some $\mu_0 > 0$. Then the semiiterative method is a regularization method of optimal order for $T^\dagger y \in \mathcal{X}_\mu$ with $0 < \mu \leq \mu_0 - 1/2$ provided the iteration is stopped with $k = k(\delta, y^\delta)$ according to the discrepancy principle with fixed $\tau > \sup_{k \in \mathbb{N}} \|r_k\|_{C[0,1]}$. In this case we have $k(\delta, y^\delta) = O(\delta^{-\frac{1}{2\mu+1}})$.*

Proof. By virtue of Proposition 6.10 the residual polynomials are uniformly bounded over $[0,1]$, hence we can find $\tau < \infty$ as required. As in the proof of Theorem 6.8 we obtain the following bound for the iteration polynomial g_k from Markov's inequality:

$$|g_k(\lambda)| \leq 2\tau k^2, \qquad 0 \leq \lambda \leq 1, \quad k \in \mathbb{N}. \tag{6.18}$$

The assertion now follows from Theorem 4.17 and Remark 4.18 with the same straightforward modifications as in the proof of Theorem 6.5 by making the identification $\alpha = k^{-2}$. ∎

This result should be compared with Theorem 6.5. It shows that semiiterative methods may yield order-optimal regularized solutions with (asymptotically) much fewer iterations than the classical Landweber iteration. Consequently, semiiterative methods with optimal rate of convergence of the approximation error are much more efficient candidates for regularizing inverse problems.

166   6. Iterative Regularization Methods

We like to use the occasion and emphasize once more the difference between continuous regularization methods as considered in Chapter 4, and semiiterative methods. For the latter we only need to stipulate the decay rate (6.17) of the modulus of convergence $\omega_\mu(k)$: the Bernstein and Markov inequalities for polynomials automatically imply (6.18) which may be used to bound the norm of the regularization operators R_k. For continuous regularization methods based on more general (piecewise continuous) functions, on the other hand, no such general inequalities exist and we therefore have to preassume (6.18) explicitly, cf. assumption (4.54) for Theorem 4.17. The important conclusion that we may draw from this section, and from Theorem 6.11 in particular, is that in designing iterative regularization methods with order-optimal accuracy we only have to take care that (6.17) is satisfied, which means that the approximation error decays with optimal rate of convergence. The regularizing properties then drop out as inherent by-product.

The applicability of other stopping rules than the discrepancy principle depends on further properties of the residual polynomials. In the Landweber iteration, for example, the values of the residual polynomials $\{r_k(\lambda)\}$ are decreasing in k for fixed $\lambda \in (0, 1]$, and hence, a stopping rule similar to the one from Theorem 4.27 applies: it would correspond to multiples of

$$\frac{r_k^2(\lambda) - r_{k+1}^2(\lambda)}{\lambda} \tag{6.19}$$

as "trial functions" s_k in (4.98); cf. [70] for more details. The criterion (4.98) with s_k as in (4.112), however, will be inefficient in general. The reason is that it requires the evaluation of

$$\langle y^\delta - Tx_k^\delta, r_k^{\frac{1}{\mu}}(TT^*)(y^\delta - Tx_k^\delta) \rangle \qquad \text{for some } \mu > 0.$$

In most cases this requires explicit knowledge of the spectral family of TT^* which, in general, can only be obtained by computing a singular value expansion or alike; the purpose of iterative regularization methods, however, is to avoid such costly routines.

6.3. The ν-Methods

An example for semiiterative methods with optimal rate of convergence are the ν-methods. Originally, they were introduced by Brakhage [34] to obtain theoretical estimates for the performance of the conjugate gradient method, cf. Section 7.4. In the meantime they have been established as a promising alternative to the conjugate gradient method. The ν-methods come with a parameter $\nu > 0$ to be chosen in advance. Occasionally, we shall assume that this parameter takes a specific value, in which case we shall speak about the ν-method in singular. We point out that the (1/2)-method, also known as *Chebyshev method* in the Russian literature, was considered somewhat earlier by Nemirovskii and Polyak [209].

6.3. The ν-Methods

The motivation behind the ν-method (ν fixed !) originates from minimization of the weighted Chebyshev norm on the left-hand side of (6.17) (with μ replaced by ν) over all possible residual polynomials of degree k. As is explained in Appendix A.1 and A.2 this leads quite naturally to a consideration of residual polynomials that are mutually orthogonal with respect to the weight function

$$w_\nu(\lambda) = \frac{\lambda^{2\nu}}{\lambda^{\frac{1}{2}}(1-\lambda)^{\frac{1}{2}}}, \qquad 0 < \lambda < 1. \tag{6.20}$$

Since w_ν is a shifted Jacobi weight, the corresponding residual polynomials are shifted copies of *Jacobi polynomials* with all their many well-known properties. For the reader's convenience we have gathered the most important properties in Appendix A.2; note that r_k equals $p_k^{(\nu)}$ in the notation of the appendix.

Theorem 6.12. *The residual polynomials $\{r_k\}$ of the ν-method ($\nu > 0$ fixed) are uniformly bounded for all $k \in \mathbb{N}$,*

$$|r_k(\lambda)| \leq 1, \qquad \lambda \in [0,1]; \tag{6.21}$$

they further satisfy

$$|\lambda^\nu r_k(\lambda)| \leq c_\nu k^{-2\nu}, \qquad \lambda \in [0,1], \tag{6.22}$$

with appropriate constants $c_\nu > 0$.

Proof. As stated in Appendix A.2, the maximum of $|r_k(\lambda)|$ over $[0,1]$ is attained at $\lambda = 0$, which yields (6.21). Estimate (6.22) is a consequence of (A.12). ∎

By virtue of Theorem 6.12 the results of the foregoing section are now applicable: in particular, it follows that the approximation error of the ν-method has optimal convergence rate when $x^\dagger = T^\dagger y \in \mathcal{X}_\nu$, and the regularized approximations obtained with the discrepancy principle (6.7) are order-optimal when $y = Tx^\dagger$ and $x^\dagger \in \mathcal{X}_\mu$, $0 < \mu \leq \nu - 1/2$; for an implementation of (6.7) any $\tau > 1$ is appropriate according to (6.21).

The ν-methods are easy to implement since their residual polynomials are mutually orthogonal. Using the connection of the residual polynomials with the Gegenbauer polynomials, and their three-term recursion, cf. (A.7), we obtain after some computations

$$x_k^\delta = x_{k-1}^\delta + \mu_k(x_{k-1}^\delta - x_{k-2}^\delta) + \omega_k T^*(y^\delta - Tx_{k-1}^\delta) \tag{6.23}$$

with $\mu_1 = 0$, $\omega_1 = (4\nu+2)/(4\nu+1)$ and

$$\begin{aligned} \mu_k &= \frac{(k-1)(2k-3)(2k+2\nu-1)}{(k+2\nu-1)(2k+4\nu-1)(2k+2\nu-3)}, \\ \omega_k &= 4\frac{(2k+2\nu-1)(k+\nu-1)}{(k+2\nu-1)(2k+4\nu-1)}, \qquad k > 1. \end{aligned} \tag{6.24}$$

168   6. Iterative Regularization Methods

For a design of sophisticated stopping rules, it has turned out more favorable to base the computation of x_k^δ on intermediate quantities z_k^δ via $x_k^\delta = T^* z_k^\delta$ (or if an initial guess $x^* \neq 0$ is used, $x_k^\delta = x^* + T^* z_k^\delta$), rather than computing x_k^δ directly via (6.23). This is the same idea as in the implementation (5.36) of iterated Tikhonov regularization. The sequence $\{z_k^\delta\}$ can be computed by a similar recursion which leads to the following algorithm:

Algorithm 6.13. (ν-Method)

- $z_0^\delta = 0$, $x_0^\delta = 0$;

- $z_1^\delta = \dfrac{4\nu + 2}{4\nu + 1} y^\delta$, $x_1^\delta = T^* z_1^\delta$;

- for $k = 2, 3, \ldots$, compute μ_k and ω_k from (6.24), and determine

$$z_k^\delta = z_{k-1}^\delta + \mu_k(z_{k-1}^\delta - z_{k-2}^\delta) + \omega_k(y^\delta - T x_{k-1}^\delta),$$
$$x_k^\delta = T^* z_k^\delta.$$

The ν-method exhibits a saturation phenomenon that has not been present in the Landweber iteration, but which we have seen before in Tikhonov regularization. If the exact solution $x^\dagger$ belongs to $\mathcal{X}_\mu$ with $\mu > \nu$, then the ν-method cannot be made an order-optimal regularization method: this is due to the fact that the ν-method has a finite qualification $\mu_0 = \nu$, i.e., the approximation error does no longer decrease with optimal rate when $\mu > \nu$, a fact first observed in the case $\nu = 1/2$ in [210], and later generalized to arbitrary $\nu > 0$ in [111]. The precise statement is as follows:

Theorem 6.14. *Let the parameter $\nu > 0$ be fixed, and consider the iterates $\{x_k\}$ of the ν-method with right-hand side y. If, for some $x \in \mathcal{X}$,*

$$\|x - x_k\| = O(k^{-2\nu}) \quad as \quad k \to \infty,$$

then $x = T^\dagger y$ and $x \in \mathcal{X}_\nu$. If $\|x - x_k\| = o(k^{-2\nu})$, then $x = 0$.

Proof. By Theorem 6.9, $y \in \mathcal{D}(T^\dagger)$ and $x = T^\dagger y$. Thus, we can express the error $x - x_k$ according to (6.12) and use the assumption to obtain

$$\|x - x_k\|^2 = \int_0^{1+} r_k^2(\lambda) \, d\|E_\lambda x\|^2 = O(k^{-4\nu}) \quad as \quad k \to \infty.$$

Unfortunately, the method of proof used in Theorem 4.11 does not apply here, since no inequality of the form $\lambda^\nu |r_k(\lambda)| \geq c k^{-2\nu}$ is valid for $\lambda \in [ck^{-2}, 1]$. This is obvious from the fact that the orthogonal polynomials r_k have all their zeros in $[0, 1]$. Therefore, consider the Christoffel function $\Lambda_n(\,\cdot\,; w_\nu)$, cf. Appendix A.3, associated with the weight function w_ν of (6.20): introducing the orthonormal multiples of r_k, i.e., $\check{r}_k^2 = \pi_k^{(\nu)} r_k^2$ (the normalizing factors $\pi_k^{(\nu)}$ are determined in (A.11)), we have

$$\Lambda_n^{-1}(\lambda; w_\nu) = \sum_{k=0}^{n-1} \check{r}_k^2(\lambda).$$

Since $\pi_k^{(\nu)} \sim k^{4\nu}$ as $k \to \infty$, there exists $c > 0$ with

$$\int_0^{1+} \check{r}_k^2(\lambda) \, d\|E_\lambda x\|^2 = \pi_k^{(\nu)} \int_0^{1+} r_k^2(\lambda) \, d\|E_\lambda x\|^2 \leq c, \quad (6.25)$$

valid for all $k \in \mathbb{N}$, hence

$$\int_0^{1+} \frac{1}{n} \Lambda_n^{-1}(\lambda; w_\nu) \, d\|E_\lambda x\|^2 = \frac{1}{n} \sum_{k=0}^{n-1} \int_0^{1+} \check{r}_k^2(\lambda) \, d\|E_\lambda x\|^2 \leq \frac{1}{n} \sum_{k=0}^{n-1} c = c.$$

According to (A.23),

$$\frac{1}{n} \Lambda_n^{-1}(\lambda; w_\mu) \sim \lambda^{-2\nu}, \quad n^{-2} \leq \lambda \leq 1,$$

and hence

$$\int_{n^{-2}}^{1+} \lambda^{-2\nu} \, d\|E_\lambda x\|^2 = O(1) \quad \text{as} \quad n \to \infty.$$

Passing to the limit as $n \to \infty$ yields $x \in \mathcal{X}_\nu$, since this is equivalent to

$$\int_0^{1+} \lambda^{-2\nu} \, d\|E_\lambda x\|^2 < \infty.$$

On the other hand, if $\|x - x_k\| = o(k^{-2\nu})$, then the left-hand side of (6.25) goes to zero as $k \to \infty$ so that

$$\int_0^{1+} \frac{1}{n} \Lambda_n^{-1}(\lambda; w_\mu) \, d\|E_\lambda x\|^2 = o(1) \quad \text{as} \quad n \to \infty.$$

This is only possible when $x = 0$ because it follows from (A.23) that $\Lambda_n^{-1}(\,\cdot\,; w_\nu)/n$ is bounded from below by a strictly positive number, uniformly over $[0, 1]$. ∎

We emphasize the difference between Theorem 6.9 and Theorem 6.14. In the latter, the rate of convergence $O(k^{-2\nu})$ is connected to the value of the parameter ν in the ν-method; it is only for this particular rate that the result of Theorem 6.9 can be strengthened in this manner.

Recall that a result similar to Theorem 6.14 holds true for iterated Tikhonov regularization of order n, when $\nu = n$, cf. Section 5.1. In the following, our major intention is to clarify that this analogy is not by accident but rather the natural effect of a much more general connection between the two methods. To this end, we compare iterated Tikhonov regularization of integer order ν with the corresponding ν-method.

170   6. Iterative Regularization Methods

Theorem 6.15. *Let $k \in \mathbb{N}_0$, $\nu \in \mathbb{N}$ be fixed, and further denote by x_k^δ (x_k) the kth iterate of the ν-method and by $x_{\alpha,\nu}^\delta$ ($x_{\alpha,\nu}$) the iterated Tikhonov approximation of order ν with regularization parameter α. We assume that both methods use the same a-priori guess $x_0 = x_0^\delta = x_{\infty,\nu} = x_{\infty,\nu}^\delta = 0$, and we set $x^\dagger = T^\dagger y$. Then there exists some $c > 0$ depending only on ν but not on k with*

$$\|x^\dagger - x_k\| \leq c \min\{\|x^\dagger - x_{\alpha,\nu}\| \mid (k+1)^{-2} \leq \alpha \leq k^{-2}\},$$

$$\|x_k - x_k^\delta\| \leq c \min\{\|x_{\alpha,\nu} - x_{\alpha,\nu}^\delta\| \mid (k+1)^{-2} \leq \alpha \leq k^{-2}\}. \quad (6.26)$$

Hereby, we understand k^{-2} as $+\infty$ when $k = 0$.

Proof. According to Section 5.1 the approximation error of iterated Tikhonov regularization is given by

$$x^\dagger - x_{\alpha,\nu} = r_{\alpha,\nu}(T^*T)T^*y,$$

where $r_{\infty,\nu} = 1$ and

$$r_{\alpha,\nu}(\lambda) = \left(\frac{\alpha}{\lambda + \alpha}\right)^\nu, \qquad 0 < \alpha < \infty.$$

$r_{\alpha,\nu}(\lambda)$ is nondecreasing in α for fixed $\lambda \in [0,1]$, hence

$$\min_{(k+1)^{-2} \leq \alpha \leq k^{-2}} r_{\alpha,\nu}(\lambda) = r_{(k+1)^{-2},\nu}(\lambda) = (1 + (k+1)^2\lambda)^{-\nu}.$$

In the following we search for a number $c > 0$ such that

$$|r_k(\lambda)| \leq c r_{(k+1)^{-2},\nu}(\lambda), \qquad 0 \leq \lambda \leq 1; \quad (6.27)$$

when this is established, the first assertion of the theorem is a straightforward consequence of (6.12) and spectral theory:

$$\|x^\dagger - x_k\| = \|r_k(T^*T)x^\dagger\| \leq c\|r_{\alpha,\nu}(T^*T)x^\dagger\| \leq c\|x^\dagger - x_{\alpha,\nu}\|,$$

valid for all $\alpha \in [(k+1)^{-2}, k^{-2}]$. Note that (6.27) is easily seen to be true when $k = 0$ since we have $r_0(\lambda) = 1$, whereas $r_{1,\nu}(\lambda) \geq 2^{-\nu}$ for $0 \leq \lambda \leq 1$.

We can thus restrict our attention to $k > 0$ in which case we split the λ-interval $[0, 1]$ at $\lambda_k = (k+1)^{-2}$ and verify (6.27) in both intervals separately. The key tool hereby is Theorem 6.12: if $\lambda \in [0, \lambda_k]$, then $1 + (k+1)^2\lambda \leq 2$, hence

$$r_{(k+1)^{-2},\nu}(\lambda) \geq 2^{-\nu}, \qquad \text{whereas} \qquad |r_k(\lambda)| \leq 1;$$

if $\lambda \in [\lambda_k, 1]$, then

$$r_{(k+1)^{-2},\nu}(\lambda) = (1 + (k+1)^2\lambda)^{-\nu} \geq (2(k+1)^2\lambda)^{-\nu},$$

whereas, with c_ν as in Theorem 6.12,

$$|r_k(\lambda)| \leq c_\nu(k^2\lambda)^{-\nu} \leq 4^\nu c_\nu((k+1)^2\lambda)^{-\nu};$$

6.3. The ν-Methods

hence (6.27) has been verified for $c = \max\{2^\nu, 8^\nu c_\nu\}$.

The proof of (6.26) follows similar lines: here we have

$$x_{\alpha,\nu} - x_{\alpha,\nu}^\delta = g_{\alpha,\nu}(T^*T)T^*(y - y^\delta).$$

with $g_{\infty,\nu} \equiv 0$ and

$$g_{\alpha,\nu}(\lambda) = \frac{1 - r_{\alpha,\nu}(\lambda)}{\lambda}, \qquad 0 < \alpha < \infty.$$

Since $g_{\alpha,\nu}(\lambda)$ is nonincreasing as function of α we now search for $c > 0$ with

$$0 \leq g_k(\lambda) \leq c g_{k-2,\nu}(\lambda), \qquad 0 \leq \lambda \leq 1, \tag{6.28}$$

and then use spectral theory as above. First, we observe that for $k = 0$ (6.28) is trivially fulfilled as $g_0 = g_{\infty,\nu} = 0$. For $k > 0$ we again consider two cases according to whether λ is less or greater than $\lambda_k = k^{-2}$. In the second case, namely, when $\lambda \in [\lambda_k, 1]$ we obtain

$$1 - r_{k-2,\nu}(\lambda) \geq 1 - r_{k-2,\nu}(\lambda_k) = 1 - 2^{-\nu},$$

whereas $0 \leq 1 - r_k(\lambda) \leq 2$ throughout the entire interval. Dividing by λ yields

$$0 \leq g_k(\lambda) = \frac{1 - r_k(\lambda)}{\lambda} \leq \frac{2}{1 - 2^{-\nu}} g_{k-2,\nu}(\lambda), \qquad \lambda_k \leq \lambda \leq 1.$$

In the remaining case, i.e., $0 \leq \lambda \leq \lambda_k$, we use a different argument based on the Mean Value Theorem: since $r_k(0) = 1$ we can find $\tilde{\lambda}$ with

$$g_k(\lambda) = \frac{1 - r_k(\lambda)}{\lambda} = |r'_k(\tilde{\lambda})|,$$

where the right-hand side is bounded by $2k^2$ according to Markov's inequality (6.16); another application of the Mean Value Theorem provides a (possibly different) $\tilde{\lambda} \in [0, \lambda_k]$ for which we obtain from the monotonicity of $r'_{\alpha,\nu}$:

$$g_{k-2,\nu}(\lambda) = |r'_{k-2,\nu}(\tilde{\lambda})| \geq |r'_{k-2,\nu}(\lambda_k)| = \frac{\nu}{2^{\nu+1}} k^2.$$

Combining all estimates, (6.28) follows with $c = \max\{2/(1 - 2^{-\nu}), 2^{\nu+2}/\nu\}$. ∎

The above result is taken from [116], where a converse of (6.26) is also proved, i.e.,

$$\|x_k - x_k^\delta\| \sim \min_{(k+1)^{-2} \leq \alpha \leq k^{-2}} \|x_{\alpha,\nu} - x_{\alpha,\nu}^\dagger\| \quad \text{as} \quad k \to \infty.$$

Thus, the data error components of the two methods are essentially equivalent. On the other hand, the iteration error $x^\dagger - x_k$ can actually vanish in the course of the iteration whereas, for $\alpha > 0$, $x^\dagger = x_{\alpha,\nu}$ can only hold if $x^\dagger = 0$ (or $x^* = x^\dagger$). This is due to the fact that r_k has zeros in $(0, 1)$ whereas $r_{\alpha,\nu}$ is strictly positive. As an

172   6. Iterative Regularization Methods

illustrative (but academic) example, consider the extreme case where the operator T is such that one of the zeros of r_k (k fixed) is an eigenvalue of TT^*: if y is a corresponding eigenvector (and $x^* = 0$), then $x_k = x^\dagger$.

Now, let $Tx = y$ be the given problem, and y^δ be a perturbation of y; consider the optimal regularization parameter α_* for iterated Tikhonov regularization of order ν, i.e., the regularization parameter which minimizes $\|x^\dagger - x_{\alpha,\nu}^\delta\|$ over all $\alpha \in \mathbb{R}^+ \cup \{\infty\}$. Then, if k is the smallest integer not less than $\alpha_*^{-\frac{1}{2}}$, Theorem 6.15 states that the propagated data error $x_k - x_k^\delta$ is of the same order of magnitude as is the corresponding data error component of $x_{\alpha_*,\nu}^\delta$. Furthermore, the approximation error $x^\dagger - x_k$ is at least as small as the corresponding approximation error of the iterated Tikhonov method.

This connection between Tikhonov regularization and the ν-method with $\nu = 1$ also clarifies why Theorem 6.14 is similar to the aforementioned converse result for Tikhonov regularization.

If the exact solution $x^\dagger$ belongs to $\mathcal{X}_\nu$, then the ν-method provides order-optimal approximations $\{x_k^\delta\}$ of $x^\dagger$, e.g., when $k \sim \delta^{-\frac{1}{2\nu+1}}$ (cf. Corollary 4.4 with the identification $\alpha = k^{-2}$). The saturation of the ν-method, however, causes a loss of accuracy when following the discrepancy principle. This is the same phenomenon that we have already observed for Tikhonov regularization.

In the following we derive a stopping rule, taken from [116], which does not share this disadvantage. We follow the general approach of Theorem 4.33 by computing numbers $\eta_k = \langle y^\delta, s_k(TT^*)y^\delta \rangle$, where the trial functions $\{s_k\}$ have a prescribed asymptotic behaviour, namely

$$s_k(\lambda) \sim (1 + k^2\lambda)^{-2\nu-1}, \quad \lambda \in [0,1], \quad k \in \mathbb{N}.$$

As mentioned earlier, a convenient choice for s_k would be a polynomial in order to avoid the computation of the spectral family of TT^*.

As we will see below, such polynomials can be obtained via the Christoffel functions $\Lambda_k(\,\cdot\,;w_{\nu+\frac{1}{2}})$ corresponding to the weight function

$$w_{\nu+\frac{1}{2}}(\lambda) = \lambda^{2\nu+\frac{1}{2}}(1-\lambda)^{-\frac{1}{2}}, \quad 0 \leq \lambda \leq 1. \tag{6.29}$$

Note that $w_{\nu+\frac{1}{2}}$ differs from the weight function w_ν of (6.20) associated with the ν-method by an additional factor λ. Nevertheless, there is an important intrinsic connection between the iterates of the ν-method and the weight function $w_{\nu+\frac{1}{2}}$ as the following result shows:

Lemma 6.16. *If α is a nondecreasing distribution function over $\mathbb{R}^+$ with infinitely many points of increase and if the residual polynomials $\{r_k\}$ are orthogonal with respect to $d\alpha(\lambda)$, then the polynomials*

$$u_k(\lambda) = \frac{r_{k-1}(\lambda) - r_k(\lambda)}{\lambda}, \quad k \in \mathbb{N},$$

are mutually orthogonal with respect to $\lambda\,d\alpha(\lambda)$.

Proof. Let $k \in \mathbb{N}$. By assumption, u_k is a well-defined polynomial of degree $k-1$, since $r_{k-1} - r_k$ has a root at $\lambda = 0$ as r_k and r_{k-1} are both residual polynomials. The orthogonality relation follows immediately: we consider the inner product of the jth monomial λ^j with u_k in the appropriate inner product space, i.e.,

$$\int_{\mathbb{R}^+} \lambda^j u_k(\lambda)\,\lambda d\alpha(\lambda) = \int_{\mathbb{R}^+} \lambda^j (r_{k-1}(\lambda) - r_k(\lambda))\,d\alpha(\lambda)$$
$$= \int_{\mathbb{R}^+} \lambda^j r_{k-1}(\lambda)\,d\alpha(\lambda) - \int_{\mathbb{R}^+} \lambda^j r_k(\lambda)\,d\alpha(\lambda)\,;$$

for $0 \leq j < k-1$ the right-hand side vanishes due to the orthogonality of $\{r_k\}$. Thus, u_k is orthogonal to the set of all polynomials of degree at most $k-2$ with respect to $\lambda\,d\alpha(\lambda)$. Since α has infinitely many points of increase, we have $r_{k-1} \neq r_k$, and hence $u_k \neq 0$ has precise degree $k-1$. It follows that $\{u_k\}$ is a sequence of orthogonal polynomials with respect to $\lambda d\alpha(\lambda)$. ∎

We emphasize that u_k is a polynomial of degree $k-1$. It may alternatively be expressed in terms of the iteration polynomials $\{g_k\}$, cf. (6.11):

$$u_k(\lambda) = \frac{r_{k-1}(\lambda) - r_k(\lambda)}{\lambda} = g_k(\lambda) - g_{k-1}(\lambda)\,.$$

Hence,

$$u_k(TT^*)y^\delta = g_k(TT^*)y^\delta - g_{k-1}(TT^*)y^\delta = z_k^\delta - z_{k-1}^\delta\,,$$

where z_k^δ has the same meaning as in Algorithm 6.13. Since $x_k^\delta = T^* z_k^\delta$, the polynomial u_k also determines the update from x_{k-1}^δ to x_k^δ, and is therefore called *update polynomial*.

Let us return to the ν-methods. By Lemma 6.16, $\{u_k\}$ is a sequence of orthogonal polynomials with respect to the weight function $w_{\nu+1/2}$ of (6.29); although this is the weight function of the $(\nu + 1/2)$-method, u_k is different from the corresponding residual polynomial of the $(\nu + 1/2)$-method, since

$$1 \neq u_k(0) = r'_{k-1}(0) - r'_k(0) = \frac{2k + 2\nu - 1}{2\nu + \frac{1}{2}}\,, \tag{6.30}$$

cf. (A.13).

As mentioned before, we construct the order-optimal stopping rule for the ν-method by means of the Christoffel functions associated with $w_{\nu+\frac{1}{2}}$, namely we define

$$s_k(\lambda) = \frac{\Lambda_{k+1}(0; w_{\nu+\frac{1}{2}})}{\Lambda_{k+1}(\lambda; w_{\nu+\frac{1}{2}})}\,, \qquad k \in \mathbb{N}_0\,, \tag{6.31}$$

and correspondingly

$$\eta_k = \langle y^\delta, s_k(TT^*)y^\delta \rangle\,, \qquad k \in \mathbb{N}_0\,.$$

174   6. Iterative Regularization Methods

We show first how to evaluate $\{\eta_k\}$ recursively. To this end, observe from (A.16) that the reciprocal of the kth Christoffel function is the sum of squares of the first k orthonormal polynomials $\check{u}_j$, $j = 1, 2, \ldots, k$, associated with $w_{\nu+\frac{1}{2}}$, and hence, for $k > 0$,

$$\Lambda_{k+1}^{-1}(\lambda; w_{\nu+\frac{1}{2}}) = \Lambda_k^{-1}(\lambda; w_{\nu+\frac{1}{2}}) + \check{u}_{k+1}^2(\lambda),$$

$$\Lambda_{k+1}^{-1}(0; w_{\nu+\frac{1}{2}}) = \Lambda_k^{-1}(0; w_{\nu+\frac{1}{2}}) + \check{u}_{k+1}^2(0).$$

Consequently,

$$s_k(\lambda) = \frac{\Lambda_k^{-1}(\lambda; w_{\nu+\frac{1}{2}}) + \check{u}_{k+1}^2(\lambda)}{\Lambda_k^{-1}(0; w_{\nu+\frac{1}{2}}) + \check{u}_{k+1}^2(0)}$$

$$= \left(\frac{\Lambda_k^{-1}(\lambda; w_{\nu+\frac{1}{2}})}{\Lambda_k^{-1}(0; w_{\nu+\frac{1}{2}})} + \frac{\check{u}_{k+1}^2(\lambda)}{\Lambda_k^{-1}(0; w_{\nu+\frac{1}{2}})} \right) \frac{\Lambda_k^{-1}(0; w_{\nu+\frac{1}{2}})}{\Lambda_k^{-1}(0; w_{\nu+\frac{1}{2}}) + \check{u}_{k+1}^2(0)}$$

$$= \frac{\Lambda_k^{-1}(0; w_{\nu+\frac{1}{2}})}{\Lambda_k^{-1}(0; w_{\nu+\frac{1}{2}}) + \check{u}_{k+1}^2(0)} s_{k-1}(\lambda) + \frac{1}{\Lambda_k^{-1}(0; w_{\nu+\frac{1}{2}}) + \check{u}_{k+1}^2(0)} \check{u}_{k+1}^2(\lambda).$$

Introducing

$$\alpha_k := \frac{\check{u}_{k+1}^2(0)}{\Lambda_k^{-1}(0; w_{\nu+\frac{1}{2}}) + \check{u}_{k+1}^2(0)} = \check{u}_{k+1}^2(0) \Lambda_{k+1}(0; w_{\nu+\frac{1}{2}}), \qquad (6.32)$$

we obtain the following expression for η_k, $k \in \mathbb{N}$:

$$\eta_k = \langle y^\delta, s_k(TT^*) y^\delta \rangle$$

$$= (1 - \alpha_k) \langle y^\delta, s_{k-1}(TT^*) y^\delta \rangle$$

$$\quad + \frac{\alpha_k}{u_{k+1}^2(0)} \langle y^\delta, u_{k+1}^2(TT^*) y^\delta \rangle$$

$$= (1 - \alpha_k) \eta_{k-1} + \frac{\alpha_k}{u_{k+1}^2(0)} \| z_{k+1}^\delta - z_k^\delta \|^2.$$

Note that $s_0 = 1$, and hence, $\eta_0 = \|y^\delta\|^2$.

It remains to compute α_k, $k \in \mathbb{N}$, which can also be done recursively using (6.32):

$$\alpha_k^{-1} = 1 + \frac{\Lambda_k^{-1}(0; w_{\nu+\frac{1}{2}})}{\check{u}_{k+1}^2(0)} = 1 + \frac{\check{u}_k^2(0)}{\check{u}_{k+1}^2(0)} \alpha_{k-1}^{-1}, \qquad \alpha_0 = 1.$$

Since the polynomials $\{\check{u}_k\}$ are orthonormal with respect to $w_{\nu+\frac{1}{2}}$, they coincide with the polynomials $\check{p}_{k-1}^{(\nu+\frac{1}{2})}$ from Appendix A.2, and hence, $\check{u}_k^2(0) = \pi_{k-1}^{(\nu+\frac{1}{2})}$ as given in (A.11). From this we obtain

$$\alpha_k^{-1} = 1 + \frac{\pi_{k-1}^{(\nu+\frac{1}{2})}}{\pi_k^{(\nu+\frac{1}{2})}} \alpha_{k-1}^{-1} = 1 + \frac{k(2k-1)(2k+2\nu-1)}{(k+2\nu)(2k+4\nu+1)(2k+2\nu+1)} \alpha_{k-1}^{-1}.$$

6.3. The ν-Methods

By using (6.30) this leads to the following stopping rule:

Algorithm 6.17. (Stopping rule for the ν-method)

- Set $\eta_0 = \|y^\delta\|^2$, $\alpha_0 = 1$, and select $\tau > 1$.

- Within the $(k+1)$st iteration of Algorithm 6.13, $k = 1, 2, \ldots$, compute

$$\alpha_k^{-1} = 1 + \frac{k(2k-1)(2k+2\nu-1)}{(k+2\nu)(2k+4\nu+1)(2k+2\nu+1)} \alpha_{k-1}^{-1},$$

$$\beta_k = \frac{(2\nu + \frac{1}{2})^2}{(2k+2\nu+1)^2} \alpha_k,$$

and determine

$$\eta_k = (1 - \alpha_k)\,\eta_{k-1} + \beta_k \|z_{k+1}^\delta - z_k^\delta\|^2.$$

- Stop iterating with $x_{k(\delta,y^\delta)}^\delta$ after $k(\delta, y^\delta)$ iterations when, for the first time, $\eta_{k(\delta,y^\delta)} \leq \tau \delta^2$.

Note that z_{k+1}^δ is required to evaluate η_k, that is, to check the stopping criterion after k iterations. Thus, Algorithm 6.13 always has to be ahead in time of Algorithm 6.17 by half an iteration. This is similar to the order-optimal parameter choice rule in Section 5.1 for (iterated) Tikhonov regularization.

The properties of this stopping rule are summarized in the following theorem:

Theorem 6.18. *Given $\nu > 0$, Algorithm 6.17 is an order-optimal regularization method whenever $y \in \mathcal{R}(T)$ and $T^\dagger y \in \mathcal{X}_\mu$, $0 < \mu \leq \nu$; in this case, $k(\delta, y^\delta) = O(\delta^{-\frac{1}{2\mu+1}})$.*

Proof. Since $\{u_k\}$ are orthogonal polynomials with respect to the shifted Jacobi weight $w_{\nu+\frac{1}{2}}$, they attain their maximum absolute value over $[0, 1]$ at $\lambda = 0$, cf. Appendix A.2. From this we conclude that

$$\Lambda_k^{-1}(\lambda; w_{\nu+\frac{1}{2}}) \leq \Lambda_k^{-1}(0; w_{\nu+\frac{1}{2}}), \qquad 0 \leq \lambda \leq 1.$$

As a matter of fact, all functions $\{s_k\}$ as defined by (6.31) are uniformly bounded,

$$0 \leq s_k(\lambda) \leq 1, \qquad 0 \leq \lambda \leq 1, \quad k \in \mathbb{N}_0.$$

From the asymptotic behaviour (A.23) for the Christoffel functions of Jacobi polynomials we further obtain

$$s_k(\lambda) \sim (1 + k^2 \lambda)^{-2\nu-1}, \qquad 0 \leq \lambda \leq 1.$$

Thus, the assertions follow from Theorem 4.33 by making the identification $\alpha = k^{-2}$, with similar modifications as in the proofs of Theorems 6.5 and 6.11. ∎

176   6. Iterative Regularization Methods

We mention in passing that the first approach from Section 4.4 to obtain order-optimal a-posteriori parameter choice rules would be based on functions $\{s_k\}$ in the definition of η_k, which are multiples of the polynomials (6.19). For the ν-methods, however, this parameter choice fails in general because of the oscillation of the residual polynomials p_k. In fact, the corresponding numbers η_k need not even be positive, and it is impossible to predict when η_k (or $|\eta_k|$) will drop below the tolerance. We refer to [116, Example 4.1] for an illustrative example. The crucial advantage of the particular sequence $\{\eta_k\}$ defined by Algorithm 6.17 is that it is based on averages of the squared orthogonal polynomials, which essentially cancels their oscillations. As a consequence, the sequence $\{\eta_k\}$ of Algorithm 6.17 decays more smoothly, until reaching the tolerance $\tau\delta$ just in time.

7. The Conjugate Gradient Method

7.1. Basic Properties

The conjugate gradient method is known as one of the most powerful algorithms for the solution of selfadjoint, positive (semi)definite well-posed linear equations. On the basis of our results for iterative methods from the foregoing chapter we are therefore led to investigate its performance when applied to the normal equation

$$T^*Tx = T^*y^\delta \tag{7.1}$$

of an ill-posed problem. We shall call this the CGNE method (conjugate gradients for the normal equation). In contrast to other iterative regularization methods like Landweber iteration or the ν-methods, the conjugate gradient algorithm is not based on a fixed sequence of polynomials $\{g_k\}$ and $\{r_k\}$. Instead, these polynomials depend on the given right-hand side. This has the advantage of a greater flexibility of the conjugate gradient iteration, but on the price that $\{x_k^\delta\}$ depends *nonlinearly* on the data y^δ as we shall explain below. In particular, this entails notably more complicated theoretical tools for the mathematical analysis since the general framework of Chapter 4 is no longer adequate.

As we will see in Theorem 7.3 the iterates $\{x_k^\delta\}$ of the conjugate gradient iteration minimize the residual in the corresponding Krylov subspace shifted by the initial guess x^*, i.e.,

$$\|y^\delta - Tx_k^\delta\| = \min\{\|y^\delta - Tx\| \mid x - x^* \in \mathcal{K}_k(T^*(y^\delta - Tx^*), T^*T)\}. \tag{7.2}$$

This has the important consequence that CGNE requires fewest iterations among all semiiterative methods if the discrepancy principle is the chosen stopping rule; in Section 7.3 we will see that the corresponding approximations are order-optimal.

CGNE is defined as follows:

Algorithm 7.1. (CGNE)

- $x_0^\delta = x^*$, $d_0 = y^\delta - Tx_0^\delta$, $p_1 = s_0 = T^*d_0$;
- for $k = 1, 2, \ldots$, unless $s_{k-1} = 0$, compute

$$\begin{aligned}
q_k &= Tp_k, \\
\alpha_k &= \|s_{k-1}\|^2 / \|q_k\|^2, \\
x_k^\delta &= x_{k-1}^\delta + \alpha_k p_k, \\
d_k &= d_{k-1} - \alpha_k q_k, \\
s_k &= T^*d_k, \\
\beta_k &= \|s_k\|^2 / \|s_{k-1}\|^2, \\
p_{k+1} &= s_k + \beta_k p_k.
\end{aligned}$$

178   7. The Conjugate Gradient Method

It can be seen from Algorithm 7.1 that $x_k^\delta - x^*$ is the same as would be obtained with right-hand side $y^\delta - Tx^*$ and zero initial guess. Hence, without loss of generality, we assume $x^* = 0$ in the sequel.

Although the original derivation of the conjugate gradient method by Hestenes and Stiefel [128] is finite-dimensional, it applies almost word by word to continuous operators with closed range in infinite-dimensional Hilbert spaces, cf., e.g., [224]. The question whether the iteration converges for ill-posed equations was posed and partially resolved by Kammerer and Nashed [148] until it was completely settled to the affirmative in case $y^\delta = y \in \mathcal{D}(T^\dagger)$ by Gilyazov [93], see also [34, 182] for estimates for the rate of convergence. In [113] it has been shown that the discrepancy principle may fail if the conjugate gradient iteration is applied straight to $Tx = y$ in the case that T is selfadjoint, positive semidefinite. Instead one has to use a variant which is called minimal residual method, cf. [113, 208], and which can also be extended to cope with indefinite selfadjoint problems (cf. [113]). We also mention the minimal error method by King [151] as an interesting variant of conjugate gradients; see [114] for its analysis as a regularization method.

In Section 7.4 we discuss the number of iterations for the conjugate gradient iteration until the stopping criterion of the discrepancy principle is met. It turns out, that under appropriate assumptions the conjugate gradient iteration requires significantly less iterations than the Krylov methods considered in the previous chapter, cf. Theorem 6.9, which justifies the interest in this algorithm.

We now turn to a verification of (7.2), and to some fundamental properties of Algorithm 7.1. It is obvious from 7.1 that

$$
\begin{aligned}
d_k &= y^\delta - Tx_k^\delta, \\
s_k &= T^*(y^\delta - Tx_k^\delta),
\end{aligned}
\tag{7.3}
$$

for all k. Furthermore, s_k and q_k satisfy the following orthogonality relations:

Lemma 7.2. *As long as $q_k \neq 0$ and $s_k \neq 0$, we have*

(i) $\quad \langle s_k, s_j \rangle = 0, \quad 0 \leq j < k$,

(ii) $\quad \langle q_k, q_j \rangle = 0, \quad 1 \leq j < k$.

Proof. We set for convenience $\beta_0 = 0$, $p_0 = q_0 = 0$, thus extending the validity of the recursion $s_k = p_{k+1} - \beta_k p_k$ to all $k \geq 0$. The proof now runs by simultaneous induction on k. For $k = 1$ we have

$$\langle s_1, s_0 \rangle = \langle T^* d_1, s_0 \rangle = \langle s_0, s_0 \rangle - \alpha_1 \langle T^* q_1, s_0 \rangle = \|s_0\|^2 - \frac{\|s_0\|^2}{\|q_1\|^2} \langle q_1, Ts_0 \rangle = 0$$

since $Ts_0 = q_1$. Assume next that both assertions of the lemma hold for some $k \geq 1$, and that q_{k+1} and s_{k+1} are different from zero. Then α_{k+1} and β_{k+1} are well-defined nonzero numbers and from Algorithm 7.1 we obtain for $1 \leq j \leq k$:

$$
\begin{aligned}
\langle q_{k+1}, q_j \rangle &= \langle Tp_{k+1}, q_j \rangle = \langle s_k, T^*q_j \rangle + \beta_k \langle q_k, q_j \rangle \\
&= \alpha_j^{-1} \langle s_k, T^*(d_{j-1} - d_j) \rangle + \beta_k \langle q_k, q_j \rangle \\
&= \alpha_j^{-1} \langle s_k, s_{j-1} \rangle - \alpha_j^{-1} \langle s_k, s_j \rangle + \beta_k \langle q_k, q_j \rangle \\
&= \begin{cases} -\dfrac{\|q_k\|^2}{\|s_{k-1}\|^2} \|s_k\|^2 + \dfrac{\|s_k\|^2}{\|s_{k-1}\|^2} \|q_k\|^2, & j = k, \\ 0, & j < k \end{cases} \\
&= 0.
\end{aligned}
$$

With this we then obtain for $0 < j \leq k$:

$$
\begin{aligned}
\langle s_{k+1}, s_j \rangle &= \langle T^*d_{k+1}, s_j \rangle = \langle T^*d_k, s_j \rangle - \alpha_{k+1} \langle T^*q_{k+1}, s_j \rangle \\
&= \langle s_k, s_j \rangle - \alpha_{k+1} \langle T^*q_{k+1}, p_{j+1} - \beta_j p_j \rangle \\
&= \langle s_k, s_j \rangle - \alpha_{k+1} \langle q_{k+1}, q_{j+1} \rangle + \alpha_{k+1} \beta_j \langle q_{k+1}, q_j \rangle \\
&= \begin{cases} \|s_k\|^2 - \dfrac{\|s_k\|^2}{\|q_{k+1}\|^2} \|q_{k+1}\|^2, & j = k, \\ 0, & j < k \end{cases} \\
&= 0,
\end{aligned}
$$

completing the proof. ∎

Note that $p_k \in \mathcal{R}(T^*)$ for all $k \geq 0$; hence, if $q_{\kappa+1} = Tp_{\kappa+1} = 0$ for some $\kappa > 0$, then this implies $p_{\kappa+1} = 0$ and we obtain from Algorithm 7.1:

$$s_\kappa = -\beta_\kappa p_\kappa \in \mathrm{span}\{s_0, \ldots, s_{\kappa-1}\}.$$

By virtue of Lemma 7.2 this is only possible if $s_\kappa = 0$, too. On the other hand, if $s_\kappa = 0$, then we observe from (7.3) that $x_\kappa^\delta = T^\dagger y^\delta$. We conclude that Algorithm 7.1 can only *break down* with division by zero if x_κ^δ has become the (least-squares) solution of the perturbed problem; in this case we set for convenience $x_k^\delta = x_\kappa^\delta$ for all $k > \kappa$ extending by this assertion (i) of Lemma 7.2 to all $k \in \mathbb{N}$, cf. (7.3). If no such finite κ with $s_\kappa = 0$ exists, we let $\kappa = \infty$, which means that we can consider κ as the "ultimate termination index" of CGNE further on.

We now verify (7.2):

Theorem 7.3. *The iterates x_k^δ of Algorithm 7.1 satisfy (7.2).*

Proof. Recall that we have assumed that $x^* = 0$ throughout the remainder of this chapter. Let $z_k \in \mathcal{K}_k(T^*y^\delta, T^*T)$. Since $\mathcal{K}_k(T^*y^\delta, T^*T) = \mathrm{span}\{s_0, \ldots, s_{k-1}\}$ with s_j given by (7.3), we can write

$$z_k - x_k^\delta = \sum_{j=0}^{k-1} \zeta_j s_j, \qquad \zeta_j \in \mathbb{R},$$

180   7. The Conjugate Gradient Method

Hence,

$$\begin{aligned}
\|y^\delta - Tz_k\|^2 &= \|y^\delta - Tx_k^\delta\|^2 - 2\sum_{j=0}^{k-1}\langle \zeta_j Ts_j, y^\delta - Tx_k^\delta\rangle + \|T\sum_{j=0}^{k-1}\zeta_j s_j\|^2 \\
&\geq \|y^\delta - Tx_k^\delta\|^2 - 2\sum_{j=0}^{k-1}\zeta_j\langle s_j, s_k\rangle = \|y^\delta - Tx_k^\delta\|^2,
\end{aligned}$$

by virtue of Lemma 7.2. ∎

For later use we reformulate Lemma 7.2 and Theorem 7.3 in terms of the polynomials associated with Algorithm 7.1. Recall from Chapter 6 that every Krylov subspace iterative method generates a sequence of iteration polynomials $\{g_k\}$ such that

$$x_k^\delta = g_k(T^*T)T^*y^\delta, \qquad g_k \in \Pi_{k-1},$$

and associated with these iteration polynomials are the residual polynomials

$$r_k(\lambda) = 1 - \lambda g_k(\lambda).$$

Corollary 7.4. *The polynomials $\{r_k\}$ generated by Algorithm 7.1 are orthogonal with respect to the inner product*

$$[\varphi, \psi] := \int_0^{\|T\|^2+} \varphi(\lambda)\psi(\lambda)\lambda\, d\|F_\lambda y^\delta\|^2.$$

Among all polynomials $\varphi \in \Pi_k$ with $\varphi(0) = 1$, r_k minimizes the quadratic functional

$$\Phi[\varphi] := \int_0^{\|T\|^2+} \varphi^2(\lambda)\, d\|F_\lambda y^\delta\|^2.$$

Proof. The proof follows readily from the identities

$$d_k = r_k(TT^*)y^\delta, \qquad s_k = r_k(T^*T)T^*y^\delta,$$

cf. (7.3), together with Lemma 7.2 and Theorem 7.3. Note that $s_\kappa = 0$ implies $\Phi[r_j] = 0$ for all $j \geq \kappa$. ∎

Remark 7.5. As can be seen from Corollary 7.4 the residual polynomials depend on y^δ via the associated orthogonality measure. Different right-hand sides y^δ yield, in general, different sequences of residual polynomials $\{r_k(\,\cdot\,;y^\delta)\}$ and iteration polynomials $\{g_k(\,\cdot\,;y^\delta)\}$, respectively. In other words, if $\{R_k\}$ denotes the regularization operator induced by the conjugate gradient iteration, namely

$$x_k^\delta = R_k y^\delta = g_k(T^*T; y^\delta)T^*y^\delta,$$

then $\{R_k\}$ is a sequence of *nonlinear* operators.

7.2. Stability and Convergence

In this section we will discuss the two features *stability* and *convergence* (for exact data $y \in \mathcal{D}(T^\dagger)$) which have turned out sufficient for regularization in the case of *linear* regularization operators, cf. Proposition 3.4.

Concerning stability, linear iterative methods as considered in the previous chapter always lead to continuous regularization operators as the kth residual polynomial is fixed, cf. Theorem 6.8. For conjugate gradients, however, the polynomials vary in dependence on the right-hand side and need no longer be uniformly bounded. Indeed, the following result by Eicke, Louis and Plato [62] shows that the kth conjugate gradient iterate does not always depend continuously on perturbations in the right-hand side.

Theorem 7.6. *Let $T = K$ be compact and non-degenerate. Then, for any $k \in \mathbb{N}$, the operator R_k which maps the right-hand side y onto the kth conjugate gradient iterate x_k is discontinuous.*

Proof. Let $(\sigma_n; v_n, u_n)$ be a singular system of K. Without loss of generality, we assume that the singular values are mutually distinct. We choose non-zero ξ_n, $1 \le n < k$, to define

$$x = \sum_{n=1}^{k-1} \xi_n v_n, \qquad y = Kx,$$

and we set

$$y^\delta = y + \delta u_m, \qquad \delta > 0, \quad m \ge k.$$

As K is non-degenerate the above construction is well-defined with $y, y^\delta \in \mathcal{R}(K)$.

By virtue of Corollary 7.4, after k steps of CGNE with right-hand sides y and y^δ, respectively, the corresponding kth residual polynomials are

$$r_k(\lambda; y) = r_{k-1}(\lambda; y) = \prod_{n=1}^{k-1}(1 - \frac{\lambda}{\sigma_n^2}),$$

$$r_k(\lambda; y^\delta) = (1 - \frac{\lambda}{\sigma_m^2})\prod_{n=1}^{k-1}(1 - \frac{\lambda}{\sigma_n^2}),$$

and hence,

$$x_k = R_k y = K^\dagger y = x, \qquad x_k^\delta = R_k y^\delta = K^\dagger y^\delta = x + \frac{\delta}{\sigma_m} v_m.$$

Choosing $m = m(\delta)$ such that $\delta/\sigma_{m(\delta)}$ diverges to infinity as $\delta \to 0$, we find

$$\|R_k y^\delta - R_k y\| \ge \frac{\delta}{\sigma_{m(\delta)}} \to \infty \quad \text{as} \quad \delta \to 0,$$

proving thus the discontinuity of R_k. ∎

182   7. The Conjugate Gradient Method

Every stopping rule must ultimately take care of this phenomenon. In particular, it follows from Theorem 7.6 that no a-priori choice of $k = k(\delta)$ can make conjugate gradients a regularization method in the sense of Definition 3.1.

Corollary 7.7. *If $T = K$ is compact and non-degenerate, then the conjugate gradient iteration with any a-priori choice $k = k(\delta)$, independent of the right-hand side y^δ, is no regularization.*

Proof. As before, let $(\sigma_n; v_n, u_n)$ be a singular system for K. Assume without loss of generality that $\liminf_{\delta \to 0} k(\delta) \geq 1$ and choose $y^\delta = \delta u_m$. As shown in the proof of Theorem 7.6,

$$x_{k(\delta)} = R_{k(\delta)} y^\delta = \frac{\delta}{\sigma_m} v_m \to \infty,$$

provided $m = m(\delta)$ is chosen accordingly. On the other hand, $y^\delta \to 0$ showing that the exact solution is $x = 0$. ∎

As the discontinuity of R_k is such a delicate matter we supply a characterization of all points of discontinuity of R_k, valid for general (not necessarily compact) operators T:

Theorem 7.8. *The operator R_k, $k \in \mathbb{N}$, is discontinuous at y if and only if Qy can be written as the sum of at most $k - 1$ eigenvectors for TT^*.*

Proof. Here we prove only the "only-if-part" of the Theorem. The "if-part" follows from Theorem 7.6 for non-degenerate, compact operators T; for general operators the proof becomes more involved, using the concept of approximate eigenfunctions (see [113] for the details).

We consider approximations y^δ of y and we set $y^0 := y$ for notational convenience. Further on, we rewrite the kth residual polynomial $r_k(\,\cdot\,; y^\delta)$ as

$$r_k(\lambda; y^\delta) = 1 - \rho_1^\delta \lambda - \ldots - \rho_k^\delta \lambda^k;$$

note that this yields

$$x_k^\delta = R_k y^\delta = \sum_{j=1}^{k} \rho_j^\delta (T^*T)^{j-1} T^* y^\delta.$$

Our aim is to show that $\rho_j^\delta \to \rho_j^0$, $1 \leq j \leq n$, as $\delta \to 0$, which readily implies $x_k^\delta \to x_k = R_k y^0$.

Denoting by μ_n^δ the moments of the inner product $[\cdot,\cdot]$ defined in Corollary 7.4, i.e.,

$$\mu_n^\delta = \langle T^* y^\delta, (T^*T)^n T^* y^\delta \rangle = [1, \lambda^n], \qquad n \in \mathbb{N}_0,$$

we conclude from Corollary 7.4 for $0 \leq n < k$:

$$0 = [\lambda^n, r_k(\lambda; y^\delta)] = \langle (T^*T)^n T^* y^\delta, r_k(T^*T; y^\delta) T^* y^\delta \rangle$$

7.2. Stability and Convergence

$$= \langle T^*y^\delta, (T^*T)^n T^*y^\delta \rangle - \sum_{j=1}^{k} \rho_j^\delta \langle T^*y^\delta, (T^*T)^{n+j} T^*y^\delta \rangle$$

$$= \mu_n^\delta - \sum_{j=1}^{k} \mu_{n+j}^\delta \rho_j^\delta.$$

Hence, we have in matrix notation:

$$M_k^\delta \mathbf{r}^\delta \equiv \begin{pmatrix} \mu_1^\delta & \mu_2^\delta & \cdots & \mu_k^\delta \\ \mu_2^\delta & \mu_3^\delta & & \mu_{k+1}^\delta \\ \vdots & & \ddots & \\ \mu_k^\delta & \mu_{k+1}^\delta & & \mu_{2k-1}^\delta \end{pmatrix} \begin{pmatrix} \rho_1^\delta \\ \rho_2^\delta \\ \vdots \\ \rho_k^\delta \end{pmatrix} = \begin{pmatrix} \mu_0^\delta \\ \mu_1^\delta \\ \vdots \\ \mu_{k-1}^\delta \end{pmatrix}. \tag{7.4}$$

If $y^\delta \to y^0 = y$ as $\delta \to 0$ then $\mu_n^\delta \to \mu_n^0$, $0 \le n \le 2k-1$, and

$$M_k^\delta \to M_k = \begin{pmatrix} \mu_1^0 & \mu_2^0 & \cdots & \mu_k^0 \\ \mu_2^0 & \mu_3^0 & & \mu_{k+1}^0 \\ \vdots & & \ddots & \\ \mu_k^0 & \mu_{k+1}^0 & & \mu_{2k-1}^0 \end{pmatrix} \quad \text{as} \quad \delta \to 0,$$

uniformly for all $\{y^\delta\}$ with $\|y^\delta - y\| \le \delta$. For $\mathbf{z} = (\zeta_1, \cdots, \zeta_k)^T$ we have

$$\mathbf{z}^* M_k \mathbf{z} = \|\sum_{j=1}^{k} \zeta_j (T^*T)^j y\|^2,$$

and hence, $\mathbf{z}^* M_k \mathbf{z} = 0$ if and only if

$$\left(\sum_{j=1}^{k} \zeta_j (T^*T)^{j-1}\right) TT^*y = 0,$$

that is, if and only if TT^*y is the sum of at most $k-1$ eigenvectors of TT^*, i.e., Qy is the sum of at most $k-1$ eigenvectors of TT^*. In other words, if Qy cannot be represented in this form, then M_k is positive definite, and it follows from (7.4) that

$$\mathbf{r}^\delta \to \mathbf{r}^0 = M_k^{-1} \begin{pmatrix} \mu_1^0 \\ \vdots \\ \mu_{k-1}^0 \end{pmatrix},$$

as was to be shown. Note that the rate of convergence is $O(\delta)$, uniformly for all $\{y^\delta\}$ with $\|y^\delta - y\| \le \delta$. ∎

Now we return to the situation we met in the proof of Theorem 7.6, namely,

$$x = \sum_{n=1}^{k-1} \xi_n v_n, \qquad y = Kx,$$

184   7. The Conjugate Gradient Method

where K is a compact operator and $(\sigma_n; v_n, u_n)$ is its singular system. As we have seen, $x_k^\delta = R_k y^\delta$ may diverge to infinity as $y^\delta \to y$ if K is non-degenerate and the perturbations $\{y^\delta\}$ are suitably chosen. By Theorem 7.8, however, if $\xi_n \neq 0$, $1 \leq n \leq k-1$, and if all singular values are mutually different, then

$$x_{k-1}^\delta = R_{k-1} y^\delta \to R_{k-1} y = x \quad \text{as} \quad y^\delta \to y,$$

with a rate of convergence depending only on y and on $\|y^\delta - y\|$. In other words, the $(k-1)$st iterate of conjugate gradients is a regularized approximation of x in this particular situation. Any stopping rule which makes the conjugate gradient iteration a regularizing algorithm *must* terminate the iteration with $k(\delta) = k-1$ if y is as above and δ is sufficiently small.

After this discussion of the stability of the conjugate gradient method, we now turn to the question whether the iteration converges for $y \in \mathcal{D}(T^\dagger)$. The first positive result in this direction has been obtained by Kammerer and Nashed [148]. They proved that the iteration converges if $y \in \mathcal{D}(T^\dagger)$ and $T^\dagger y \in \mathcal{R}(T^*)$. Gilyazov [93] was the first to settle the general case $y \in \mathcal{D}(T^\dagger)$ (see [113] for the details of his argument). The following proof is due to Nemirovskii and Polyak [209].

Theorem 7.9.   *The conjugate gradient iterates x_k converge to $T^\dagger y$ for all $y \in \mathcal{D}(T^\dagger)$.*

Proof. If the iteration terminates after a finite number of steps then the corresponding iterate coincides with $T^\dagger y$ as mentioned before. Hence, in the sequel we assume that the iteration does not terminate. In this case, every polynomial r_k with $k > 0$ has simple real zeros $\lambda_{j,k}$, $j = 1, \ldots, k$, with

$$0 < \lambda_{1,k} < \lambda_{2,k} < \ldots < \lambda_{k,k} \leq \|T\|^2$$

because of their orthogonality, cf. Appendix A.2. Moreover, due to the normalization $r_k(0) = 1$ we have the representation

$$r_k(\lambda) = \prod_{j=1}^{k} \left(1 - \frac{\lambda}{\lambda_{j,k}}\right), \tag{7.5}$$

and

$$|r_k'(0)| = \sum_{j=1}^{k} \frac{1}{\lambda_{j,k}}. \tag{7.6}$$

Since $r_k(\lambda)/(\lambda_{1,k} - \lambda)$ is a polynomial of degree $k-1$, the orthogonality relation in Corollary 7.4 yields

$$0 = \int_0^{\|T\|^2+} r_k(\lambda) \frac{r_k(\lambda)}{\lambda - \lambda_{1,k}} \lambda\, d\|F_\lambda y\|^2,$$

or equivalently,

$$\int_0^{\lambda_{1,k}} r_k^2(\lambda) \frac{\lambda}{\lambda_{1,k} - \lambda} d\|F_\lambda y\|^2 = \int_{\lambda_{1,k}}^{\|T\|^2+} r_k^2(\lambda) \frac{\lambda}{\lambda - \lambda_{1,k}} d\|F_\lambda y\|^2.$$

7.2. Stability and Convergence

Since $\lambda/(\lambda - \lambda_{1,k}) \geq 1$ for $\lambda \geq \lambda_{1,k}$ we obtain

$$\int_0^{\lambda_{1,k}} r_k^2(\lambda) \frac{\lambda}{\lambda_{1,k} - \lambda} d\|F_\lambda y\|^2 \geq \int_{\lambda_{1,k}}^{\|T\|^2+} r_k^2(\lambda) d\|F_\lambda y\|^2.$$

Consequently,

$$\begin{aligned}\|y - Tx_k\|^2 &= \int_0^{\lambda_{1,k}} r_k^2(\lambda) d\|F_\lambda y\|^2 + \int_{\lambda_{1,k}}^{\|T\|^2+} r_k^2(\lambda) d\|F_\lambda y\|^2 \\ &\leq \int_0^{\lambda_{1,k}} r_k^2(\lambda) \left(1 + \frac{\lambda}{\lambda_{1,k} - \lambda}\right) d\|F_\lambda y\|^2.\end{aligned}$$

Introducing

$$\varphi_k(\lambda) := r_k(\lambda) \left(\frac{\lambda_{1,k}}{\lambda_{1,k} - \lambda}\right)^{\frac{1}{2}}, \qquad 0 \leq \lambda \leq \lambda_{1,k},$$

the previous inequality can be rewritten as

$$\|y - Tx_k\| \leq \|F_{\lambda_{1,k}} \varphi_k(TT^*) y\|. \tag{7.7}$$

We need to derive upper bounds for φ_k in $[0, \lambda_{1,k}]$. First, as we observe from (7.5),

$$0 \leq \varphi_k(\lambda) \leq 1, \qquad 0 \leq \lambda \leq \lambda_{1,k}.$$

Second, by elementary calculus we can see that, for a given $\nu > 0$, the maximum of $\lambda^\nu \varphi_k^2(\lambda)$ in $[0, \lambda_{1,k}]$ is attained at $\lambda = \lambda_*$, determined as the unique solution of

$$\nu + \lambda_* \left(\frac{1}{\lambda_{1,k} - \lambda_*} - \sum_{j=1}^k \frac{2}{\lambda_{j,k} - \lambda_*}\right) = 0$$

in $[0, \lambda_{1,k}]$, and hence, cf. (7.6),

$$\nu \geq \lambda_* \sum_{j=1}^k \frac{1}{\lambda_{j,k} - \lambda_*} \geq \lambda_* \sum_{j=1}^k \frac{1}{\lambda_{j,k}} = \lambda_* |r_k'(0)|.$$

Thus, $\lambda_* \leq \nu |r_k'(0)|^{-1}$, and therefore (and because $\varphi_k^2(\lambda^*) \leq 1$), we obtain

$$\lambda^\nu \varphi_k^2(\lambda) \leq \lambda_*^\nu \varphi_k^2(\lambda_*) \leq \nu^\nu |r_k'(0)|^{-\nu}, \qquad \nu > 0, \; 0 \leq \lambda \leq \lambda_{1,k}. \tag{7.8}$$

We emphasize that up to now we have not used that $y \in \mathcal{D}(T^\dagger)$. This will be done next: rewriting $Qy = Tx^\dagger$ with $x^\dagger = T^\dagger y$, we apply (7.8) with $\nu = 1$ to obtain from (7.7) that

$$\|Qy - Tx_k\|^2 \leq \|F_{\lambda_{1,k}} \varphi_k(TT^*) Tx^\dagger\|^2 \leq |r_k'(0)|^{-1} \|E_{\lambda_{1,k}} x^\dagger\|^2.$$

Now we fix $0 < \varepsilon \leq \lambda_{1,k}$ and estimate the iteration error as follows:

$$\begin{aligned}\|x^\dagger - x_k\| &= \|r_k(T^*T) x^\dagger\| \leq \|E_\varepsilon r_k(T^*T) x^\dagger\| + \|(I - E_\varepsilon) r_k(T^*T) x^\dagger\| \\ &\leq \|E_\varepsilon r_k(T^*T) x^\dagger\| + \varepsilon^{-\frac{1}{2}} \|(I - F_\varepsilon) r_k(TT^*) Qy\| \\ &\leq \|E_\varepsilon x^\dagger\| + \varepsilon^{-\frac{1}{2}} \|Qy - Tx_k\| \\ &\leq \|E_\varepsilon x^\dagger\| + \left(\frac{|r_k'(0)|^{-1}}{\varepsilon}\right)^{\frac{1}{2}} \|E_{\lambda_{1,k}} x^\dagger\|.\end{aligned}$$

186   7. The Conjugate Gradient Method

Two cases have to be considered: if $\lambda_{1,k} \to 0$ as $k \to \infty$, then we choose $\varepsilon = \varepsilon_k = \lambda_{1,k}$ and obtain, since $|r'_k(0)| \geq \lambda_{1,k}^{-1}$,

$$\|x^\dagger - x_k\| \leq 2\|E_{\lambda_{1,k}}x^\dagger\| \to 0 \quad \text{as} \quad k \to \infty.$$

On the other hand, if $\lambda_{1,k} \to \lambda_1 > 0$ as $k \to \infty$ (note that the smallest root of an orthogonal polynomial sequence is strictly decreasing as a consequence of the interlacing property), then we choose $\varepsilon_k = |r'_k(0)|^{-\frac{1}{2}}$. Since $|r'_k(0)| \geq k\|T\|^{-2}$, cf. (7.6), $\varepsilon_k \to 0$ as $k \to \infty$, and hence, $\varepsilon_k < \lambda_{1,k}$ for k sufficiently large. In this case we obtain

$$\|x^\dagger - x_k\| \leq \|E_{\varepsilon_k}x^\dagger\| + \varepsilon_k^{\frac{1}{2}}\|x^\dagger\| \to 0 \quad \text{as} \quad k \to \infty.$$

Consequently, in any case, $\|x^\dagger - x_k\| \to 0$. ∎

We will modify the above proof in the following section to obtain convergence rates for the regularized solutions.

7.3. The Discrepancy Principle

We now turn to the regularizing properties of CGNE with a properly chosen stopping rule, namely with the discrepancy principle. Throughout we require attainability of the unperturbed problem,

$$y \in \mathcal{R}(T), \qquad \|y^\delta - y\| \leq \delta,$$

where the noise level $\delta > 0$ is known. As in Section 6.1, the discrepancy principle requires an a-priori chosen $\tau > 1$, and then Algorithm 7.1 is terminated with $k = k(\delta, y^\delta)$ when

$$\|y^\delta - Tx^\delta_{k(\delta,y^\delta)}\| \leq \tau\delta < \|y^\delta - Tx^\delta_{k(\delta,y^\delta)-1}\|. \tag{7.9}$$

By virtue of Theorem 7.3 the residual of the kth CGNE iterate is always smaller than the corresponding residual of the Landweber iterate defined in Section 6.1; Proposition 6.4 thus guarantees a finite termination index $k(\delta, y^\delta)$ for CGNE. Moreover, as an immediate consequence of Theorem 7.3 the residuals of CGNE are nonincreasing, and hence $k(\delta, y^\delta)$ is uniquely determined by (7.9). We further note that $s_k \neq 0$ for all $k < k(\delta, y^\delta)$: for, if there is a finite ultimate termination index $\kappa \in \mathbb{N}$ with $s_\kappa = 0$, then this implies that $x^\delta_\kappa = T^\dagger y^\delta$, in which case

$$\|y^\delta - Tx^\delta_\kappa\| = \|(I-Q)y^\delta\| = \|(I-Q)(y^\delta - y)\| \leq \delta,$$

hence $k(\delta, y^\delta) \leq \kappa$.

Our aim is to prove Nemirovskii's theorem [208] that CGNE combined with the discrepancy principle is an order-optimal regularization method if the exact solution belongs to one of the sets $\mathcal{X}_{\mu,\rho}$ of (3.28). In doing so we observe for the first time the

7.3. The Discrepancy Principle

intrinsic difficulties arising from the nonlinearity of the algorithm because it prohibits a separate treatment of approximation and data error. Moreover, compared to the previous chapters, the present situation is further complicated by the fact that there is no general bound for $\{r_k\}$ over the spectrum of T.

Nevertheless, there are still a few analogies to the proofs in the previous chapters. For example, in Section 3.1 we have observed that the data error in the regularized approximations is amplified by $\|R_\alpha\|$, essentially. In all examples from Chapters 4 to 6 we have had

$$\|R_\alpha\| \leq C\sqrt{g_\alpha(0)} = C\sqrt{|r'_\alpha(0)|},$$

and the right-hand side has not only been a measure for the growth of the data error, but also determined the decrease of approximation error and residual norm, respectively. Here, $\sqrt{|r'_k(0)|}$ plays a similar role.

Lemma 7.10. *Let $y = Tx^\dagger$ with $x^\dagger \in \mathcal{X}_{\mu,\rho}$. Then, for $0 < k \leq \kappa$,*

$$\|y^\delta - Tx_k^\delta\| \leq \delta + c\,|r'_k(0)|^{-\mu-\frac{1}{2}}\rho.$$

Proof. In the proof of Theorem 7.9 it has been shown, cf. (7.7), that

$$\|y^\delta - Tx_k^\delta\| \leq \|F_{\lambda_{1,k}}\varphi_k(TT^*)y^\delta\|,$$

where φ_k is bounded by 1 in $[0, \lambda_{1,k}]$, and satisfies (7.8) with $\nu = 2\mu + 1$, i.e.,

$$\lambda^{2\mu+1}\varphi_k^2(\lambda) \leq (2\mu+1)^{2\mu+1}|r'_k(0)|^{-2\mu-1}, \qquad 0 \leq \lambda \leq \lambda_{1,k}.$$

Rewriting $y = T(T^*T)^\mu w$ with $\|w\| \leq \rho$, and inserting these estimates for φ_k, we obtain

$$\begin{aligned}\|y^\delta - Tx_k^\delta\| &\leq \|F_{\lambda_{1,k}}\varphi_k(TT^*)(y^\delta - y)\| + \|F_{\lambda_{1,k}}\varphi_k(TT^*)y\| \\ &\leq \delta + \|E_{\lambda_{1,k}}\varphi_k(T^*T)(T^*T)^{\mu+\frac{1}{2}}w\| \\ &\leq \delta + (2\mu+1)^{\mu+\frac{1}{2}}|r'_k(0)|^{-\mu-\frac{1}{2}}\rho,\end{aligned}$$

which yields the assertion. ∎

On the other hand, we have the following estimate for the iteration error:

Lemma 7.11. *Assume that $y = Tx^\dagger$ with $x^\dagger \in \mathcal{X}_{\mu,\rho}$. Then, for $0 < k \leq \kappa$,*

$$\|x_k^\delta - x^\dagger\| \leq c(\rho^{\frac{1}{2\mu+1}}\delta_k^{\frac{2\mu}{2\mu+1}} + \sqrt{|r'_k(0)|}\,\delta_k),$$

where

$$\delta_k := \max\{\|Tx_k^\delta - y^\delta\|, \delta\}. \tag{7.10}$$

Proof. By the interpolation inequality

$$\|T^\dagger y\| \leq c\rho^{\frac{1}{2\mu+1}}\|y\|^{\frac{2\mu}{2\mu+1}} \leq c\rho^{\frac{1}{2\mu+1}}(\|y^\delta\| + \|y - y^\delta\|)^{\frac{2\mu}{2\mu+1}},$$

188   7. The Conjugate Gradient Method

hence the assertion of the lemma is true for $k = 0$. Now let $0 < k \leq \kappa$. By assumption we have

$$x^\dagger = T^\dagger y = (T^*T)^\mu w,$$

and we choose a positive ε such that

$$0 < \varepsilon \leq |r_k'(0)|^{-1}, \qquad (7.11)$$

which in particular implies that ε is smaller than $\lambda_{1,k}$, cf. (7.6). Next, we introduce

$$\tilde{x}_k = g_k(T^*T; y^\delta)T^*y,$$

where we recollect the full notation for the iteration polynomial to emphasize that $\tilde{x}_k$ is *not* the iterate that would be computed by CGNE with exact right-hand side y. Below we return to the more compact notation for g_k again and obtain

$$\begin{aligned}
\|x^\dagger - x_k^\delta\| &\leq \|E_\varepsilon(x^\dagger - x_k^\delta)\| + \|(I - E_\varepsilon)(x^\dagger - x_k^\delta)\| \\
&\leq \|E_\varepsilon(x^\dagger - \tilde{x}_k)\| + \|E_\varepsilon(\tilde{x}_k - x_k^\delta)\| + \varepsilon^{-\frac{1}{2}}\|(I - F_\varepsilon)(y - Tx_k^\delta)\| \\
&\leq \|E_\varepsilon r_k(T^*T)(T^*T)^\mu w\| + \|E_\varepsilon g_k(T^*T)T^*(y - y^\delta)\| + \varepsilon^{-\frac{1}{2}}\|y - Tx_k^\delta\| \\
&\leq \|\lambda^\mu r_k(\lambda)\|_{C[0,\varepsilon]}\rho + \|\lambda^{\frac{1}{2}} g_k(\lambda)\|_{C[0,\varepsilon]}\delta + \varepsilon^{-\frac{1}{2}}(\|y^\delta - Tx_k^\delta\| + \delta).
\end{aligned}$$

Since $\varepsilon < \lambda_{1,k}$, r_k is convex in $[0, \varepsilon]$ by (7.5), and hence,

$$0 \leq \lambda g_k^2(\lambda) = \frac{1 - r_k(\lambda)}{\lambda}(1 - r_k(\lambda)) \leq |r_k'(0)|, \qquad 0 \leq \lambda \leq \varepsilon.$$

Furthermore, we obviously have

$$0 \leq \lambda^\mu r_k(\lambda) \leq \varepsilon^\mu, \qquad 0 \leq \lambda \leq \varepsilon.$$

Therefore, we conclude that

$$\|x^\dagger - x_k^\delta\| \leq \varepsilon^\mu \rho + 2\varepsilon^{-\frac{1}{2}}\delta_k + |r_k'(0)|^{\frac{1}{2}}\delta. \qquad (7.12)$$

Note that the right-hand side of (7.12) is a decreasing function of ε in $(0, \varepsilon_*)$ and increasing in (ε_*, ∞), where ε_* is determined through

$$\varepsilon_*^{\mu + \frac{1}{2}} = \frac{1}{\mu}\frac{\delta_k}{\rho}.$$

Taking (7.11) into account the estimation of (7.12) splits into two different cases: if $\varepsilon_* < |r_k'(0)|^{-1}$, then we may choose $\varepsilon = \varepsilon_*$ and obtain

$$\begin{aligned}
\|x^\dagger - x_k^\delta\| &\leq \varepsilon_*^\mu \rho + 2\varepsilon_*^{-\frac{1}{2}}\delta_k + |r_k'(0)|^{\frac{1}{2}}\delta \\
&= c\rho^{\frac{1}{2\mu+1}}\delta_k^{\frac{2\mu}{2\mu+1}} + |r_k'(0)|^{\frac{1}{2}}\delta;
\end{aligned}$$

7.3. The Discrepancy Principle

if $\varepsilon_* \geq |r'_k(0)|^{-1}$, then we set $\varepsilon = |r'_k(0)|^{-1}$ in (7.12) which yields

$$\begin{aligned} \|x^\dagger - x_k^\delta\| &\leq |r'_k(0)|^{-\mu}\rho + 2|r'_k(0)|^{\frac{1}{2}}\delta_k + |r'_k(0)|^{\frac{1}{2}}\delta \\ &\leq \varepsilon_*^\mu \rho + 3|r'_k(0)|^{\frac{1}{2}}\delta_k \\ &= c\rho^{\frac{1}{2\mu+1}}\delta_k^{\frac{2\mu}{2\mu+1}} + 3|r'_k(0)|^{\frac{1}{2}}\delta_k . \end{aligned}$$

Thus, we always have

$$\|x^\dagger - x_k^\delta\| \leq c\,(\rho^{\frac{1}{2\mu+1}}\delta_k^{\frac{2\mu}{2\mu+1}} + |r'_k(0)|^{\frac{1}{2}}\delta_k), \qquad (7.13)$$

as was to be shown. ∎

The combination of the previous two lemmata now yields the desired result.

Theorem 7.12. *If $y \in \mathcal{R}(T)$ and if CGNE is stopped according to the discrepancy principle (7.9) with $k(\delta, y^\delta)$, then CGNE is an order-optimal regularization method, i.e., if $T^\dagger y \in \mathcal{X}_{\mu,\rho}$, then*

$$\|T^\dagger y - x_{k(\delta,y^\delta)}^\delta\| \leq c\rho^{\frac{1}{2\mu+1}}\delta^{\frac{2\mu}{2\mu+1}} .$$

Proof. By the definition of the stopping index, the number $\delta_{k(\delta,y^\delta)}$ as introduced in (7.10) satisfies

$$\delta_{k(\delta,y^\delta)} \leq \tau\delta .$$

Hence, according to Lemma 7.11 it remains to estimate $|r'_{k(\delta,y^\delta)}|$ at the stopping index. For the sake of notational convenience, we briefly write k for $k(\delta, y^\delta)$ in the sequel. We may assume, without loss of generality, that $k \geq 1$, since $r'_0 = 0$. Using Lemma 7.10 we conclude that

$$\tau\delta < \|y^\delta - Tx_{k-1}^\delta\| \leq \delta + c|r'_{k-1}(0)|^{-\mu-\frac{1}{2}}\rho ,$$

where the right-hand side has to be understood as $+\infty$ for $k = 1$. Since $\tau > 1$ this implies that

$$|r'_{k-1}(0)| \leq c\left(\frac{\rho}{\delta}\right)^{\frac{2}{2\mu+1}} . \qquad (7.14)$$

It remains to estimate

$$\pi_k := r'_{k-1}(0) - r'_k(0) .$$

Recall from Lemma 6.16 that the update polynomial

$$u_k(\lambda) = \frac{r_{k-1}(\lambda) - r_k(\lambda)}{\lambda}$$

has degree $k - 1$, and

$$[u_k, \lambda\varphi] = 0 \qquad \text{for every } \varphi \in \Pi_{k-2} .$$

190   7. The Conjugate Gradient Method

Moreover, by definition,
$$u_k(0) = \pi_k. \tag{7.15}$$

Substituting $u_k = \pi_k + \lambda \varphi$, then we have $\varphi \in \Pi_{k-2}$ and

$$[u_k, u_k] = \pi_k[u_k, 1] + [u_k, \lambda \varphi] = \pi_k[u_k, 1] = \pi_k[r_{k-1} - r_k, \frac{1}{\lambda}].$$

Using the orthogonality of $\{r_j\}$ we obtain

$$[r_j, \frac{1}{\lambda}] = [r_j, -g_j + \frac{1}{\lambda}] = [r_j, \frac{1}{\lambda} r_j], \quad j = k-1, k,$$

and hence,

$$[u_k, u_k] = \pi_k[r_{k-1}, \frac{1}{\lambda} r_{k-1}] - \pi_k[r_k, \frac{1}{\lambda} r_k]. \tag{7.16}$$

Note that the zeros of r_k and r_{k-1} are interlacing, cf. Appendix A.2; hence, by studying the sign of $r_{k-1} - r_k$ at the zeros of r_{k-1}, it becomes obvious that the $k-1$ roots $\tilde{\lambda}_{j,k-1}$ of u_k are interlacing with the roots $\lambda_{j,k-1}$ of r_{k-1} in the following way:

$$0 < \lambda_{1,k-1} < \tilde{\lambda}_{1,k-1} < \lambda_{2,k-1} < \tilde{\lambda}_{2,k-1} < \cdots < \lambda_{k-1,k-1} < \tilde{\lambda}_{k-1,k-1}.$$

Together with (7.15) this shows that

$$0 \leq u_k(\lambda) \leq \pi_k, \quad 0 \leq \lambda \leq \varepsilon \leq \lambda_{1,k-1}; \tag{7.17}$$

in particular, choosing

$$\varepsilon = \left(c_\varepsilon \frac{\delta}{\rho}\right)^{\frac{2}{2\mu+1}} \tag{7.18}$$

with c_ε sufficiently small, (7.17) holds true by virtue of (7.14) and the fact that $|r'_{k-1}(0)|^{-1} \leq \lambda_{1,k-1}$.

Since $u_k/\pi_k \in \Pi_{k-1}$ has value 1 in the origin, the optimality property of Corollary 7.4 of r_{k-1} implies

$$\pi_k[r_{k-1}, \frac{1}{\lambda} r_{k-1}]^{\frac{1}{2}} = \pi_k \|r_{k-1}(TT^*)y^\delta\|$$
$$\leq \|u_k(TT^*)y^\delta\|$$
$$\leq \|E_\varepsilon u_k(TT^*)y^\delta\| + \|(I - E_\varepsilon)u_k(TT^*)y^\delta\|$$
$$\leq \|E_\varepsilon u_k(TT^*)y^\delta\| + \varepsilon^{-\frac{1}{2}}\|(I - F_\varepsilon)T^* u_k(TT^*)y^\delta\|.$$

Using (7.17) this yields the inequality

$$\pi_k[r_{k-1}, \frac{1}{\lambda} r_{k-1}]^{\frac{1}{2}} \leq \pi_k \|E_\varepsilon y^\delta\| + \varepsilon^{-\frac{1}{2}} [u_k, u_k]^{\frac{1}{2}}. \tag{7.19}$$

The two terms on the right-hand side of (7.19) can be estimated as follows (note that $T^\dagger y = (T^*T)^\mu w$ with $\|w\| \leq \rho$). Using (7.18) and the definition of the stopping rule, we obtain that

$$\begin{aligned} \|E_\varepsilon y^\delta\| &\le \|E_\varepsilon(y^\delta - y)\| + \|E_\varepsilon T(T^*T)^\mu w\| \\ &\le \delta + \varepsilon^{\mu+\frac{1}{2}} \rho = (1 + c_\varepsilon)\delta \\ &\le \frac{1+c_\varepsilon}{\tau} [r_{k-1}, \frac{1}{\lambda} r_{k-1}]^{\frac{1}{2}} ; \end{aligned}$$

on the other hand, using (7.16), we have

$$[u_k, u_k]^{\frac{1}{2}} \le \pi_k^{\frac{1}{2}} [r_{k-1}, \frac{1}{\lambda} r_{k-1}]^{\frac{1}{2}} .$$

Inserting this into (7.19) leads to

$$\left(1 - \frac{1+c_\varepsilon}{\tau}\right) \pi_k^{\frac{1}{2}} \le \varepsilon^{-\frac{1}{2}} ,$$

and, by choosing $c_\varepsilon < \tau - 1$, it follows from (7.18) that

$$\pi_k \le c \left(\frac{\rho}{\delta}\right)^{\frac{2}{2\mu+1}} .$$

Since $r'_k(0) = r'_{k-1}(0) - \pi_k$, (7.14) and the triangle inequality finally yield that

$$|r'_{k(\delta, y^\delta)}(0)| \le c \left(\frac{\rho}{\delta}\right)^{\frac{2}{2\mu+1}} . \tag{7.20}$$

Inserting this into Lemma 7.11 completes the proof. ∎

We emphasize that (in the attainable case) CGNE with the discrepancy principle is an order-optimal regularization method *for all $\mathcal{X}_\mu$* with $\mu > 0$: there is no saturation as in Tikhonov's method or in the ν-methods. Concerning the non-attainable case where $y \in \mathcal{D}(T^\dagger) \setminus \mathcal{R}(T)$, we observe that for right-hand sides y and Qy, respectively, Algorithm 7.1 generates the same sequence of iterates. Consequently, we can obtain order-optimal accuracy in this case, too, if we are able to replace y^δ by Qy^δ in the stopping criterion (7.9).

7.4. The Number of Iterations

The efficiency of iterative regularization methods – besides their accuracy – depends on the number $k(\delta, y^\delta)$ of iterations to meet the stopping criterion. With this in mind, and because of Theorem 7.3, one cannot do better than using CGNE, in particular if the discrepancy principle is the chosen stopping rule. We now investigate how many iterations CGNE will require.

As we will see, this mainly depends on the spectral family $\{F_\lambda\}$ of TT^* and on the smoothness of the solution. We will give two different estimates for $k(\delta, y^\delta)$:

192   7. The Conjugate Gradient Method

the first one concerning the worst-case error is due to Nemirovskii [208], see also [34]; the second estimate, which is significantly better than the worst-case bound in appropriate circumstances, is due to Nemirovskii and Polyak [209].

Theorem 7.13. *If $y \in \mathcal{R}(T)$ and $T^\dagger y \in \mathcal{X}_{\mu,\rho}$ then*

$$k(\delta, y^\delta) \leq c \left(\frac{\rho}{\delta}\right)^{\frac{1}{2\mu+1}}, \qquad (7.21)$$

and this estimate is sharp in the sense that the exponent cannot be replaced by a smaller one.

Proof. We write $T^\dagger y = (T^*T)^\mu w$ with $\|w\| \leq \rho$, and consider the Jacobi polynomial

$$\varphi_k(\lambda) := p_k^{(\mu+\frac{1}{2})}\left(\frac{\lambda}{\|T\|^2}\right), \qquad (7.22)$$

cf. Appendix A.2. Note that φ_k is the kth residual polynomial of the $(\mu + 1/2)$-method. Thus, we have, cf. Theorem 6.12,

$$|\varphi_k(\lambda)| \leq 1, \qquad |\lambda^{\mu+\frac{1}{2}}\varphi_k(\lambda)| \leq ck^{-2\mu-1},$$

for $0 \leq \lambda \leq \|T\|^2$ and some constant $c > 0$. Corollary 7.4 yields

$$\begin{aligned}
\|y^\delta - Tx_k^\delta\| &\leq \|\varphi_k(TT^*)y^\delta\| \\
&\leq \|\varphi_k(TT^*)(y^\delta - y)\| + \|\varphi_k(TT^*)TT^\dagger y\| \\
&\leq \|y^\delta - y\| + \|\lambda^{\mu+\frac{1}{2}}\varphi_k(\lambda)\|_{C[0,\|T\|^2]}\|w\| \\
&\leq \delta + ck^{-2\mu-1}\rho.
\end{aligned}$$

Thus, $\|y^\delta - Tx_k^\delta\| \leq \tau\delta$ if

$$k^{-2\mu-1} \leq \frac{\tau-1}{c}\frac{\delta}{\rho},$$

i.e., if k is given by the right-hand side of (7.21) with modified $c > 0$. By (7.9) this implies $k(\delta, y^\delta) \leq k$ as was to be shown.

Now, let $\varepsilon > 0$ be arbitrarily fixed and choose T and y_ε such that, cf. (A.9),

$$d\|F_\lambda y_\varepsilon\|^2 = w_{\mu+\frac{1}{4}+\varepsilon}(\lambda)\,d\lambda = \lambda^{2(\mu+\varepsilon)}(1-\lambda)^{-\frac{1}{2}}\,d\lambda, \qquad 0 < \lambda < 1.$$

It follows that

$$\int_0^{1+} \lambda^{-2\mu-1}\,d\|F_\lambda y_\varepsilon\|^2 = \int_0^1 \lambda^{2\varepsilon-1}(1-\lambda)^{-\frac{1}{2}}\,d\lambda < \infty,$$

and hence $T^\dagger y_\varepsilon \in \mathcal{X}_\mu$ for all $\varepsilon > 0$. With y_ε as right-hand side CGNE chooses the kth residual polynomial r_k such that

$$\|y_\varepsilon - Tx_k\|^2 = \int_0^1 r_k^2(\lambda)w_{\mu+\frac{1}{4}+\varepsilon}(\lambda)\,d\lambda = \Phi[r_k]$$

7.4. The Number of Iterations   193

is minimized among all possible residual polynomials of degree k. Thus, cf. (A.20), $\|y_\varepsilon - Tx_k\|^2$ is given by the kth Christoffel function associated with the Jacobi weight $w_{\mu+\frac{1}{4}+\varepsilon}$, i.e.,

$$\|y_\varepsilon - Tx_k\|^2 = \Lambda_{k+1}(0; w_{\mu+\frac{1}{4}+\varepsilon}) \sim k^{-4(\mu+\varepsilon)-2} \quad \text{as} \quad k \to \infty,$$

where the latter asymptotic behaviour follows from (A.23).

If $\delta > 0$ is the available bound for the noise level in the (nevertheless precise) data $y_\varepsilon^\delta = y_\varepsilon$, then the stopping criterion (7.9) will consequently be satisfied with

$$k(\delta, y^\delta) \sim \delta^{-\frac{1}{2(\mu+\varepsilon)+1}} \quad \text{as} \quad \delta \to 0.$$

By letting $\varepsilon \to 0$ we conclude that the exponent in (7.21) is best possible. ∎

The estimate of Theorem 7.13 is the best possible *uniform* bound if no further information on the spectral family $\{F_\lambda\}$ associated with T is available. With more such information better bounds may be obtained as the following result shows:

Theorem 7.14. *Let $T = K$ be a compact operator, $K^\dagger y \in \mathcal{X}_{\mu,\rho}$.*

(i) If the singular values σ_n of K decay like $O(n^{-\alpha})$ as $n \to \infty$ with some $\alpha > 0$, then

$$k(\delta, y^\delta) \leq c \left(\frac{\rho}{\delta}\right)^{\frac{1}{(2\mu+1)(\alpha+1)}}.$$

(ii) If the singular values of K decay like $O(q^n)$ as $n \to \infty$ with some $q < 1$, then

$$k(\delta, y^\delta) \leq c \left(1 + \log^+ \frac{\rho}{\delta}\right),$$

where $\log^+ t = \log t$ for $t \geq 1$, and $\log^+ t = 0$ otherwise.

Proof. As before, denote by $\{r_k\}$ the residual polynomials of CGNE for this particular problem; we assume that $\{\sigma_n\}$ is in non-increasing order. Choose $m \in \mathbb{N}$, $0 \leq m \leq k$, and define

$$\psi_m(\lambda) := \prod_{j=1}^m \left(1 - \frac{\lambda}{\sigma_j^2}\right).$$

Taking φ_{k-m} to be the shifted Jacobi polynomial of (7.22) again, this time with degree $k - m$, we set

$$p_k(\lambda) := \psi(\lambda)\varphi_{k-m}\left(\frac{\|K\|^2}{\sigma_{m+1}^2}\lambda\right).$$

Since $p_k \in \Pi_k$ and $p_k(0) = 1$, Corollary 7.4 implies that

$$\|y^\delta - Kx_k^\delta\| \leq \|p_k(KK^*)y^\delta\|.$$

7. The Conjugate Gradient Method

Since p_k vanishes for all $\lambda \in \sigma(KK^*) \cap (\sigma_{m+1}^2, \|K\|^2]$, and since we have that $p_k(\lambda) \leq \varphi_{k-m}(\|K\|^2 \lambda / \sigma_{m+1}^2) \leq 1$ in the remaining interval $[0, \sigma_{m+1}^2]$, this implies

$$
\begin{aligned}
\|y^\delta - Kx_k^\delta\| &\leq \|F_{\sigma_m^2} p_k(KK^*) y^\delta\| \\
&\leq \|F_{\sigma_m^2} p_k(KK^*)(y^\delta - y)\| + \|F_{\sigma_m^2} p_k(KK^*) KK^\dagger y\| \\
&\leq \|y^\delta - y\| + \|\lambda^{\mu+\frac{1}{2}} \varphi_{k-m}\left(\frac{\|K\|^2}{\sigma_{m+1}^2} \lambda\right)\|_{C[0,\sigma_{m+1}^2]} \|w\| \\
&\leq \delta + c\sigma_{m+1}^{2\mu+1}(k-m)^{-2\mu-1} \rho.
\end{aligned}
$$

In the case that $\sigma_n = O(n^{-\alpha})$ the choice $m \approx k/2$ now yields

$$\|y^\delta - Kx_k^\delta\| \leq \delta + ck^{-(2\mu+1)(1+\alpha)} \rho,$$

and hence $\|y^\delta - Kx_k^\delta\| \leq \tau \delta$ if

$$k^{-(2\mu+1)(1+\alpha)} \leq \frac{\tau - 1}{c} \frac{\delta}{\rho}.$$

This is assertion (i). If the singular values σ_n decay like $O(q^n)$, then we choose $m = k$ and obtain

$$\|y^\delta - Kx_k^\delta\| \leq \delta + c\sigma_{k+1}^{2\mu+1} \rho \leq \delta + c\tilde{q}^{k+1} \rho,$$

where $\tilde{q} = q^{2\mu+1} < 1$. This implies the second assertion. ∎

We emphasize that this result is always better than the corresponding worst case bound from Theorem 7.13. For Fredholm integral equations with smooth and non-degenerate $\mathcal{L}^2$-kernels the difference between the two bounds can be enormous, showing that CGNE will require significantly less iterations for the same order of accuracy than, for example, the ν-methods, cf. Theorem 6.18.

To conclude this section we give a third estimate for $k(\delta, y^\delta)$ that might be of interest in connection with *preconditioning*: roughly spoken the following theorem suggests to transform the normal equations (7.1) into an equivalent (still ill-posed) problem with clustered spectrum. Whether and how this is actually possible, strongly depends on the particular situation, see the concluding discussion.

Theorem 7.15. *Let p be a positive integer, and assume*

$$\sigma(TT^*) \subset [0, \varepsilon_0] \cup [1 - \varepsilon_1, 1 + \varepsilon_1] \cup \{\lambda_1, \ldots, \lambda_p\},$$

where ε_0 and ε_1 are small positive numbers with $\varepsilon_0 < 1 - \varepsilon_1$ and $\lambda_j > 1 + \varepsilon_1$, $1 \leq j \leq p$. Then, if $T^\dagger y \in \mathcal{X}_{\mu,\rho}$ and

$$\varepsilon_0^{\mu+\frac{1}{2}} \leq (\tau - 1) \frac{\delta}{\rho}, \tag{7.23}$$

then there is some $c(\varepsilon_1) > 0$, independent of ε_0 and p such that

$$k(\delta, y^\delta) \leq p + c(\varepsilon_1)\left(1 + \log^+ \frac{\rho}{\delta}\right).$$

Moreover, $c(\varepsilon_1) \to 0$ as $\varepsilon_1 \to 0$.

Proof. Let T_k denote the kth Chebyshev polynomial of the first kind and consider

$$\varphi_{p+k}(\lambda) = T_k\left(\frac{1-\lambda}{\varepsilon_1}\right) T_k^{-1}\left(\frac{1}{\varepsilon_1}\right) \prod_{j=1}^{p}\left(1 - \frac{\lambda}{\lambda_j}\right) \in \Pi_{p+k}$$

with $\varphi_{p+k}(0) = 1$. Then

$$\begin{aligned}\varphi_{p+k}(\lambda_j) &= 0, & 1 \leq j \leq p, \\ |\varphi_{p+k}(\lambda)| &\leq 1, & 0 \leq \lambda \leq \varepsilon_0,\end{aligned}$$

and, cf. Appendix A.1,

$$|\varphi_{p+k}(\lambda)| \leq T_k^{-1}\left(\frac{1}{\varepsilon_1}\right) \leq 2\varepsilon_1^k, \qquad 1 - \varepsilon_1 \leq \lambda \leq 1 + \varepsilon_1.$$

Therefore, and because of our assumptions, we observe the following: if

$$k|\log \varepsilon_1| \geq c_0 \left(1 + \log^+ \frac{\rho}{\delta}\right),$$

with c_0 sufficiently large (depending only on τ and μ), then

$$|\varphi_{p+k}(\lambda)| \leq \min\{1, (\tau-1)2^{-\mu-\frac{1}{2}}\frac{\delta}{\rho}\}, \qquad 1 - \varepsilon_1 \leq \lambda \leq 1 + \varepsilon_1.$$

Together with (7.23) this yields

$$\begin{aligned}|\varphi_{p+k}(\lambda)| &\leq 1, \\ |\lambda^{\mu+\frac{1}{2}}\varphi_{p+k}(\lambda)| &\leq (\tau-1)\frac{\delta}{\rho},\end{aligned}$$

for all $\lambda \in \sigma(TT^*)$. Rewriting $T^\dagger y = (T^*T)^\mu w$ with $\|w\| \leq \rho$ and using the optimality of the $(p+k)$th residual polynomial of CGNE, cf. Corollary 7.4, we finally obtain

$$\begin{aligned}\|y^\delta - Tx_{p+k}^\delta\| &\leq \|\varphi_{p+k}(TT^*)y^\delta\| \\ &\leq \|\varphi_{p+k}(TT^*)(y^\delta - y)\| + \|\varphi_{p+k}(TT^*)T(T^*T)^\mu w\| \\ &\leq \|y^\delta - y\| + (\tau-1)\frac{\delta}{\rho}\|w\| \\ &\leq \tau\delta.\end{aligned}$$

7. The Conjugate Gradient Method

As a consequence we have from (7.9) that

$$k(\delta, y^\delta) \leq p + \frac{c_0}{|\log \varepsilon_1|} \left(1 + \log^+ \frac{\rho}{\delta}\right),$$

and the assertion of the theorem holds with $c(\varepsilon_1) = c_0/|\log \varepsilon_1|$ which goes to zero as $\varepsilon_1 \to 0$. ∎

Well-posed problems may be preconditioned by a postmultiplication of T with some *preconditioner* U which is chosen so as to approximate the (generalized) inverse of T. Since $T^\dagger$ is unbounded for ill-posed problems, a preconditioner chosen in this way will be unbounded or will have large norm, and each iteration with the preconditioned problem will be unstable. Instead, reasonable preconditioners should be bounded, in which case they will not affect the essential singularity of the resolvent of TT^* near $\lambda = 0$; the preconditioned problem will still have a spectral cluster in $[0, \varepsilon_0]$. As can be seen from Theorem 7.15, a good preconditioner should cluster the remaining part of the spectrum around $\lambda = 1$ except for possibly finitely many outliers λ_1 through λ_p.

The actual magnitude of ε_0 reflects the *degree of preconditioning*: large ε_0 means less preconditioning; small ε_0 indicates more preconditioning. With "enough" preconditioning, namely with (7.23) satisfied, only about $p + \log(\rho/\delta)$ iterations are required to reach order-optimal accuracy. However, as $\delta \to 0$, (7.23) will eventually be violated. Then the error will decrease as follows: replacing δ by $c\varepsilon_0^{\mu+\frac{1}{2}}$ in Theorem 7.15 we conclude that the first $p + |\log \varepsilon_0|$ iterations rapidly reduce the error-level to about ε_0^μ; after that, the further decrease of the error depends on the spectral distribution in the interval $[0, \varepsilon_0]$ and will in general be governed by the worst-case bound from Theorem 7.13. We refer to [119] for a specific example where preconditioning with the above philosophy has been applied successfully to an ill-posed image restoration problem.

8. Regularization With Differential Operators

8.1. Weighted Generalized Inverses

In the previous chapters we have been working with the space $\mathcal{X}$ equipped with its natural norm $\|\cdot\|$. This has led to the Moore-Penrose generalized inverse for an appropriate *notion of solution* of
$$Tx = y. \tag{8.1}$$
In the case when $\mathcal{N}(T) \neq \{0\}$ this is of course not the only possibility we have. Instead, we may also consider the problem of approximating the least-squares solution $x_L^\dagger$ of (8.1), which minimizes a different (semi)norm, namely
$$\|Lx_L^\dagger\| = \inf\{\|Lz\| \mid z \text{ is least-squares solution of } Tx = y\}.$$
This leads to the notion of weighted generalized inverses of T. Yet, to guarantee that such a solution (uniquely) exists, we will have to stipulate conditions on L, e.g., those given below.

Even in the case where $\mathcal{N}(T) = \{0\}$, and hence the uniqueness of a least-squares solution is guaranteed, one may alter the norm in $\mathcal{X}$ for the regularization process in order to enforce special features of the regularized approximations. For instance, in his original papers [269, 270], Tikhonov considered his regularization method with the more general functional
$$\|Tx - y\|^2 + \alpha \|Lx\|^2 \to \min, \qquad x \in \mathcal{D}(L). \tag{8.2}$$
Important choices include those where L has a nontrivial nullspace, i.e., when L is a *differential operator*: in many practical examples, e.g., in spline smoothing, cf. [286], L is taken to be the second derivative operator.

By virtue of (8.2), the regularized solutions automatically belong to $\mathcal{D}(L)$. As we will see below, these approximations can only converge if the solution belongs to $\mathcal{D}(L)$, too; in other words, the use of (8.2) is formally only appropriate with proper a-priori knowledge about the solution. However, we will be able to relax this condition later on in the section on Hilbert scales.

To be specific, let L further on be a closed, densely defined operator mapping $\mathcal{D}(L) \subset \mathcal{X}$ onto a Hilbert space $\mathcal{Z}$ with closed range $\mathcal{R}(L)$. A necessary condition for the following analysis is the so-called *complementation condition*, cf. Morozov [194, p. 33], which requires the existence of $\gamma > 0$ with
$$\|Tx\|^2 + \|Lx\|^2 \geq \gamma \|x\|^2, \qquad x \in \mathcal{D}(L). \tag{8.3}$$
It follows that a necessary condition for (8.3) to hold is that the nullspaces of T and L intersect only trivially. On the other hand, a sufficient condition which guarantees (8.3) is that
$$\dim \mathcal{N}(L) < \infty, \qquad \mathcal{N}(L) \cap \mathcal{N}(T) = \{0\}.$$

8. Regularization With Differential Operators

In fact, in this case $\|Tx\|/\|x\|$ has a positive minimum for $x \in \mathcal{N}(L) \setminus \{0\}$, whereas $\|Lx\|/\|x\|$ remains above some positive number for $x \in \mathcal{N}(L)^\perp$, since L has closed range. We emphasize that this sufficient condition is very important in practice as it applies in the case where L is a differential operator in $\mathbb{R}$.

From (8.3) follows that the restriction

$$T_0 := T|_{\mathcal{N}(L)}$$

of T to the nullspace of L has closed range, and therefore, T_0 has a continuous Moore-Penrose inverse $T_0^\dagger$ with $\|T_0^\dagger\| \leq \gamma^{-1}$, cf. Proposition 2.4.

Following Locker and Prenter [181], we now continue by defining a second inner product and norm on $\mathcal{D}(L)$, namely

$$\langle x, \tilde{x} \rangle_* := \langle Tx, T\tilde{x} \rangle + \langle Lx, L\tilde{x} \rangle, \qquad \|x\|_* := \langle x, x \rangle_*^{\frac{1}{2}}; \tag{8.4}$$

notice that $\|x\|_*^2 \geq \gamma \|x\|^2 > 0$ for $x \neq 0$ by virtue of (8.3), and hence defines a norm. In the sequel, we will make use of *both* topologies in $\mathcal{D}(L)$, namely the $*$-topology induced by (8.4) and the topology induced by the inner product of $\mathcal{X}$. To avoid confusion, we will mention explicitly when the $*$-topology is concerned; the notations T^* for the adjoint of T and $\mathcal{L}^\perp$ for the orthogonal complement of $\mathcal{L}$ *always* correspond to the original inner product in $\mathcal{X}$.

Lemma 8.1. *The subspace $\mathcal{D}(L)$ equipped with the inner product $\langle \cdot, \cdot \rangle_*$ is a Hilbert space. With respect to this inner product,*

$$\mathcal{L} := \mathcal{R}(T^*T_0)^\perp \cap \mathcal{D}(L) = \{x \in \mathcal{D}(L) \mid T^*Tx \perp \mathcal{N}(L)\} \tag{8.5}$$

is the orthogonal complement of $\mathcal{N}(L)$, and

$$\mathcal{T} := \{x \in \mathcal{D}(L) \mid \langle Lx, L\tilde{x} \rangle = 0 \text{ for all } \tilde{x} \in \mathcal{N}(T) \cap \mathcal{D}(L)\} \tag{8.6}$$

is the orthogonal complement of $\mathcal{N}(T) \cap \mathcal{D}(L)$ in $\mathcal{D}(L)$.

Proof. Let $\{x_n\}$ be a Cauchy sequence in $\mathcal{D}(L)$ with respect to the norm $\|\cdot\|_*$, i.e.,

$$\|x_n - x_m\|_*^2 = \|Tx_n - Tx_m\|^2 + \|Lx_n - Lx_m\|^2 \to 0$$

as $n, m \to \infty$. Consequently, $\{Tx_n\}$ is a Cauchy sequence in $\mathcal{Y}$ and $\{Lx_n\}$ is a Cauchy sequence in $\mathcal{Z}$, which provides elements $y \in \mathcal{Y}$ and $z \in \mathcal{Z}$ with

$$Tx_n \to y, \quad Lx_n \to z \quad \text{as} \quad n \to \infty. \tag{8.7}$$

Now we split $x_n = u_n + v_n$ into its components $u_n \in \mathcal{N}(L)$ and $v_n \in \mathcal{N}(L)^\perp$: since the range of L is closed, the Moore-Penrose generalized inverse $L^\dagger$ is continuous, and hence

$$v_n = L^\dagger L x_n \to v := L^\dagger z \quad \text{as} \quad n \to \infty.$$

8.1. Weighted Generalized Inverses   199

It follows that

$$T_0 u_n = T u_n = T x_n - T v_n \to y - T L^\dagger z \quad \text{as} \quad n \to \infty,$$

and since the range of T_0 is closed this implies that

$$u_n \to u := T_0^\dagger (y - T L^\dagger z) \in \mathcal{N}(L) \quad \text{as} \quad n \to \infty.$$

In other words, in the original topology of $\mathcal{X}$, the sequence $\{x_n\}$ converges to $x := u + v \in \mathcal{D}(L)$. Since $L x = L v = L L^\dagger z = z$, (8.7) implies convergence of $\{x_n\}$ to x also with respect to $\langle \cdot, \cdot \rangle_*$, and hence $\mathcal{D}(L)$ is complete in this topology.

An element $x \in \mathcal{D}(L)$ is orthogonal to $\mathcal{N}(L)$ with respect to (8.4) if and only if, for every $\tilde{x} \in \mathcal{N}(L)$,

$$\langle x, \tilde{x} \rangle_* = \langle T x, T \tilde{x} \rangle + \langle L x, L \tilde{x} \rangle = \langle T^* T x, \tilde{x} \rangle = 0.$$

Therefore, the orthogonal complement of $\mathcal{N}(L)$ in $\mathcal{D}(L)$ with respect to (8.4) is given by $\mathcal{L}$.

On the other hand, $x \in \mathcal{D}(L)$ is orthogonal to $\mathcal{N}(T) \cap \mathcal{D}(L)$ if and only if, for every $\tilde{x} \in \mathcal{N}(T) \cap \mathcal{D}(L)$,

$$\langle x, \tilde{x} \rangle_* = \langle T x, T \tilde{x} \rangle + \langle L x, L \tilde{x} \rangle = \langle L x, L \tilde{x} \rangle = 0,$$

that is, if and only if $x \in \mathcal{T}$. ∎

Equipped with the *-inner product in $\mathcal{D}(L)$, there are unique Moore-Penrose generalized inverses of $L : \mathcal{D}(L) \to \mathcal{Z}$ and of the restriction $T : \mathcal{D}(L) \to \mathcal{Y}$, denoted by $L_T^\dagger$ and $T_L^\dagger$, respectively. Note that L and T are bounded operators with respect to the norm $\| \cdot \|_*$; this is obvious from the proof of Lemma 8.1 It is important to realize that $T_L^\dagger$ and $L_T^\dagger$ are different from the (standard) Moore-Penrose inverses $T^\dagger$ and $L^\dagger$ which result from the original inner product in $\mathcal{X}$; rather, one can think of them as *weighted Moore-Penrose generalized inverses*, since they are obtained via the weighted inner product (8.4):

Theorem 8.2. *For $y \in \mathcal{D}(T_L^\dagger)$, $x = T_L^\dagger y$ is a least-squares solution of (8.1). For any other least-squares solution $\tilde{x} \in \mathcal{D}(L)$ of (8.1),*

$$\|L x\| < \|L \tilde{x}\|. \tag{8.8}$$

If the range of T is non-closed, then $T_L^\dagger$ is unbounded.

Proof. Since $T_L^\dagger$ is the Moore-Penrose generalized inverse of $T : \mathcal{D}(L) \to \mathcal{Y}$, with $\mathcal{D}(L)$ being the Hilbert space with inner product (8.4), it follows from Section 2.1 that $x = T_L^\dagger y$ is a least-squares solution of (8.1), and

$$\langle T_L^\dagger y, T_L^\dagger y \rangle_* < \langle \tilde{x}, \tilde{x} \rangle_*$$

200   8. Regularization With Differential Operators

for any other least-squares solution $\tilde{x} \in \mathcal{D}(L)$ of (8.1). Since

$$\langle \tilde{x}, \tilde{x} \rangle_* = \|T\tilde{x}\|^2 + \|L\tilde{x}\|^2 = \|Qy\|^2 + \|L\tilde{x}\|^2$$

for any least-squares solution $\tilde{x} \in \mathcal{D}(L)$ of (8.1), assertion (8.8) follows.

If $\mathcal{R}(T)$ is non-closed, $\mathcal{R}(T|_{\mathcal{D}(L)})$ cannot be closed, either, since $\mathcal{D}(L)$ is a dense subspace of $\mathcal{X}$. As in the proof of Proposition 2.4 it therefore follows that $T_L^\dagger$ is unbounded. ∎

Remark 8.3. It is now clear why the nullspaces of T and L must have no points in common besides the origin. If this were not the case, the solution $T_L^\dagger y$ as characterized in the above theorem would no longer be unique; any element in $\mathcal{N}(T) \cap \mathcal{N}(L)$ could be added to $T_L^\dagger y$, still leading to a least-squares solution of (8.1) with the same (semi)norm $\|L \cdot \|$.

It should be noted that $T_L^\dagger$ and $L_T^\dagger$ belong to the general class of so-called *inner and outer generalized inverses* (or *algebraic generalized inverses*), since they satisfy the first two Moore-Penrose conditions as given in Proposition 2.3. We refer to the survey paper by Nashed [203] as a general reference on this subject.

Opposed to the result on $T_L^\dagger$ as given in Theorem 8.2, $L_T^\dagger$ is bounded as operator from $\mathcal{Z}$ onto $\mathcal{D}(L)$. We summarize the mapping properties of $T_L^\dagger$ and $L_T^\dagger$, which follow from Definition 2.2 and Lemma 8.1:

$$\begin{aligned}
\mathcal{D}(T_L^\dagger) &= \mathcal{R}(T|_{\mathcal{D}(L)}) + \mathcal{R}(T)^\perp, & \mathcal{D}(L_T^\dagger) &= \mathcal{Z}, \\
\mathcal{N}(T_L^\dagger) &= \mathcal{R}(T)^\perp, & \mathcal{N}(L_T^\dagger) &= \{0\}, \\
\mathcal{R}(T_L^\dagger) &= \mathcal{T}, & \mathcal{R}(L_T^\dagger) &= \mathcal{L}.
\end{aligned}$$

The following lemma studies $\mathcal{D}(T_L^\dagger)$ and $\mathcal{D}(L)$ in greater detail.

Lemma 8.4. $\mathcal{D}(T_L^\dagger)$ *can be decomposed into three pairwise orthogonal subspaces,*

$$\mathcal{D}_1 = \mathcal{R}(TL_T^\dagger), \quad \mathcal{D}_2 = \mathcal{R}(T_0) \quad \text{and} \quad \mathcal{D}_3 = \mathcal{R}(T)^\perp.$$

Furthermore, $\overline{\mathcal{D}_1 + \mathcal{D}_2} + \mathcal{D}_3$ *is a decomposition of* $\mathcal{Y}$. *In* $\mathcal{D}(L)$ *we have the (oblique) decomposition*

$$\mathcal{D}(L) = \mathcal{N}(L) + (\mathcal{L} \cap \mathcal{T}) + (\mathcal{N}(T) \cap \mathcal{D}(L)).$$

Proof. Since $\mathcal{D}_3$ is the orthogonal complement of $\overline{\mathcal{R}(T)} = \overline{\mathcal{R}(T|_{\mathcal{D}(L)})}$, it has to be shown that $\mathcal{D}_1 + \mathcal{D}_2$ is an orthogonal decomposition of the entire range of $T|_{\mathcal{D}(L)}$. By definition, for $y_1 \in \mathcal{D}_1$ and $y_2 \in \mathcal{D}_2$ there exist $x_1 \in \mathcal{R}(L_T^\dagger)$ and $x_2 \in \mathcal{N}(L)$ with $y_1 = Tx_1$ and $y_2 = Tx_2$, hence

$$\langle y_1, y_2 \rangle = \langle Tx_1, Tx_2 \rangle = \langle x_1, T^*Tx_2 \rangle.$$

Since $x_1 \in \mathcal{R}(L_T^\dagger) \subset \mathcal{L}$ the right-hand side vanishes, cf. (8.5), showing that $\mathcal{D}_1 \perp \mathcal{D}_2$. Finally, since $\mathcal{N}(L) + \mathcal{R}(L_T^\dagger)$ span $\mathcal{D}(L)$ it is clear that $\mathcal{D}_1 + \mathcal{D}_2$ span $\mathcal{R}(T|_{\mathcal{D}(L)})$.

As $\mathcal{Y}$ is the closure of $\mathcal{D}(T_L^\dagger)$, we also have

$$\mathcal{Y} = \overline{\mathcal{D}_1} + \overline{\mathcal{D}_2} + \overline{\mathcal{D}_3}.$$

Since $\mathcal{D}_2$ and $\mathcal{D}_3$ are closed subspaces (recall that the range of T_0 is closed), the second assertion of the lemma follows.

Turning to $\mathcal{D}(L)$ we first recall from Lemma 8.1 that $\mathcal{L}$ and $\mathcal{N}(L)$, and $\mathcal{T}$ and $\mathcal{N}(T) \cap \mathcal{D}(L)$ are complementary subspaces of $\mathcal{D}(L)$, respectively. By definition (8.5) of $\mathcal{L}$, we have $\mathcal{N}(T) \cap \mathcal{D}(L) \subset \mathcal{L}$, hence

$$\begin{aligned}\mathcal{D}(L) &= \mathcal{N}(L) + \mathcal{L} = \mathcal{N}(L) + \mathcal{L} \cap (\mathcal{T} + \mathcal{N}(T) \cap \mathcal{D}(L)) \\ &= \mathcal{N}(L) + (\mathcal{L} \cap \mathcal{T}) + (\mathcal{N}(T) \cap \mathcal{D}(L)),\end{aligned}$$

and the proof is complete. ∎

Finally, consider the adjoint operator L^* of L which is closed, densely defined in $\mathcal{Z}$ with closed range $\mathcal{R}(L^*) = \mathcal{N}(L)^\perp$. Further on, denote by $(L_T^\dagger)^*$ the adjoint of $L_T^\dagger$; $(L_T^\dagger)^*$ is defined in $\mathcal{X}$ and maps into $\mathcal{Z}$. Recall that these adjoint operators are always defined via the original inner product in $\mathcal{X}$. The following properties of L^* and $(L_T^\dagger)^*$ will be required in the following section.

Lemma 8.5. $\mathcal{D}_2 + \mathcal{D}_3$ *is contained in the nullspace of* $(L_T^\dagger)^*T^*$, *while the range of* $(L_T^\dagger)^*T^*$ *is contained in* $\mathcal{D}(L^*)$. *Furthermore, if* $y \in \mathcal{Y}$ *is split into* $y_1 \in \overline{\mathcal{D}_1}$ *and* $y_{2,3} \in \mathcal{D}_2 + \mathcal{D}_3$, *then*

$$L^*(L_T^\dagger)^*T^*y = T^*y_1. \tag{8.9}$$

Proof. Since $\mathcal{D}_3$ is perpendicular to $\mathcal{R}(T)$, $\mathcal{D}_3$ belongs to $\mathcal{N}(T^*) \subset \mathcal{N}((L_T^\dagger)^*T^*)$. If $y \in \mathcal{D}_2$, then $y = Tx$ for some $x \in \mathcal{N}(L)$ and $T^*y = T^*Tx \in \mathcal{L}^\perp$, cf. (8.5); since $\mathcal{L}^\perp$ is in the orthogonal complement of $\mathcal{R}(L_T^\dagger)$ this implies that $(L_T^\dagger)^*T^*y = 0$. Now let $y \in \mathcal{Y}$ be split into $y_1 \in \overline{\mathcal{D}_1}$ and $y_{2,3} \in \mathcal{D}_2 + \mathcal{D}_3$; then, by the previous assertion,

$$(L_T^\dagger)^*T^*y = (L_T^\dagger)^*T^*y_1.$$

Given $x \in \mathcal{L}$, we have $L_T^\dagger Lx = x$, and therefore

$$\langle Lx, (L_T^\dagger)^*T^*y \rangle = \langle Lx, (L_T^\dagger)^*T^*y_1 \rangle = \langle L_T^\dagger Lx, T^*y_1 \rangle = \langle x, T^*y_1 \rangle. \tag{8.10}$$

On the other hand, if $x \in \mathcal{N}(L)$, then we approximate y_1 by elements in $\mathcal{D}_1$, i.e., we let $\{x_n\} \in \mathcal{L}$ be such that

$$y_1 = \lim_{n \to \infty} Tx_n.$$

Using the continuity of T^*, the definition (8.5) of $\mathcal{L}$ implies

$$\langle x, T^*y_1 \rangle = \lim_{n \to \infty} \langle x, T^*Tx_n \rangle = 0.$$

Therefore, we have $0 = \langle Lx, (L_T^\dagger)^*T^*y \rangle = \langle x, T^*y_1 \rangle$ for $x \in \mathcal{N}(L)$. Since (8.10) has been established for $x \in \mathcal{L}$ before, we conclude that

$$\langle Lx, (L_T^\dagger)^*T^*y \rangle = \langle x, T^*y_1 \rangle$$

8. Regularization With Differential Operators

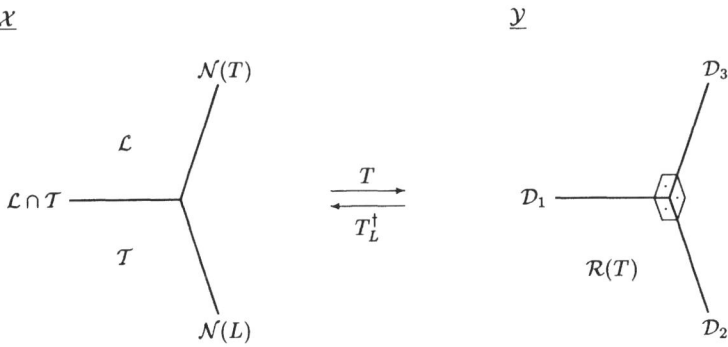

Figure 8.1: Decompositions of $\mathcal{X}$ and $\mathcal{Y}$

holds for all $x \in \mathcal{D}(L)$. This proves (8.9). ∎

We emphasize that we need to be able to apply Lemma 8.5 to perturbed data; hence, we have to consider $(L_T^\dagger)^* T^*$ as operator from $\mathcal{Y}$ to $\mathcal{Z}$, which makes the final part of the above proof somewhat technical.

We conclude this section with a graphical illustration of the algebraic mapping structure of T and $T_L^\dagger$, cf. Figure 8.1; corresponding subspaces of $\mathcal{X}$ and $\mathcal{Y}$ (i.e., parallel lines in Figure 8.1) are mapped onto each other; $\mathcal{N}(T)$ and $\mathcal{D}_3$ are the nullspaces of T and $T_L^\dagger$, respectively. The three subspaces of $\mathcal{Y}$ (but not those of $\mathcal{X}$) are mutually orthogonal.

8.2. Regularization with Seminorms

To approximate $T_L^\dagger y$, one can in principle use all regularization methods from the previous chapters, by considering T as an operator from $\mathcal{D}(L)$ to $\mathcal{Y}$, where $\mathcal{D}(L)$ is equipped with the inner product (8.4). This is the approach taken by Locker and Prenter [181] concerning Tikhonov regularization. From a computational point of view, this has certain drawbacks, since it requires the computation of the adjoint $T^\sharp$ of T with respect to the inner product (8.4), i.e., cf. [181, Lemma 4.1],

$$T^\sharp = (T^*T + L^*L)^{-1} T^* \ : \ \mathcal{Y} \to \mathcal{D}(L).$$

In other words, any application of $T^\sharp$ requires the solution of a linear equation with $T^*T + L^*L$. Another approach has been taken in [112]. Although this approach also requires the solution of linear equations since $L_T^\dagger$ and $(L_T^\dagger)^*$ have to be implemented, such systems can often be solved much more efficiently in the case when L is a differential operator, cf. Section 9.2.

8.2. Regularization with Seminorms

To this end we introduce

$$B := TL_T^\dagger, \qquad (8.11)$$

and observe that $B : \mathcal{Z} \to \mathcal{Y}$ is continuous under the conditions on L from the previous section. We begin with an algebraic proposition (due to Eldén [64]), which is of interest in itself, since it states some kind of *reverse order law* for the generalized inverse of an operator product.

Proposition 8.6. *The Moore-Penrose generalized inverse of B is given by $B^\dagger = LT_L^\dagger$.*

Proof. We make use of the definitions in Lemma 8.4. There it has been shown that

$$\mathcal{R}(B) = \mathcal{D}_1,$$

and it also follows from Lemma 8.4 that

$$\mathcal{D}(T_L^\dagger) = \mathcal{D}_1 + \mathcal{D}_1^\perp = \mathcal{D}(B^\dagger).$$

Consider now $B(LT_L^\dagger) : \mathcal{D}(T_L^\dagger) \to \mathcal{Y}$. Decomposing $y \in \mathcal{D}(T_L^\dagger)$ according to Lemma 8.4 into $y = y_1 + y_2 + y_3$ we obtain

$$BLT_L^\dagger y = BL(T_L^\dagger y_1 + T_L^\dagger y_2) = BLT_L^\dagger y_1,$$

since $T_L^\dagger y_2 \in \mathcal{N}(L)$. As $T_L^\dagger y_1 \in \mathcal{L}$, and $L_T^\dagger L$ is an (oblique) projector onto $\mathcal{L}$, this implies

$$BLT_L^\dagger y = BLT_L^\dagger y_1 = T(L_T^\dagger L)T_L^\dagger y_1 = TT_L^\dagger y_1 = y_1.$$

In other words, $B(LT_L^\dagger)$ is the orthogonal projector onto $\mathcal{D}_1 = \mathcal{R}(B)$.

Next we consider $(LT_L^\dagger)B : \mathcal{Z} \to \mathcal{Z}$. Given $z \in \mathcal{Z}$ we have $L_T^\dagger z \in \mathcal{L}$, and we can decompose

$$L_T^\dagger z = x_1 + x_2, \qquad x_1 \in \mathcal{L} \cap \mathcal{T}, \quad x_2 \in \mathcal{N}(T),$$

by virtue of Lemma 8.4. It follows that

$$(LT_L^\dagger)Bz = LT_L^\dagger T(L_T^\dagger z) = LT_L^\dagger T(x_1 + x_2) = Lx_1$$

because $T_L^\dagger T$ is an (oblique) projector with range $\mathcal{T}$. Since L is onto, $z = LL_T^\dagger z = Lx_1 + Lx_2$, and hence $(LT_L^\dagger)B$ is the projector with range

$$\mathcal{R}(L|_{\mathcal{L} \cap \mathcal{T}}) = \mathcal{R}(L|_\mathcal{T})$$

and nullspace $\mathcal{R}(L|_{\mathcal{N}(T) \cap \mathcal{D}(L)})$. The latter coincides with the nullspace of B, and the former is orthogonal to this space by virtue of Lemma 8.1. From Proposition 2.3 and the subsequent remark it therefore follows that $LT_L^\dagger$ is the Moore-Penrose generalized inverse of B. ∎

8. Regularization With Differential Operators

For the definition of regularization methods we choose functions $g_\alpha(\lambda)$ approximating $1/\lambda$ as in Chapter 4. Since the Moore-Penrose generalized inverse of $T_0 = T|_{\mathcal{N}(L)}$ is a continuous mapping, one can determine $T_0^\dagger$ in a stable way. The result

$$x_0^\dagger := T_0^\dagger y \tag{8.12}$$

is the component of $T_L^\dagger y$ in $\mathcal{N}(L)$ associated with the decomposition shown in Figure 8.1. Now we define the regularization $\{R_\alpha\}$ as

$$R_\alpha = T_0^\dagger + L_T^\dagger g_\alpha(B^*B)B^* \tag{8.13}$$

Since $L_T^\dagger g_\alpha(B^*B)B^*$ belongs to $\mathcal{R}(L_T^\dagger) = \mathcal{L}$, the decomposition in (8.13) is an orthogonal one with respect to the inner product $\langle \cdot, \cdot \rangle_*$ according to Lemma 8.1.

Theorem 8.7. *Let $\{g_\alpha\}$ be a family of piecewise continuous functions over $[0, \|B\|^2]$ which fulfills the assumptions of Theorem 4.1. Then, for all $y \in \mathcal{D}(T_L^\dagger)$,*

$$R_\alpha y \to T_L^\dagger y, \quad LR_\alpha y \to LT_L^\dagger y, \quad TR_\alpha y \to Qy,$$

as $\alpha \to 0$. If $y \notin \mathcal{D}(T_L^\dagger)$, then $\lim_{\alpha \to 0} \|LR_\alpha y\| = \infty$.

Proof. Let $z_\alpha := g_\alpha(B^*B)B^*y$: since $z_\alpha \in \mathcal{Z} = \mathcal{R}(L)$,

$$LR_\alpha y = LL_T^\dagger z_\alpha = z_\alpha,$$

and hence, if $y \in \mathcal{D}(B^\dagger)$, Theorem 4.1 and Proposition 8.6 yield

$$LR_\alpha y = g_\alpha(B^*B)B^*y \to B^\dagger y = LT_L^\dagger y \quad \text{as} \quad \alpha \to 0. \tag{8.14}$$

On the other hand, by virtue of the same theorem, if $y \notin \mathcal{D}(B^\dagger) = \mathcal{D}(T_L^\dagger)$, then $\|LR_\alpha y\| \to \infty$ as $\alpha \to 0$.

In the remaining part of the proof, we restrict our attention to $y \in \mathcal{D}(T_L^\dagger)$. By definition (8.12) of $x_0^\dagger$,

$$TR_\alpha y = Tx_0^\dagger + Bg_\alpha(B^*B)B^*y = T_0T_0^\dagger y + Bg_\alpha(B^*B)B^*y,$$

and Theorem 4.1 yields

$$TR_\alpha y \to (T_0T_0^\dagger + BB^\dagger)y \quad \text{as} \quad \alpha \to 0.$$

Since $T_0T_0^\dagger$ is the orthogonal projector onto $\mathcal{D}_2$ and $BB^\dagger$ is the orthogonal projector onto $\mathcal{D}_1$ it follows that $\lim_{\alpha \to 0} TR_\alpha y = Qy$. Finally, using the complementation condition (8.3),

$$\|R_\alpha y - T_L^\dagger y\|^2 \leq \frac{1}{\gamma}(\|TR_\alpha y - Qy\|^2 + \|LR_\alpha y - LT_L^\dagger y\|^2),$$

hence $R_\alpha y \to T_L^\dagger y$ as $\alpha \to 0$. ∎

Remark 8.8. The approximations given by (8.13) belong to $T \cap D(L^*L)$; the latter is obvious from (8.14), since $LR_\alpha y \in \mathcal{R}(B^*)$, which is a subset of $\mathcal{D}(L^*)$ by Lemma 8.5. Moreover, Lemma 8.5 yields for all $x \in \mathcal{N}(T) \cap \mathcal{D}(L)$,

$$\langle LR_\alpha y, Lx \rangle = \langle L^*LR_\alpha y, x \rangle = \langle T^*y_1, x \rangle = \langle y_1, Tx \rangle$$

with some element $y_1 \in \overline{\mathcal{D}_1}$. Since $x \in \mathcal{N}(T)$, the right-hand side vanishes, establishing that $R_\alpha y \in T$, cf. (8.6).

Since the regularized approximations belong to $\mathcal{D}(L^*L)$, they have additional smoothness properties as compared to the exact solution which may only belong to $\mathcal{D}(L)$. This is similar to the $L = I$ case, where we compute regularized approximations in $\mathcal{R}(T^*)$ even though the exact solution may have no similar smoothness property.

The convergence result of Theorem 8.7 is equivalent to convergence with respect to the so-called *graph norm* $\|\cdot\|_L$ given by

$$\|x\|_L^2 := \|x\|^2 + \|Lx\|^2, \qquad x \in \mathcal{D}(L). \tag{8.15}$$

In spite of the divergence result for $y \notin \mathcal{D}(T_L^\dagger)$ with respect to the graph norm, we will see in Section 8.5 that the approximations $R_\alpha y$ may nevertheless converge in $\mathcal{X}$ to a reasonable solution of (8.1) if $y \in \mathcal{D}(T^\dagger)$, provided we have certain relations between L and T.

We also remark that it is again possible to supply an initial guess x^* for the exact solution. In this case, the regularization scheme is applied to the right-hand side $y - Tx^*$, and then the regularized solution is $x^* + R_\alpha(y - Tx^*)$.

Remark 8.9. It can be seen from the proof of Theorem 8.7 that for $y \in \mathcal{D}(T_L^\dagger)$ the rate of convergence of $R_\alpha y \to T_L^\dagger y$ is bounded by the rate of convergence of $LR_\alpha y \to B^\dagger y$. Because of (8.14) the latter can be estimated by means of the results from the foregoing chapters, cf., e.g., Theorem 4.3. These results apply if

$$B^\dagger y \in \mathcal{R}((B^*B)^\mu) \qquad \text{for some } \mu > 0. \tag{8.16}$$

It is difficult to give an interpretation of this condition for arbitrary $\mu > 0$ without further information about L and T. We will come back to this question in the Hilbert scale context of Section 8.5. However, for the important special cases $\mu = 1/2$ and $\mu = 1$, respectively, the condition (8.16) has the following illuminating interpretation.

Proposition 8.10. *Let* $y \in \mathcal{D}(T_L^\dagger) = \mathcal{R}(T|_{\mathcal{D}(L)}) + \mathcal{R}(T)^\perp$. *Then the following holds:*

(i) $B^\dagger y \in \mathcal{R}((B^*B)^{\frac{1}{2}})$ *if and only if* $T_L^\dagger y \in \mathcal{D}(L^*L)$ *and* $(L^*L)T_L^\dagger y \in \mathcal{R}(T^*)$.

(ii) $B^\dagger y \in \mathcal{R}(B^*B)$ *if and only if* $T_L^\dagger y \in \mathcal{D}(L^*L)$ *and* $(L^*L)T_L^\dagger y \in \mathcal{R}(T^*T|_{\mathcal{D}(L)})$.

8. Regularization With Differential Operators

Proof. (i) Recall that $B^\dagger y \in \mathcal{R}((B^*B)^{\frac{1}{2}})$ if and only if $B^\dagger y \in \mathcal{R}(B^*) = \mathcal{R}((L_T^\dagger)^*T^*)$. Because of the representation for $B^\dagger$ derived in Proposition 8.6, this is the case if and only if

$$LT_L^\dagger y = (L_T^\dagger)^* T^* w \tag{8.17}$$

for some $w \in \mathcal{Y}$. Assume (8.17) holds for some $w \in \mathcal{Y}$. Then, Lemma 8.5 shows that the right-hand side belongs to $\mathcal{D}(L^*)$, and hence $T_L^\dagger y \in \mathcal{D}(L^*L)$. Moreover, (8.9) yields some $w_1 \in \overline{\mathcal{D}_1}$ with

$$L^* L T_L^\dagger y = T^* w_1 \in \mathcal{R}(T^*). \tag{8.18}$$

On the other hand, if $T_L^\dagger y \in \mathcal{D}(L^*L)$, then we make use of the fact that $LL_T^\dagger = I$, since L is onto, and hence,

$$B^\dagger y = LT_L^\dagger y = (LL_T^\dagger)^* LT_L^\dagger y = (L_T^\dagger)^*(L^* LT_L^\dagger y).$$

In other words, assuming $L^* LT_L^\dagger y \in \mathcal{R}(T^*)$, $B^\dagger y$ has a representation as in (8.17), and hence $B^\dagger y \in \mathcal{R}(B^*)$.

(ii) If $B^\dagger y \in \mathcal{R}(B^*B)$, then w in (8.17) can be chosen from $\mathcal{R}(B) = \mathcal{D}_1$, cf. Lemma 8.4. Following the proof of part (i) we, therefore, arrive at (8.18) with $w_1 = w$ according to Lemma 8.5. Since $w \in \mathcal{D}_1 \subset \mathcal{R}(T|_{\mathcal{D}(L)})$ the assertion

$$L^* LT_L^\dagger y \in \mathcal{R}(T^*T|_{\mathcal{D}(L)})$$

follows. In turn, assuming the latter relation, we obtain as in the proof of case (i) that

$$B^\dagger y = (L_T^\dagger)^*(L^* LT_L^\dagger y) = (L_T^\dagger)^* T^* T x$$

for some $x \in \mathcal{D}(L)$. Since any $x \in \mathcal{D}(L)$ can be rewritten as $L_T^\dagger z + \tilde{x}$ with $z \in \mathcal{Z}$ and $\tilde{x} \in \mathcal{N}(L)$ we have

$$B^\dagger y = B^* B z + (L_T^\dagger)^* T^* T \tilde{x}.$$

The last term, however, vanishes since $T\tilde{x} = T_0 \tilde{x} \in \mathcal{N}((L_T^\dagger)^* T^*)$, cf. Lemma 8.5. This proves that $B^\dagger y \in \mathcal{R}(B^*B)$. ∎

This result states that the classical smoothness assumptions $T^\dagger y \in \mathcal{R}(T^*)$ and $T^\dagger y \in \mathcal{R}(T^*T)$ in the case $L = I$ have to be replaced by the *same* smoothness assumptions for $L^* LT_L^\dagger y$. For example, if L is the second derivative operator in $\mathcal{L}^2[0,1]$, then

$$L^* u = u'', \quad u \in \mathcal{D}(L^*) = \{u \in \mathcal{H}^2[0,1] \mid u(0) = u(1) = u'(0) = u'(1) = 0\}.$$

The smoothness condition $B^\dagger y \in \mathcal{R}(B^*B)$ therefore means in this case that $T_L^\dagger y$ has four derivatives in $\mathcal{L}^2[0,1]$, with the second and third derivative satisfying appropriate boundary conditions, and the *fourth* derivative of $T_L^\dagger y$ satisfying the "classical" smoothness assumption to be in $\mathcal{R}(T^*T)$.

8.3. Examples

We exemplify the scheme of the previous section for three regularization methods, namely Tikhonov regularization, Landweber iteration, and the conjugate gradient iteration.

Example 8.11. In Chapter 5 we have seen that *Tikhonov regularization* can be described by the function $g_\alpha(\lambda) = (\lambda + \alpha)^{-1}$. Now we will show that (8.13) with this particular function g_α yields precisely the solution of (8.2). To this end, redefine x_α via (8.13), i.e.,

$$x_\alpha = R_\alpha y = x_0^\dagger + L_T^\dagger g_\alpha(B^*B) B^* y.$$

As in the proof of Theorem 8.7, cf. (8.14),

$$L^* L x_\alpha = L^* g_\alpha(B^*B) B^* y;$$

recall that $x_\alpha \in \mathcal{D}(L^*L)$ has been established in Remark 8.8. Using (8.9),

$$T^*T x_\alpha = T^*T_0 T_0^\dagger y + T^* T L_T^\dagger g_\alpha(B^*B) B^* y = T^* T_0^* T_0^\dagger y + L^*(L_T^\dagger)^* T^* T L_T^\dagger g_\alpha(B^*B) B^* y,$$

since $\mathcal{R}(T L_T^\dagger) = \mathcal{D}_1$. Hence, we obtain

$$\begin{aligned}(T^*T + \alpha L^*L)x_\alpha &= T^* T_0^* T_0^\dagger y + (L^*(L_T^\dagger)^* T^* T L_T^\dagger + \alpha L^*) g_\alpha(B^*B) B^* y \\ &= T^* T_0^* T_0^\dagger y + L^*(B^*B + \alpha I)(B^*B + \alpha I)^{-1} B^* y \\ &= T^* T_0^* T_0^\dagger y + L^*(L_T^\dagger)^* T^* y.\end{aligned}$$

Decomposing y according to Lemma 8.4 into

$$y = y_1 + y_2 + y_3, \qquad y_1 \in \overline{\mathcal{D}_1},\ y_2 \in \mathcal{D}_2,\ y_3 \in \mathcal{D}_3,$$

then we have $T_0 T_0^\dagger y = y_2$, and a final application of Lemma 8.5 yields

$$(T^*T + \alpha L^*L)x_\alpha = T^* y_2 + T^* y_1 = T^* y.$$

Since this is the Euler equation for (8.2), x_α is the minimizer of the Tikhonov functional in general form.

As $g_\alpha(\lambda) = (\lambda + \alpha)^{-1}$ satisfies the conditions of Theorem 8.7, it follows that the regularized solution x_α of the Tikhonov method in general form converges to $T_L^\dagger y$ as $\alpha \to 0$ if $y \in \mathcal{D}(T_L^\dagger)$. This special instance of Theorem 8.7 has been obtained by Locker and Prenter [181], see also Morozov [194].

Example 8.12. The functions $\{g_k\}$ corresponding to the *Landweber iteration* are $g_k(\lambda) = \sum_{j=0}^{k-1}(1-\lambda)^j$, $k \in \mathbb{N}_0$, cf. Chapter 6. The regularization operators

$$R_k = T_0^\dagger + L_T^\dagger g_k(B^*B) B^*, \qquad k \in \mathbb{N}_0,$$

208   8. Regularization With Differential Operators

can be evaluated recursively, since

$$g_k(\lambda) = 1 + (1 - \lambda)g_{k-1}(\lambda), \qquad k \in \mathbb{N}.$$

It follows that

$$\begin{aligned}
R_k y &= T_0^\dagger y + L_T^\dagger B^* y + L_T^\dagger (I - B^* B) g_{k-1}(B^* B) B^* y \\
&= R_{k-1} y + L_T^\dagger B^* (y - T L_T^\dagger g_{k-1}(B^* B) B^* y) \\
&= R_{k-1} y + L_T^\dagger (L_T^\dagger)^* T^* (y - T R_{k-1} y + T T_0^\dagger y).
\end{aligned}$$

Because of Lemma 8.5 we have

$$TT_0^\dagger y \in \mathcal{D}_2 \subset \mathcal{N}((L_T^\dagger)^* T^*),$$

so that we finally obtain the following iterative scheme for $x_k = R_k y$:

$$x_0 = x_0^\dagger, \qquad x_k = x_{k-1} + L_T^\dagger (L_T^\dagger)^* T^* (y - T x_{k-1}), \qquad k \in \mathbb{N}.$$

Again, $\{g_k\}$ satisfies the conditions of Theorem 8.7, so that the sequence $\{x_k\}$ converges to $T_L^\dagger y$ if $y \in \mathcal{D}(T_L^\dagger)$.

Example 8.13. As final example, consider the *conjugate gradient iteration* in $\mathcal{D}(L)$ with respect to the modified inner product (8.4):

Algorithm 8.14.

- $x_0 = x_0^\dagger, \quad d_0 = y - T x_0, \quad s_0 = (L_T^\dagger)^* T^* d_0, \quad p_1 = L_T^\dagger s_0;$

- for $k = 1, 2, \ldots,$ unless $s_{k-1} = 0$, compute

$$\begin{aligned}
q_k &= T p_k, \\
\alpha_k &= \|s_{k-1}\|^2 / \|q_k\|^2, \\
x_k &= x_{k-1} + \alpha_k p_k, \\
d_k &= d_{k-1} - \alpha_k q_k, \\
s_k &= (L_T^\dagger)^* T^* d_k, \\
\beta_k &= \|s_k\|^2 / \|s_{k-1}\|^2, \\
p_{k+1} &= L_T^\dagger s_k + \beta_k p_k.
\end{aligned}$$

It follows by induction (note that $B^* y = B^*(y - Tx_0)$ by Lemma 8.5) that

$$p_k \in L_T^\dagger (\mathcal{K}_k(B^* y, B^* B)),$$

where $\mathcal{K}_k(B^* y, B^* B)$ is the corresponding Krylov subspace, and hence x_k is of the form (8.13) with a certain polynomial g_k of degree $k - 1$. The basic properties of this variant of CGNE are summarized in the following result, valid for all $y \in \mathcal{Y}$:

8.3. Examples

Theorem 8.15. *The kth iterate x_k of Algorithm 8.14 minimizes $\|y - Tx_k\|$ in the augmented Krylov subspace*

$$x_0^\dagger + L_T^\dagger(\mathcal{K}_k(B^*y, B^*B)).$$

The corresponding residual polynomials $\{r_k\}$ associated with this algorithm are orthogonal with respect to the inner product

$$[\varphi, \psi] := \int_0^{\|B\|^2+} \varphi(\lambda)\psi(\lambda)\lambda\, d\|H_\lambda y\|^2;$$

here, $\{H_\lambda\}$ is the spectral family of BB^*.

Proof. Comparing Algorithms 7.1 and 8.14 one observes that the kth iterate x_k of Algorithm 8.14 can be rewritten as

$$x_k = x_0^\dagger + L_T^\dagger z_k,$$

where z_k is the kth iterate obtained by the classical conjugate gradient method CGNE applied to the problem

$$Bz = y.$$

From (8.13) we conclude that $z_k = g_k(B^*B)B^*y$, and hence the residual polynomials of the two algorithms coincide. The orthogonality assertion now follows from Corollary 7.4. Furthermore, it follows from Theorem 7.3 that $\|y - Bz_k\|$ is minimal among all elements $z \in \mathcal{K}_k(B^*y, B^*B)$, so that, by Lemma 8.4,

$$\begin{aligned}\|y - Tx_k\|^2 &= \|y - TL_T^\dagger z_k - T_0 x_0^\dagger\|^2 = \|y - Bz_k\|^2 + \|T_0 T_0^\dagger y\|^2 \\ &\leq \|y - Tx\|^2\end{aligned}$$

for any other element $x \in x_0^\dagger + L_T^\dagger(\mathcal{K}_k(B^*y, B^*B))$. ∎

The above theorem is the analogue of Corollary 7.4 and therefore forms the basis for generalizing the properties of the conjugate gradient method as determined in Chapter 7 to the case of a modified inner product (8.4) in $\mathcal{D}(L)$. As an example we obtain immediately the following convergence result.

Theorem 8.16. *If $y \in \mathcal{D}(T_L^\dagger)$, then the iterates of Algorithm 8.14 converge to $T_L^\dagger y$ as $k \to \infty$ with respect to the graph norm (8.15).*

Proof. By Theorem 7.9 the iterates z_k introduced in the proof of Theorem 8.15 converge to $B^\dagger y$; furthermore, Bz_k converges to the projection of y onto $\mathcal{D}_1$. As in the proof of Theorem 8.7 it, therefore, follows that

$$Lx_k = z_k \to B^\dagger y = LT_L^\dagger y, \qquad Tx_k \to Qy.$$

The complete assertion follows from the complementation condition (8.3) as in the proof of Theorem 8.7. ∎

8.4. Hilbert Scales

In Section 8.2 convergence has not been analyzed in the norm of $\mathcal{X}$, but in the stronger graph norm $\|\cdot\|_L$ of (8.15). An essential requirement for this approach was that the solution is an element of $\mathcal{D}(L)$. The question arises, whether convergence of these regularized approximations can be achieved in the original norm $\|\cdot\|$ of $\mathcal{X}$ even when no solution of (8.1) is an element of $\mathcal{D}(L)$. If this happens to be the case, then this is called *oversmoothing*. The convergence analysis of this case will be presented in the general framework of Hilbert scales.

We first present a short introduction into Hilbert scales and prove those results which will be essential for the analysis of regularization methods in the next section. For a rather complete theory on Hilbert scales see Krein and Petunin [164].

Further on, let L be a densely defined unbounded selfadjoint strictly positive operator in the Hilbert space $\mathcal{X}$, i.e., L is a closed operator in $\mathcal{X}$ satisfying:

$$\mathcal{D}(L) = \mathcal{D}(L^*) \quad \text{is dense in } \mathcal{X},$$

$$\langle Lx, y \rangle = \langle x, Ly \rangle \quad \text{for all } x, y \in \mathcal{D}(L),$$

and there exists $\gamma > 0$ such that

$$\langle Lx, x \rangle \geq \gamma \|x\|^2 \quad \text{for all } x \in \mathcal{D}(L). \tag{8.19}$$

We consider the set $\mathcal{M}$ of all elements x for which all the powers of L are defined, i.e.,

$$\mathcal{M} := \bigcap_{k=0}^{\infty} \mathcal{D}(L^k). \tag{8.20}$$

Lemma 8.17. *The set $\mathcal{M}$ defined by (8.20) is dense in $\mathcal{X}$.*

Proof. Let $\{E_\lambda\}$ denote the spectral family of L (cf. Section 2.3) and let $x \in \mathcal{X}$ be an arbitrary, but fixed element. We consider the sequence $\{x_n\}$ defined by

$$x_n := E_n x, \quad n \in \mathbb{N}.$$

Due to (8.19) and the fact that

$$\int_\gamma^\infty \lambda^{2k} \, d\|E_\lambda x_n\|^2 = \int_\gamma^{n+} \lambda^{2k} \, d\|E_\lambda x\|^2 \leq n^{2k} \|x\|^2 < \infty,$$

$x_n \in \mathcal{D}(L^k)$ for all $k, n \in \mathbb{N}$. Hence

$$\{x_n\} \subset \mathcal{M}.$$

Since by the definition of $\{E_\lambda\}$

$$\lim_{n \to \infty} x_n = \lim_{n \to \infty} E_n x = x,$$

and since $x \in \mathcal{X}$ was arbitrary, $\mathcal{M}$ is dense in $\mathcal{X}$. ∎

It follows by spectral theory that L^s is defined on $\mathcal{M}$ for all $s \in \mathbb{R}$ and that

$$\mathcal{M} = \bigcap_{s \in \mathbb{R}} \mathcal{D}(L^s).$$

Definition 8.18. *In $\mathcal{M}$, defined by (8.20), we introduce for all $s \in \mathbb{R}$ the inner products and norms*

$$\langle x, y \rangle_s := \langle L^s x, L^s y \rangle, \tag{8.21}$$

$$\|x\|_s := \|L^s x\|, \tag{8.22}$$

$x, y \in \mathcal{M}$, respectively.

The Hilbert space $\mathcal{X}_s$ is defined as the completion of $\mathcal{M}$ with respect to the norm $\|\cdot\|_s$ in (8.22). $(\mathcal{X}_s)_{s \in \mathbb{R}}$ is called the Hilbert scale *induced by L.*

Proposition 8.19. *Let L be as above and let $(\mathcal{X}_s)_{s \in \mathbb{R}}$ be the Hilbert scale induced by L. Then the following assertions hold:*

(i) *Let $-\infty < s < t < \infty$. Then the space $\mathcal{X}_t$ is densely and continuously embedded in $\mathcal{X}_s$.*

(ii) *Let $s, t \in \mathbb{R}$. The operator L^{t-s}, defined on $\mathcal{M}$, has a unique extension to $\mathcal{X}_t$ which is an isomorphism from $\mathcal{X}_t$ onto $\mathcal{X}_s$. This extension, again denoted by L^{t-s}, is selfadjoint and strictly positive as restriction to $\mathcal{X}_t$ in $\mathcal{X}_s$ if $t > s$. Moreover, $L^{t-s} = L^t L^{-s}$ holds for the appropriate extensions, especially, $(L^s)^{-1} = L^{-s}$.*

(iii) *If $s \geq 0$, then $\mathcal{X}_s = \mathcal{D}(L^s)$ and $\mathcal{X}_{-s} = (\mathcal{X}_s)'$, i.e., $\mathcal{X}_{-s}$ is the dual space of $\mathcal{X}_s$.*

(iv) *Let $-\infty < q < r < s < \infty$ and let $x \in \mathcal{X}_s$. Then the* interpolation inequality

$$\|x\|_r \leq \|x\|_q^{\frac{s-r}{s-q}} \|x\|_s^{\frac{r-q}{s-q}} \tag{8.23}$$

holds.

Proof. (i) Let $\{E_\lambda\}$ be the spectral family of L and let $x \in \mathcal{M}$. Due to (8.19) and (8.22), we obtain

$$\|x\|_t^2 = \int_\gamma^\infty \lambda^{2t} \, d\|E_\lambda x\|^2$$

for all $t \in \mathbb{R}$. Hence, for $s < t$ we have the estimate

$$\|x\|_s \leq \gamma^{s-t} \|x\|_t. \tag{8.24}$$

By the definition of the spaces $\mathcal{X}_t$, this implies that $\mathcal{X}_t$ is densely and continuously embedded in $\mathcal{X}_s$.

(ii) It is an immediate consequence of the definition of the set $\mathcal{M}$ that L^u is a one-to-one mapping of $\mathcal{M}$ onto $\mathcal{M}$ for all $u \in \mathbb{R}$. For any $x \in \mathcal{M}$ and $s, t \in \mathbb{R}$

$$\begin{aligned}
L^{t-s}x &= \int_\gamma^\infty \lambda^{t-s}\, dE_\lambda x = \int_\gamma^\infty \lambda^t \lambda^{-s}\, dE_\lambda x \\
&= \int_\gamma^\infty \lambda^t\, d_\lambda \int_\gamma^\lambda \mu^{-s}\, dE_\mu x = \int_\gamma^\infty \lambda^t\, dE_\lambda L^{-s}x \\
&= L^t L^{-s}x.
\end{aligned}$$

Thus,
$$L^{t-s} = L^t L^{-s} \tag{8.25}$$

holds in $\mathcal{M}$. In addition,
$$\|x\|_t = \|L^t x\| = \|L^s L^{t-s} x\| = \|L^{t-s} x\|_s. \tag{8.26}$$

By definition of the spaces $\mathcal{X}_t$ and $\mathcal{X}_s$, this implies that L^{t-s} has a unique continuous injective extension to $\mathcal{X}_t$ with $L^{t-s}(\mathcal{X}_t) \subset \mathcal{X}_s$. The surjectivity of this mapping follows from the fact that

$$\mathcal{X}_s = \overline{\mathcal{M}} = \overline{L^{t-s}(\mathcal{M})} \subset \overline{L^{t-s}(\mathcal{X}_t)} \subset \mathcal{X}_s,$$

where the closure is taken in $\mathcal{X}_s$. Moreover, (8.26) also holds for the extension. A continuity argument now shows that (8.25) also holds for the appropriate extensions.

It remains to be shown that L^{t-s} is selfadjoint and strictly positive as restriction to $\mathcal{X}_t$ in $\mathcal{X}_s$ if $t > s$. Let $x, y \in \mathcal{M}$. Since L^t is selfadjoint in $\mathcal{D}(L^t)$ for all $t \in \mathbb{R}$ (cf. Proposition 2.16 (iii)), by (8.25) we obtain that

$$\begin{aligned}
\langle L^{t-s}x, y \rangle_s &= \langle L^s L^{t-s}x, L^s y \rangle = \langle L^t x, L^s y \rangle \\
&= \langle L^s L^t x, y \rangle = \langle L^t L^s x, y \rangle \\
&= \langle L^s x, L^t y \rangle = \langle L^s x, L^s L^{t-s} y \rangle \\
&= \langle x, L^{t-s} y \rangle_s.
\end{aligned}$$

Moreover, by (8.24) and (8.25)
$$\|L^{t-s} x\|_s = \|L^t x\| \geq \gamma^{t-s} \|x\|_s.$$

The assertion now follows by (8.24) and (8.26). If $s = t$, we obtain that $I = L^s L^{-s}$, and hence that $(L^s)^{-1} = L^{-s}$.

(iii) Let $s \geq 0$ and $x \in \mathcal{D}(L^s)$. Then $L^s x \in \mathcal{X}$ and
$$L^s E_n x = E_n L^s x \to L^s x \quad \text{as} \quad n \to \infty.$$

Since $\{E_n x\}$ is a sequence in $\mathcal{M}$ (see the proof of Lemma 8.17), this implies that $x \in \mathcal{X}_s$. $\mathcal{X}_s = \mathcal{D}(L^s)$ now follows if we are able to show that $\mathcal{D}(L^s)$ is complete with respect to $\|\cdot\|_s$. Let $\{x_n\}$ be a Cauchy sequence in $\mathcal{D}(L^s)$ with respect to $\|\cdot\|_s$. Then, by (8.24), $\{x_n\}$ and $\{L^s x_n\}$ are Cauchy sequences in $\mathcal{X}$. Since L^s is closed, there exists an element $x \in \mathcal{D}(L^s)$ such that $x_n \to x$ and $L^s x_n \to L^s x$ in $\mathcal{X}$ (here, the completeness of $\mathcal{X}$ is used). Thus, $x_n \to x$ in $\mathcal{D}(L^s)$ with respect to $\|\cdot\|_s$. The fact that $\mathcal{X}_{-s}$ is the dual space of $\mathcal{X}_s$ follows with (ii).

(iv) The validity of the interpolation inequality for all $-\infty < q < r < s < \infty$ and $x \in \mathcal{X}_s$ follows from the Hölder inequality:

$$\begin{aligned}
\|x\|_r^2 &= \int_\gamma^\infty \lambda^{2r}\, d\|E_\lambda x\|^2 \\
&= \int_\gamma^\infty \lambda^{2q \frac{s-r}{s-q}} \lambda^{2s \frac{r-q}{s-q}}\, d\|E_\lambda x\|^2 \\
&\leq \left(\int_\gamma^\infty \lambda^{2q}\, d\|E_\lambda x\|^2\right)^{\frac{s-r}{s-q}} \left(\int_\gamma^\infty \lambda^{2s}\, d\|E_\lambda x\|^2\right)^{\frac{r-q}{s-q}} \\
&= \left(\|x\|_q^2\right)^{\frac{s-r}{s-q}} \left(\|x\|_s^2\right)^{\frac{r-q}{s-q}}. \qquad \blacksquare
\end{aligned}$$

Remark 8.20. Note that the usual Sobolev spaces $(H^s(\Omega))_{s \in \mathbb{R}}$ (cf. [180]) are no Hilbert scale if Ω is an open bounded subset of $\mathbb{R}^n$, but for any fixed $m \in \mathbb{N}$ $(H^s(\Omega))_{0 \leq s \leq m}$ (with possibly equivalent norms) is part of a Hilbert scale (cf. [216]).

One special case of operators inducing a Hilbert scale is $L = (T^*T)^{-1}$, where T is an injective linear bounded operator from a Hilbert space $\mathcal{X}$ into another Hilbert space $\mathcal{Y}$ with non-closed range; if T is not injective, it has to be restricted to the orthogonal complement of its nullspace. It is then an immediate consequence of Proposition 8.19 that $\mathcal{X}_\mu = \mathcal{R}((T^*T)^\mu)$ is dense in $\mathcal{X}_\nu = \mathcal{R}((T^*T)^\nu)$ for $0 \leq \nu < \mu$.

The analysis of the next section is essentially based on Corollary 8.22 of the next proposition, the so called inequality of Heinz (cf. [125], see also [164]).

Proposition 8.21. *Let L and A be two densely defined unbounded selfadjoint strictly positive operators in $\mathcal{X}$ with $\mathcal{D}(A) \subset \mathcal{D}(L)$ and*

$$\|Lx\| \leq \|Ax\| \quad \text{for all } x \in \mathcal{D}(A).$$

Then, for all $\nu \in [0,1]$, $\mathcal{D}(A^\nu) \subset \mathcal{D}(L^\nu)$ and

$$\|L^\nu x\| \leq \|A^\nu x\| \quad \text{for all } x \in \mathcal{D}(A^\nu).$$

Proof. The assertion trivially holds for $\nu = 0$ and $\nu = 1$. Let now $0 < \nu < 1$. For L as above we show that

$$\|L^{-\nu} x\|^2 = c(\nu) \int_0^\infty t^{-\nu} \|(L^2 + tI)^{-\frac{1}{2}} x\|^2\, dt \tag{8.27}$$

holds for all $x \in \mathcal{X}$, where

$$c(\nu) := \left(\int_0^\infty \frac{t^{-\nu}}{1+t}\, dt\right)^{-1} \geq \nu(1-\nu).$$

214   8. Regularization With Differential Operators

Let $\{E_\lambda\}$ denote the spectral family of L and assume that (8.19) holds. Using Fubini's Theorem and the fact that

$$\lambda^{-2\nu} = c(\nu) \int_0^\infty \frac{t^{-\nu}}{\lambda^2 + t} dt,$$

we obtain that

$$\begin{aligned}
\|L^{-\nu}x\|^2 &= \int_\gamma^\infty \lambda^{-2\nu} d\|E_\lambda x\|^2 \\
&= \int_\gamma^\infty \left[c(\nu) \int_0^\infty \frac{t^{-\nu}}{\lambda^2 + t} dt \right] d\|E_\lambda x\|^2 \\
&= c(\nu) \int_0^\infty t^{-\nu} \left[\int_\gamma^\infty \frac{1}{\lambda^2 + t} d\|E_\lambda x\|^2 \right] dt \\
&= c(\nu) \int_0^\infty t^{-\nu} \|(L^2 + tI)^{-\frac{1}{2}} x\|^2 dt.
\end{aligned}$$

Since $\|Lx\| \leq \|Ax\|$ on $\mathcal{D}(A)$, we also have that

$$\|(L^2 + tI)^{\frac{1}{2}} x\| \leq \|(A^2 + tI)^{\frac{1}{2}} x\|$$

for all $t \geq 0$; note that $\mathcal{D}(L) = \mathcal{D}((L^2 + tI)^{\frac{1}{2}})$ and that $\mathcal{D}(A) = \mathcal{D}((A^2 + tI)^{\frac{1}{2}})$. A duality argument (cf. Proposition 8.19 (iii)) now shows that

$$\|(L^2 + tI)^{-\frac{1}{2}} x\| \geq \|(A^2 + tI)^{-\frac{1}{2}} x\| \quad \text{for all } x \in \mathcal{X}.$$

Thus, by (8.27),

$$\|L^{-\nu} x\| \geq \|A^{-\nu} x\| \quad \text{for all} \quad x \in \mathcal{X}.$$

The same duality argument therefore yields

$$\|L^\nu x\| \leq \|A^\nu x\| \quad \text{for all } x \in \mathcal{D}(A).$$

The assertion now follows by continuity. ∎

Corollary 8.22. *Let $(\mathcal{X}_s)_{s \in \mathbb{R}}$ be a Hilbert scale induced by L and let $T : \mathcal{X} \to \mathcal{Y}$ be a bounded operator satisfying*

$$\underline{m} \|x\|_{-a} \leq \|Tx\| \leq \overline{m} \|x\|_{-a} \tag{8.28}$$

on $\mathcal{X}$ for some $a > 0$ and $0 < \underline{m} \leq \overline{m} < \infty$. Then for $B := TL^{-s}$, $s \geq 0$ and $|\nu| \leq 1$

$$\underline{c}(\nu) \|x\|_{-\nu(a+s)} \leq \|(B^*B)^{\frac{\nu}{2}} x\| \leq \overline{c}(\nu) \|x\|_{-\nu(a+s)} \tag{8.29}$$

holds on $\mathcal{D}((B^*B)^{\frac{\nu}{2}})$ with $\underline{c}(\nu) = \min(\underline{m}^\nu, \overline{m}^\nu)$ and $\overline{c}(\nu) = \max(\underline{m}^\nu, \overline{m}^\nu)$. Moreover, $\mathcal{R}((B^*B)^{\frac{\nu}{2}}) = \mathcal{X}_{\nu(a+s)}$, where $(B^*B)^{\frac{\nu}{2}}$ has to be replaced by its extension to $\mathcal{X}$ if $\nu < 0$.

Proof. Since $\|(B^*B)^{\frac{1}{2}}x\| = \|Bx\| = \|TL^{-s}x\|$, (8.28) implies that

$$\underline{m}\|x\|_{-(a+s)} \leq \|(B^*B)^{\frac{1}{2}}x\| \leq \overline{m}\|x\|_{-(a+s)}$$

on $\mathcal{X}$. Proposition 8.19 (ii), (iii) and a duality argument now imply that $\mathcal{R}((B^*B)^{\frac{1}{2}}) = \mathcal{D}((B^*B)^{-\frac{1}{2}}) = \mathcal{X}_{a+s} = \mathcal{D}(L^{a+s})$ and that

$$\overline{m}^{-1}\|x\|_{a+s} \leq \|(B^*B)^{-\frac{1}{2}}x\| \leq \underline{m}^{-1}\|x\|_{a+s}$$

on $\mathcal{X}_{a+s}$. By scaling and applying Proposition 8.21 we obtain that $\mathcal{R}((B^*B)^{\frac{\nu}{2}}) = \mathcal{D}((B^*B)^{-\frac{\nu}{2}}) = \mathcal{X}_{\nu(a+s)}$ and that

$$\overline{m}^{-\nu}\|x\|_{\nu(a+s)} \leq \|(B^*B)^{-\frac{\nu}{2}}x\| \leq \underline{m}^{-\nu}\|x\|_{\nu(a+s)}$$

on $\mathcal{X}_{\nu(a+s)}$ for $0 \leq \nu \leq 1$. The rest of the assertion follows now by duality. ∎

8.5. Regularization in Hilbert Scales

Regularization in Hilbert scales was first introduced for the special case of Tikhonov regularization by Natterer [206]. He regularized the linear problem $Tx = y$ by minimizing the functional

$$\|Tx - y^\delta\|^2 + \alpha\|x\|_s^2$$

over $x \in \mathcal{X}_s$, where $\|\cdot\|_s$ denotes the corresponding norm in a Hilbert scale (cf. Section 8.4). For the theory to work he needed the assumption that the operator T satisfies condition (8.28). This restricts the class of operators to injective mappings. The number a in condition (8.28) can be interpreted as a *degree of ill-posedness*.

Natterer showed that the regularized solutions converge towards the exact solution with the rate $O(\delta^{\frac{u}{a+u}})$ in the norm of $\mathcal{X}$ if the regularization parameter α is chosen properly (depending on a and u) and if the exact solution $x^\dagger$ satisfies

$$x^\dagger \in \mathcal{X}_u \tag{8.30}$$

for some $u \geq 0$. Note that (8.30) includes the case where $\|x^\dagger\|_s$ is not well defined, as opposed to the setting of Example 8.11. Further note that this rate is optimal in the sense that it is of the same order as the best possible worst case error for recovering $x^\dagger$ under the given information. The source condition (8.30) is the natural generalization of condition $x^\dagger \in \mathcal{X}_\mu = \mathcal{R}((T^*T)^\mu)$ for the regularization in Hilbert spaces (see Chapters 3 and 4).

As in Section 4.1, we develop the convergence analysis for general regularization methods based on spectral theory. Let $(\mathcal{X}_s)_{s \in \mathbb{R}}$ be the Hilbert scale induced by a densely defined unbounded selfadjoint strictly positive operator L (cf. Section 8.4) and let B be defined as in Corollary 8.22, i.e., $B := TL^{-s} : \mathcal{X} \to \mathcal{Y}$, $s \geq 0$ (compare (8.11) for the analogous definition in the context of regularization with differential

216   8. Regularization With Differential Operators

operators). Moreover, let $g_\alpha : [0, \|B\|^2] \to \mathbb{R}$, $\alpha > 0$, be a family of piecewise continuous functions and $r_\alpha(\lambda) = 1 - \lambda g_\alpha(\lambda)$. As in Section 4.1, we assume that the following conditions hold on $[0, \|B\|^2]$:

$$\lim_{\alpha \to 0} g_\alpha(\lambda) = \frac{1}{\lambda}, \quad \lambda \neq 0, \tag{8.31}$$

$$|g_\alpha(\lambda)| \leq \hat{c}\alpha^{-1}, \tag{8.32}$$

$$\lambda^\mu |r_\alpha(\lambda)| \leq c_\mu \alpha^\mu, \quad 0 \leq \mu \leq \mu_0, \tag{8.33}$$

with $\hat{c}, c_\mu > 0$ independent of α and $\mu_0 \in [1, \infty]$. If $\mu_0 = \infty$, then $\mu \leq \mu_0$ in (8.33) has to be replaced by $\mu < \mu_0$. Note that $c_0 \geq 1$, since $r_\alpha(0) = 1$, and that (8.32) corresponds to the requirement $G_\alpha = O(1/\alpha)$ (cf. (4.14) and (4.28)).

The regularized solutions in $\mathcal{X}_s$ will be defined by (cf. (8.13))

$$x_\alpha^\delta := L^{-s} g_\alpha(B^*B) B^* y^\delta, \tag{8.34}$$

where we assume as in the sections above that $\|y - y^\delta\| \leq \delta$ and that $Tx^\dagger = Qy$. Note that in all considerations to follow it would suffice that $\|Q(y-y^\delta)\| \leq \delta$. Since $L^{-2s}T^*$ is the adjoint of the restriction of T to $\mathcal{X}_s$, (8.34) is equivalent to

$$x_\alpha^\delta = g_\alpha(L^{-2s}T^*T)L^{-2s}T^* y^\delta.$$

Hence, x_α^δ is the regularized solution of the problem $T|_{\mathcal{X}_s} x = y$ in the sense of Section 4.1. Therefore, if $x^\dagger \in \mathcal{X}_s$, all results about convergence (with respect to the norm in $\mathcal{X}_s$) from that section are applicable. In the next theorem, we will establish the aforementioned convergence results with respect to weaker norms even if $x^\dagger \notin \mathcal{X}_s$:

Theorem 8.23. *Let the assumptions (8.28) and (8.30) – (8.33) hold and let x_α^δ be the regularized solutions of $Tx = y^\delta$ defined by (8.34). If the regularization parameter α is chosen as*

$$\alpha = \bar{c} \left(\frac{\delta}{\|x^\dagger\|_u} \right)^{\frac{2(a+s)}{a+u}}, \tag{8.35}$$

$\bar{c} > 0$, *and if* $u \leq a + 2s$, *then we obtain the estimate*

$$\|x_\alpha^\delta - x^\dagger\| \leq C \delta^{\frac{u}{a+u}} \|x^\dagger\|_u^{\frac{a}{a+u}}$$

for some constant $C > 0$.

Proof. First we derive an estimate for the propagated data error $\|x_\alpha^\delta - x_\alpha\|$. Note that (8.32) and (8.33) (with $\mu = 0$) imply that

$$\lambda^t |g_\alpha(\lambda)| \leq \max\{\hat{c}, 1 + c_0\} \alpha^{t-1}, \quad \lambda \in [0, \|B\|^2], \ 0 \leq t \leq 1.$$

Using this estimate with $t = (a+2s)/(2a+2s)$, it follows with (8.34), (8.29) (with $\nu = s/(a+s)$), and $\|y - y^\delta\| \leq \delta$ that

$$\|x_\alpha^\delta - x_\alpha\| \leq c\|(B^*B)^{\frac{s}{2(a+s)}} g_\alpha(B^*B) B^*(y^\delta - y)\| \leq c\delta \alpha^{-\frac{a}{2(a+s)}}. \tag{8.36}$$

8.5. Regularization in Hilbert Scales

We use c as a generic constant. Since, by Proposition 8.19 (ii), $L^s x^\dagger \in \mathcal{X}_{u-s}$ and since $u \leq a + 2s$, Corollary 8.22 implies that there is an element $v \in \mathcal{X}$ such that

$$L^s x^\dagger = (B^* B)^{\frac{u-s}{2(a+s)}} v. \tag{8.37}$$

This together with (8.34), $T x^\dagger = Q y$, (8.29) (with $\nu = s/(a+s)$) and (8.33) (with $\mu = u/(2a+2s)$) implies that

$$\begin{aligned}
\|x_\alpha - x^\dagger\| &= \|L^{-s}[g_\alpha(B^*B)B^*B - I]L^s x^\dagger\| \\
&\leq c\|(B^*B)^{\frac{u}{2(a+s)}} r_\alpha(B^*B) v\| \\
&\leq c\alpha^{\frac{u}{2(a+s)}} \|v\|.
\end{aligned} \tag{8.38}$$

A further application of (8.29) (with $\nu = (s-u)/(2a+2s)$) yields together with (8.37) that

$$\|v\| \leq c\|x^\dagger\|_u.$$

By the triangle inequality, this together with (8.35), (8.36) and (8.38) implies the assertion. ∎

Remark 8.24. It is an immediate consequence of the proof above that we have at least convergence if $x^\dagger \in \mathcal{X} = \mathcal{X}_0$ and if $\alpha \to 0$ and $\delta \alpha^{-\frac{a}{2(a+s)}} \to 0$ as $\delta \to 0$. A careful inspection of the proof shows that for the choice $\alpha \sim \delta^{\frac{2(a+s)}{a+u}}$ we even obtain a rate for other intermediate norms, namely

$$\|x_\alpha^\delta - x^\dagger\|_r = O(\delta^{\frac{u-r}{a+u}}), \tag{8.39}$$

for $-a \leq r \leq u \leq a + 2s$.

The result of Theorem 8.23 only holds for $u \leq a + 2s$, although we have seen in Sections 4.1 and 6.1 that there are methods with qualification $\mu_0 = \infty$, i.e., where no saturation is to be expected. This restriciton is due to the fact that Corollary 8.22 is applicable only if $u \leq a + 2s$. However, it follows from the results of Section 4.1, (8.28) and the interpolation inequality (8.23) that the rate $O(\delta^{\frac{u}{a+u}})$ in Theorem 8.23 can be obtained if $u \leq 2\mu_0(a+s) - a$ and if $x^\dagger \in \mathcal{R}((L^{-2s}T^*T|_{\mathcal{X}_s})^{\frac{u-s}{2(a+s)}})$. If $\mu_0 > 1$, then u can now be larger than $a + 2s$. But, for $u > a + 2s$, $\mathcal{X}_u$ and $\mathcal{R}((L^{-2s}T^*T|_{\mathcal{X}_s})^{\frac{u-s}{2(a+s)}})$ are, in general, different subspaces.

If L^{-s} and T^*T commute, then it follows from the proof of Theorem 8.23 that the rates may even be extended to $u \leq 2\mu_0(a+s)$ if $x^\dagger \in \mathcal{R}((B^*B)^{\frac{u}{2(a+s)}})$, since in this case we have that $\mathcal{X}_u = \mathcal{R}((B^*B)^{\frac{u}{2(a+s)}})$ for all $u \geq 0$. Here, the parameter r in (8.39) is subject to the restriction

$$\max\{-a, u - 2\mu_0(a+s)\} \leq r \leq \min\{a+2s, u\}.$$

For the special case that $L := (T^*T)^{-1}$, we have that $a = 1/2$ and $\mathcal{X}_u = \mathcal{R}((T^*T)^u)$, so that we obtain results from Section 4.1 as special cases.

218   8. Regularization With Differential Operators

The parameter choice in Theorem 8.23 depends on the degree of ill-posedness a and the smoothness parameter u. We will show next that Morozov's discrepancy principle yields the rates of Theorem 8.23. Note that this a-posteriori parameter selection method only depends on available information. As in Section 1.3, we now assume that $y \in \mathcal{R}(T)$. Here, the regularization parameter is defined as

$$\alpha(\delta, y^\delta) := \sup\{\alpha \mid \|Tx_\alpha^\delta - y^\delta\| \leq \tau\delta\}, \qquad (8.40)$$

where $\tau > c_0$, with c_0 as in (8.33). Note that $\alpha(\delta, y^\delta)$ also depends on τ.

If $y \neq 0$, the parameter choice (8.40) yields a well-defined $\alpha(\delta, y^\delta) \in (0, \infty]$ and $\alpha(\delta, y^\delta) < \infty$, for $\delta > 0$ sufficiently small, since, due to (8.31) – (8.33) and (8.34),

$$\lim_{\alpha \to 0} \|Tx_\alpha^\delta - y^\delta\| = \|(I - Q)y^\delta\| \leq \delta \leq c_0 \delta < \tau\delta$$

and, by (8.32),

$$\lim_{\alpha \to \infty} \|Tx_\alpha^\delta - y^\delta\| = \|y^\delta\| \geq \|y\| - \delta.$$

We now prove that this parameter choice yields the rates of Theorem 8.23 with $O(\cdot)$ even replaced by $o(\cdot)$.

Theorem 8.25. *Let the assumptions (8.28) and (8.30) – (8.33) hold. Moreover, assume that $y \in \mathcal{R}(T)$, $y \neq 0$, and that $r_\alpha(\lambda)$ is continuous from the left for all $\lambda \in [0, \|B\|^2]$ as a function of α. For the regularized solutions x_α^δ, defined by (8.34), with $\alpha = \alpha(\delta, y^\delta)$ chosen as in (8.40), we obtain the rates*

$$\|x_{\alpha(\delta,y^\delta)}^\delta - x^\dagger\| = \begin{cases} o(\delta^{\frac{u}{a+u}}), & \text{if } u < a + 2s \text{ or } u = a + 2s, \mu_0 > 1, \\ O(\delta^{\frac{u}{a+u}}), & \text{if } u = a + 2s, \mu_0 = 1. \end{cases}$$

Proof. It follows from the proof of Theorem 8.23 that

$$\|x_\alpha^\delta - x^\dagger\| = O(\delta\alpha^{-\frac{a}{2(a+s)}} + \|(B^*B)^{\frac{u}{2(a+s)}} r_\alpha(B^*B)v\|) \qquad (8.41)$$

with v as in (8.37).

Due to (2.49), (8.37) and (8.33) we have the estimate

$$\begin{aligned}
\|(B^*B)^{\frac{u}{2(a+s)}} r_\alpha(B^*B)v\| &\leq \|(B^*B)^{\frac{a+u}{2(a+s)}} r_\alpha(B^*B)v\|^{\frac{u}{a+u}} \|r_\alpha(B^*B)v\|^{\frac{a}{a+u}} \\
&= \|r_\alpha(BB^*)y\|^{\frac{u}{a+u}} \|r_\alpha(B^*B)v\|^{\frac{a}{a+u}} \\
&\leq (\|Tx_\alpha^\delta - y^\delta\| + c_0\delta)^{\frac{u}{a+u}} \|r_\alpha(B^*B)v\|^{\frac{a}{a+u}}. \qquad (8.42)
\end{aligned}$$

Let now

$$J_\alpha^2 := \int_0^{\|B\|^2+} (\lambda\alpha^{-1})^{\frac{a+u}{a+s}} r_\alpha^2(\lambda) \, d\|E_\lambda v\|^2.$$

Then by the definition of $\alpha(\delta, y^\delta)$

$$\begin{aligned}
(2\alpha(\delta, y^\delta))^{\frac{a+u}{2(a+s)}} J_{2\alpha(\delta,y^\delta)} &= \|r_{2\alpha(\delta,y^\delta)}(BB^*)y\| \\
&\geq \|r_{2\alpha(\delta,y^\delta)}(BB^*)y^\delta\| - c_0\delta \\
&> (\tau - c_0)\delta.
\end{aligned}$$

8.5. Regularization in Hilbert Scales

Thus, we obtain that

$$\delta\alpha(\delta,y^\delta)^{-\frac{a}{2(a+s)}} = O(\delta^{\frac{u}{a+u}} \mathsf{J}_{2\alpha(\delta,y^\delta)}^{\frac{a}{a+u}}). \tag{8.43}$$

This together with (8.41), (8.42), and (8.33) implies that

$$\|x^\delta_{\alpha(\delta,y^\delta)} - x^\dagger\| = O(\delta^{\frac{u}{a+u}}).$$

To establish the $o(\cdot)$-term in the upper bound, let us first assume that $\alpha(\delta,y^\delta) \to 0$ as $\delta \to 0$. Since, by (8.31),

$$\lim_{\alpha \to 0} \|r_\alpha(B^*B)v\| = 0,$$

(8.42) implies that

$$\|(B^*B)^{\frac{u}{2(a+s)}} r_{\alpha(\delta,y^\delta)}(B^*B)v\| = o(\delta^{\frac{u}{a+u}}).$$

Due to (8.41) and (8.43), it remains to be shown that $\mathsf{J}_{2\alpha(\delta,y^\delta)} = o(1)$. Note that it is an immediate consequence of (8.31) and (8.33) that

$$\lim_{\alpha \to 0} \lambda^\mu \alpha^{-\mu} |r_\alpha(\lambda)| = 0, \quad 0 \le \mu < \mu_0, \ \lambda \in [0, \|B\|^2].$$

Thus,

$$\mathsf{J}_{2\alpha(\delta,y^\delta)} = o(1) \quad \text{for} \quad u < a + 2s \text{ or } u = a + 2s, \mu_0 > 1.$$

Finally, we have to consider the case, where $\lim_{\delta \to 0} \alpha(\delta,y^\delta) \ne 0$. Then there are $y_n = y^{\delta_n}$ with

$$\|y - y_n\| \le \delta_n \to 0 \quad \text{as} \quad n \to \infty,$$

and corresponding regularization parameters $\alpha_n := \alpha(\delta_n, y_n)$ with

$$\liminf_{n \to \infty} \alpha_n > 0.$$

Since r_α is continuous from the left with respect to α, (8.40) implies that

$$\|Tx_{\alpha_n} - y_n\| \le \tau \delta_n.$$

Because of (8.33), it follows for any $\delta > 0$ and y^δ with $\|y^\delta - y\| \le \delta$ that

$$\|r_{\alpha_n}(BB^*)y^\delta\| \le \|r_{\alpha_n}(BB^*)(y^\delta - y)\| + \|r_{\alpha_n}(BB^*)(y - y_n)\| + \|Tx_{\alpha_n} - y_n\|$$
$$\le c_0\delta + c_0\delta_n + \tau\delta_n,$$

and hence,

$$\|r_{\alpha_n}(BB^*)y^\delta\| \le \tau\delta,$$

provided n is sufficiently large. Therefore, by (8.40), we necessarily have $\alpha(\delta, y^\delta) \ge \alpha_n$ for n sufficiently large, and hence,

$$\liminf_{\delta \to 0} \alpha(\delta, y^\delta) > 0. \tag{8.44}$$

8. Regularization With Differential Operators

Moreover, due to (8.37), (8.40), and (8.33)

$$\|(B^*B)^{\frac{a+u}{2(a+s)}} r_{\alpha(\delta,y^\delta)}(B^*B)v\| = \|r_{\alpha(\delta,y^\delta)}(BB^*)y\|$$
$$= O(\|Tx^\delta_{\alpha(\delta,y^\delta)} - y^\delta\| + \delta) - O(\delta), \qquad (8.45)$$

as $\delta \to 0$. Since, due to (8.44) and (8.32), $|r_{\alpha(\delta,y^\delta)}(\lambda)| \geq 1/2$ for λ sufficiently small, (8.45) implies that a $\rho > 0$ exists such that

$$\int_0^\rho d\|E_\lambda v\|^2 = 0$$

and hence that

$$\|(B^*B)^{\frac{u}{2(a+s)}} r_\alpha(B^*B)v\| \leq \rho^{-\frac{a}{2(a+s)}} \|(B^*B)^{\frac{a+u}{2(a+s)}} r_\alpha(B^*B)v\| = O(\delta).$$

This together with (8.41) and (8.44) implies that

$$\|x^\delta_{\alpha(\delta,y^\delta)} - x^\dagger\| = O(\delta). \quad \blacksquare$$

Remark 8.26. Note that we have even proved above that we actually have the rate $O(\delta)$ if $\liminf_{\delta \to 0} \alpha(\delta, y^\delta)$ happens to be positive.

In [251] the same rates as in Theorem 8.25 were obtained under monotonicity restrictions on g_α and r_α for $s \leq u \leq a + 2s$. The restriction $s \leq u$ means, that the exact solution has to be at least an element of $\mathcal{X}_s$, i.e., that oversmoothing is not allowed.

We know from Section 4.3 that Morozov's discrepancy principle is not always optimal if $\mu_0 < \infty$ (μ_0 as in (8.33)). Therefore, for the case where L^{-s} and T^*T commute, we will see a saturation effect for $u = 2\mu_0(a+s) - a$ instead of $u = 2\mu_0(a+s)$ (cf. Remark 8.24).

For a treatment of Tikhonov regularization in Hilbert scales, combined with finite-dimensional approximation, see [213].

9. Numerical Realization

9.1. Derivation of the Discrete Problem

In this chapter we discuss efficient numerical realizations of the most popular regularization schemes, namely Tikhonov regularization and iterative regularization methods. The numerical implementation can usually be divided in three steps: discretization, transformation into standard form (cf. Section 9.2), and regularization of the standard form problem.

Note that although the process of discretization itself can often be interpreted as a regularization (cf. Section 3.3), additional regularization is usually recommended to stabilize the numerical algorithm, and to allow for the use of larger subspaces in the discretization.

We denote these subspaces by $\mathcal{X}_n \subset \mathcal{X}$ and $\mathcal{Y}_m \subset \mathcal{Y}$ with bases

$$\mathcal{X}_n = \operatorname{span}\{\varphi_1, \ldots, \varphi_n\}, \qquad \mathcal{Y}_m = \operatorname{span}\{\psi_1, \ldots, \psi_m\}, \tag{9.1}$$

and define the approximation $x_n \in \mathcal{X}_n$ as solution of the projected equation

$$Q_m T x_n = Q_m y. \tag{9.2}$$

Here, $Q_m : \mathcal{Y} \to \mathcal{Y}_m$ is an appropriate projection operator, cf. Section 3.3.

In general we will represent $x_n \in \mathcal{X}_n$ by its coordinate vector $\mathbf{x} \in \mathbb{R}^n$ corresponding to the representation

$$x_n = \xi_1 \varphi_1 + \ldots + \xi_n \varphi_n, \qquad \mathbf{x} = [\xi_1, \ldots, \xi_n]^T. \tag{9.3}$$

With this notion, the projected equation (9.2) yields a system of equations for the n unknown coordinates $\{\xi_i\}$. As said before, in most applications this system will be too ill-conditioned to be solved numerically without further regularization of some sort. To this end we have to introduce appropriate norms in the finite-dimensional spaces.

The impact of the chosen norms is most explicit in Tikhonov regularization. In the finite-dimensional setting, Tikhonov regularization amounts to finding the minimum of the quadratic functional (cf. Theorem 5.1)

$$\|\mathbf{b} - A\mathbf{x}\|_2^2 + \alpha \|B\mathbf{x}\|_2^2 = (\mathbf{b} - A\mathbf{x})^T(\mathbf{b} - A\mathbf{x}) + \alpha \mathbf{x}^T B^T B \mathbf{x}. \tag{9.4}$$

Here, $\|\cdot\|_2$ denotes the Euclidean norm, and usually the matrix B determines the way the smoothness (or roughness) of $\mathbf{x}$ must be measured so as to be consistent with $\|x\|_{\mathcal{X}}$, i.e., the norm of x in $\mathcal{X}$. Similarly, A and $\mathbf{b}$ have to be chosen in such a way that

$$\|\mathbf{b} - A\mathbf{x}\|_2 \approx \|Q_m(y - Tx)\|_{\mathcal{Y}}. \tag{9.5}$$

9. Numerical Realization

Recall from Section 5.1 that the minimizer $\mathbf{x}$ of (9.4) satisfies the $n \times n$ system of Euler equations

$$(A^T A + \alpha B^T B)\mathbf{x} = A^T \mathbf{b}. \tag{9.6}$$

The matrix B corresponding to the coefficient vector $\mathbf{x} \in \mathbb{R}^n$ is comparatively easy to derive. In view of the above consistency requirement

$$\|B\mathbf{x}\|_2 \approx \|x_n\|_{\mathcal{X}} = \|\xi_1 \varphi_1 + \ldots + \xi_n \varphi_n\|_{\mathcal{X}},$$

it is clear that the choice $B = I$ will be appropriate in exceptional instances only, e.g., when the functions φ_k are the simple indicator functions for the subset of piecewise constants. In general, the correct norm to use is defined via the Gramian matrix

$$G = [\langle \varphi_i, \varphi_j \rangle_{\mathcal{X}}], \tag{9.7}$$

since

$$\|\xi_1 \varphi_1 + \ldots + \xi_n \varphi_n\|_{\mathcal{X}}^2 = \sum_{i,j=1}^{n} \xi_i \xi_j \langle \varphi_i, \varphi_j \rangle_{\mathcal{X}} = \mathbf{x}^T G \mathbf{x}.$$

It follows that G is positive definite whenever the functions $\{\varphi_k \mid k = 1, \ldots, n\}$ form a basis of $\mathcal{X}_n$. Consequently, G can be factorized to obtain $G = B^T B$, which yields the desired identity

$$\|\xi_1 \varphi_1 + \ldots + \xi_n \varphi_n\|_{\mathcal{X}} = \|B\mathbf{x}\|_2.$$

For the algorithms in the subsequent sections it will turn out convenient to choose B upper triangular, i.e., as the Cholesky factor of G. Note that whenever $\{\varphi_k\}$ is a B-spline basis of some Sobolev space, the Gramian matrix G will be banded, since φ_i and φ_j in (9.7) have disjunct supports for $|i - j|$ sufficiently large. Consequently the Cholesky factor B is easily obtained with only $O(l^2 n)$ operations, where l is the bandwidth of G. (Here, and in the sequel, an "operation" refers to a multiplication followed by an addition.)

If $\mathcal{X}$ is equipped with some (semi)norm (induced by a differential operator as in Chapter 8), then the above approach has to be modified accordingly, by using a different bilinear form in (9.7). In most cases the Gramian matrix G will still be banded, but may be rank deficient. The corresponding (banded) Cholesky factor can be chosen to have full rank, e.g., to have less rows than columns, cf. (9.13) below.

The choice of A and $\mathbf{b}$ is more delicate, and strongly depends on the particular discretization employed. We shall exemplify this for two prominent examples.

Example 9.1. In *least-squares collocation* we consider the integral equation

$$Kx = \int_0^1 k(\cdot, t) x(t) \, dt = y,$$

with continuous kernel k and $x \in \mathcal{L}^2$, and we have measurements of y at n collocation points $\{s_i \mid i = 1, \ldots, n\}$. In Example 3.25 we have seen that the solution x_n of the

9.1. Derivation of the Discrete Problem

discretized problem has the form (9.3) with $\varphi_i = k(s_i, \cdot)$, and the vector $\mathbf{x} \in \mathbb{R}^n$ is a solution of

$$M\mathbf{x} = \mathbf{y} \quad \text{with } M = [M(s_i, s_j)], \ \mathbf{y} = [y(s_i)];$$

here, as in (3.51), $M(s,t)$ is the reproducing kernel function. Actually, we have seen more, namely that $M\mathbf{x}$ is the vector with entries $(Kx)(s_i)$, $i = 1, \ldots, n$. Therefore, assuming that the data are perturbed by random noise (e.g., white noise), the choice $A = M$ and $\mathbf{b} = \mathbf{y}$ is a good choice for the discrete problem, since in this case $\|\mathbf{b} - A\mathbf{x}\|_2$ is a reasonable way of measuring the data fit (9.5).

Example 9.2. Next, we consider the combination of regularization and projection as investigated in Section 5.2. There, we have chosen an arbitrary subspace $\mathcal{Y}_n$, and let $\mathcal{X}_n = T^*\mathcal{Y}_n$. Without regularization we would be looking for a solution of (9.2), where Q_n is the orthogonal projector onto $\mathcal{Y}_n$. As shown in Theorem 3.24, if $y \in \mathcal{D}(T^\dagger)$, then the solution of (9.2) would be the orthogonal projection of $x^\dagger$ onto $\mathcal{X}_n$. We let $\varphi_k = T^*\psi_k$ for $k = 1, \ldots, n$ (and $m = n$). It follows that x_n solves (9.2) if and only if

$$\langle Q_n T x_n, \psi_i \rangle_\mathcal{Y} = \langle T x_n, \psi_i \rangle_\mathcal{Y} = \langle y, \psi_i \rangle_\mathcal{Y}, \quad i = 1, \ldots, n.$$

Expanding x_n as before, cf. (9.3), we obtain the following linear system for $\mathbf{x}$:

$$M\mathbf{x} = \mathbf{y} \quad \text{with } M = [\langle T^*\psi_i, T^*\psi_j \rangle_\mathcal{X}], \ \mathbf{y} = [\langle y, \psi_i \rangle_\mathcal{Y}]. \tag{9.8}$$

M is symmetric and positive definite if and only if $\mathcal{Y}_n \cap \mathcal{R}(T)^\perp = \{0\}$.

We now use the approach of Section 5.2 to derive the suitable setting for computing regularized approximations of $\mathbf{x}$ in the discrete parameter space. According to this approach, cf. (5.43), Tikhonov regularization amounts to computing

$$x_{\alpha,n}^\delta = (T_n^* T_n + \alpha I)^{-1} T_n^* y = T_n^*(T_n T_n^* + \alpha I)^{-1} y =: T_n^* z_{\alpha,n}^\delta,$$

where $T_n = Q_n T$. From the last equality follows that $z_{\alpha,n}^\delta$ solves

$$\langle T_n^* z_{\alpha,n}^\delta, T_n^*\psi_i \rangle_\mathcal{X} + \alpha \langle z_{\alpha,n}^\delta, \psi_i \rangle_\mathcal{Y} = \langle y, \psi_i \rangle_\mathcal{Y}, \quad i = 1, \ldots, n. \tag{9.9}$$

Expanding the regularized approximation $x_{\alpha,n}^\delta$ like x in (9.3) before, with $\varphi_i = T^*\psi_i$, we observe that $z_{\alpha,n}^\delta = \xi_1 \psi_1 + \ldots \xi_n \psi_n$, and hence, it follows from (9.9) that the coefficient vector $\mathbf{x}$ corresponding to the regularized approximation solves the linear equation

$$(M + \alpha H)\mathbf{x} = \mathbf{y}, \tag{9.10}$$

with M and $\mathbf{y}$ as in (9.8), and with $H = [\langle \psi_i, \psi_j \rangle_\mathcal{Y}]$.

Note that (9.10) corresponds to (9.6) with $A^T A = M$ and $B^T B = H$. Therefore, we may take any A such that $A^T A = M$ (recall that M is positive definite), and then let $\mathbf{b} = A^{-T} \mathbf{y}$. However, for numerical purposes, an explicit factorization of M is not always necessary, and the vector $\mathbf{b}$ is hardly ever required. For example, iterative methods can use (9.10) as it stands, without any further preprocessing (but see also Section 9.5 below).

To efficiently implement direct methods for solving (9.10), however, an inexpensive factorization of $M = A^T A$ would be helpful, in particular, if (9.10) has to be solved for several values of α. Note that such a factorization need not be the Cholesky one. In the sequel we briefly comment on one possible way to obtain an approximate factorization without *any* computations. To this end, assume that $\mathcal{X} = \mathcal{L}^2$, so that the (i,j)th entry of M is given by

$$M_{ij} = \langle T^*\psi_i, T^*\psi_j \rangle_{\mathcal{X}} = \langle \varphi_i, \varphi_j \rangle_{\mathcal{X}} = \int \varphi_i(t)\varphi_j(t)\,dt\,.$$

For computations this integral will usually be approximated by some quadrature rule with p nodes $\{t_k\}$ and corresponding positive weights $\{\omega_k\}$, i.e.,

$$M_{ij} \approx \sum_{k=1}^{p} \omega_k \varphi_i(t_k)\varphi_j(t_k)\,.$$

If the weights $\{\omega_k\}$ are positive, then this can be rewritten as $M_{ij} \approx \mathbf{a}_i^T \mathbf{a}_j$ with

$$\mathbf{a}_i = [\sqrt{\omega_1}\varphi_i(t_1),\ldots,\sqrt{\omega_p}\varphi_i(t_p)]^T, \qquad i=1,\ldots,n\,,$$

which gives the desired factorization

$$M \approx A^T A, \quad \text{with} \quad A = [\mathbf{a}_1,\ldots,\mathbf{a}_n] \in \mathbb{R}^{p\times n}\,.$$

We mention in passing that the special structure of the matrix pencil in (9.6) allows a theoretical treatment of its dependency on α in terms of the so-called *generalized* (or *quotient*) *singular value decomposition*, cf. [29, 96]. For numerical purposes, however, an implementation via singular value decompositions is often prohibitively expensive.

9.2. Reduction to Standard Form

It has become a useful notion to distinguish between *regularization in standard form* and *regularization in general form*, depending on whether the coordinate space $\mathbb{R}^n$, in which the unknown vector $\mathbf{x}$ lives, is equipped with the Euclidean norm or not. Concerning (9.4), regularization is in standard form if $B = I$ and in general form otherwise. We shall assume throughout that $\mathcal{N}(A) \cap \mathcal{N}(B) = \{\mathbf{0}\}$, in which case the minimizer of (9.4) is unique. Note that this assumption may be seen as a reminiscent of the fundamental condition (8.3) from the continuous setting.

Basically, we treat general form problems with the techniques of Chapter 8, namely by making use of the weighted generalized inverse $B_A^\dagger$ of B, cf. Theorem 8.2. As in Section 8.2 we substitute

$$\mathbf{x} = \mathbf{x}_0^\dagger + B_A^\dagger \mathbf{z}\,, \tag{9.11}$$

9.2. Reduction to Standard Form

where $\mathbf{x}_0^\dagger$ is a "harmless" component in the nullspace of B, and $\mathbf{z}$ is approximated by
$$\mathbf{z}_\alpha = g_\alpha(\overline{A}^T\overline{A})\overline{A}^T\overline{\mathbf{b}}, \qquad \overline{A} = AB_A^\dagger, \quad \overline{\mathbf{b}} = \mathbf{b} - A\mathbf{x}_0^\dagger.$$
In other words, $\mathbf{z}_\alpha$ is a regularized solution of the standard form problem
$$\overline{A}\mathbf{z} = \overline{\mathbf{b}}. \tag{9.12}$$

We shall now comment on an efficient way of implementing this transformation. To this end, we stipulate the following assumptions on the appearance of B, namely that B has full row rank and upper diagonal bandstructure with non-vanishing diagonal entries, i.e.,

$$B = \begin{bmatrix} \begin{array}{c}\rule{1cm}{0pt}\end{array} \end{bmatrix} =: [B_1, B_2] \in \mathbb{R}^{(n-p)\times n}. \tag{9.13}$$

In other words, B_1 is the leading nonsingular upper triangular $(n-p) \times (n-p)$ submatrix of B, and p is the dimension of $\mathcal{N}(B)$. Assume that $l \ll n$ is the bandwidth of B; as suggested by (9.13) we typically have $l \geq p$. From the decomposition (9.13) we can readily compute a basis for the nullspace of B in $O(pln)$ operations by choosing the columns of
$$N = \begin{bmatrix} B_1^{-1}B_2 \\ -I \end{bmatrix}. \tag{9.14}$$

We now show how to compute $\overline{A}$ of (9.12). In the case when B is nonsingular $\overline{A}$ reduces to $\overline{A} = AB^{-1}$, i.e. $\overline{A}^T = B^{-T}A^T$, and the rows $\{\overline{\mathbf{a}}_k^T\}$ of $\overline{A}$ can be computed as the solutions of the m linear system
$$B^T\overline{\mathbf{a}}_k = \mathbf{a}_k, \qquad k = 1, \ldots, m, \tag{9.15}$$
where $\mathbf{a}_k^T$ denotes the kth row of A. Since B is banded with bandwidth l, this takes only about $(l+1)nm$ operations.

We are now looking for a similar way of computing $\overline{A}^T = (B_A^\dagger)^T A^T$ in the case when the nullspace of B is nontrivial. The generalization of (9.15) to this situation is part of the following lemma.

Lemma 9.3. *Let* $\mathcal{B} = \{\mathbf{x} \in \mathbb{R}^n \mid A^T A\mathbf{x} \perp \mathcal{N}(B)\}$, *and let* $P_\mathcal{B}$ *be the oblique projector onto* $\mathcal{B}$ *along* $\mathcal{N}(B)$.

(i) *For any vector* $\mathbf{z} \in \mathbb{R}^{n-p}$, $\mathbf{x} = B_A^\dagger\mathbf{z}$ *is the unique solution* $\mathbf{x}$ *in* $\mathcal{B}$ *of* $B\mathbf{x} = \mathbf{z}$.

(ii) *For any vector* $\mathbf{x} \in \mathbb{R}^n$, $\mathbf{z} = (B_A^\dagger)^T\mathbf{x}$ *is the (unique) solution* $\mathbf{z}$ *of* $B^T\mathbf{z} = P_\mathcal{B}^T\mathbf{x}$.

Proof. Recall from Lemma 8.1 that $\mathcal{B}$ and $\mathcal{N}(B)$ are complementary subspaces of $\mathbb{R}^n$. Hence, $P_\mathcal{B}$ is a well-defined projector in $\mathbb{R}^n$.

(i) For $z \in \mathbb{R}^{n-p}$ let $x = B_A^\dagger z$. Clearly, $x \in \mathcal{R}(B_A^\dagger) = \mathcal{B}$, cf. Section 8.1, and x solves $Bx = BB_A^\dagger z = z$ since $BB_A^\dagger = I$. Also, there is no further solution of this equation in $\mathcal{B}$, since the difference of any two such solutions would belong to $\mathcal{B} \cap \mathcal{N}(B) = \{0\}$.

(ii) From the definition of P_B follows $P_B = B_A^\dagger B$. Now, let $z = (B_A^\dagger)^T x$ for some $x \in \mathbb{R}^n$. Then we have
$$B^T z = (B_A^\dagger B)^T x = P_B^T x,$$
as has been claimed. Since B^T has full column rank, z is uniquely defined by this equation. ∎

To determine the kth row z of $\overline{A}^T$ we have to apply part (ii) of this lemma with x being the transpose of the kth row of A. Decomposing the vector $d := P_B^T x$ similar to the decomposition of B^T, cf. (9.13), we thus find z as the solution of
$$\begin{bmatrix} B_1^T \\ B_2^T \end{bmatrix} z = P_B^T x = \begin{bmatrix} d_1 \\ d_2 \end{bmatrix}.$$

Since B_1 is nonsingular, z is easily computed by forward substitution as
$$z = B_1^{-T} d_1$$
in approximately $(l+1)n$ operations.

It remains to determine $P_B^T x$ for some arbitrary $x \in \mathbb{R}^n$. To this end, we recall that the columns of N contain a basis for $\mathcal{N}(B)$, and that, cf. (8.5),
$$\mathcal{B}^\perp = A^T A(\mathcal{N}(B)).$$

Hence, $\mathcal{B}^\perp$ is spanned by the columns of $A^T A N$ and we find $d = P_B^T x$ from the condition that
$$d = x - A^T A N c \perp \mathcal{N}(B) \quad \text{for some } c \in \mathbb{R}^p.$$

It follows that d is orthogonal to the columns of N, i.e.,
$$0 = N^T d = N^T x - (AN)^T (AN) c,$$
which gives a small $p \times p$-dimensional system of equations for the unknown coefficient vector c.

We summarize the computation of $\overline{A}$ in the following algorithm:

Algorithm 9.4. (Transformation to standard form)

Determine the kth row $\overline{a}_k^T$ of $\overline{A}$ ($k = 1, \ldots, m$) from the corresponding row a_k^T of A as follows:

- solve $(AN)^T (AN) c = N^T a_k = (AN)^T e_k$ for c, i.e., $c = (AN)^\dagger e_k$ (e_k denotes the kth Cartesian basis vector),

- compute $\mathbf{d} = \mathbf{a}_k - A^T A N \mathbf{c}$,

- denote by $\mathbf{d}_1$ the first $n-p$ components of $\mathbf{d}$, and determine $\bar{\mathbf{a}}_k = B_1^{-T}\mathbf{d}_1$ by forward substitution.

This algorithm takes approximately $(l+2p+1)mn$ operations provided the matrix $A^T A N$ and the inverse of the little $p \times p$ matrix $(AN)^T(AN)$ are computed once and for all in the very beginning. Instead of inverting this cross product matrix, one could also compute $\mathbf{c}$ from a QR-decomposition of AN. Alternatively, one could choose the nullspace vectors $\mathbf{n}_i$ of B in the columns of N to be $A^T A$-orthogonal, i.e., $\mathbf{n}_i^T A^T A \mathbf{n}_j = \delta_{ij}$. Note that the only assumption imposed on N has been to provide a basis for $\mathcal{N}(B)$. In this latter variant, AN would be an orthogonal matrix, i.e., $(AN)^T(AN) = I$. The amount of work, however, does not change dramatically, since this choice of basis vectors essentially calls for a (weighted) Gram-Schmidt orthogonalization process.

Having determined $\overline{A}$ and the regularized solution $\mathbf{z}$ of (9.12), it remains to compute from (9.11) the approximation $\mathbf{x}$ of the original general form problem. To this end, recall from part (i) of Lemma 9.3 that $\mathbf{x}_B = B_A^\dagger \mathbf{z}$ is the unique solution in $\mathcal{B}$ of $B\mathbf{x} = \mathbf{z}$. Because of the special form (9.13) of B, the general solution of this equation has the form

$$\begin{bmatrix} B_1^{-1}\mathbf{z} \\ 0 \end{bmatrix} + N\mathbf{c} \qquad \text{for some } \mathbf{c} \in \mathbb{R}^p.$$

We determine $\mathbf{c}$ from the condition $\mathbf{x}_B \in \mathcal{B}$, i.e., $A^T A \mathbf{x}_B \perp \mathcal{N}(B)$, which yields

$$0 = N^T(A^T A \mathbf{x}_B) = (A^T A N)^T \begin{bmatrix} B_1^{-1}\mathbf{z} \\ 0 \end{bmatrix} + (AN)^T(AN)\mathbf{c}.$$

Finally, to compute $\mathbf{x}_0^\dagger$, we recall the identity (8.12) which gives

$$\mathbf{x}_0^\dagger = A_0^\dagger \mathbf{b},$$

where A_0 is the restriction of A to $\mathcal{N}(B)$. Writing $\mathbf{x}_0^\dagger = N\mathbf{c}^\dagger$, and identifying A_0 with the matrix AN, we obtain

$$\mathbf{c}^\dagger = (AN)^\dagger \mathbf{b}, \qquad \mathbf{x}_0^\dagger = N\mathbf{c}^\dagger.$$

Since AN has full column rank, $\mathbf{c}^\dagger$ can alternatively be written as the unique solution of the linear system

$$(AN)^T(AN)\mathbf{c}^\dagger = (AN)^T \mathbf{b}$$

with the familiar $p \times p$ matrix $(AN)^T(AN)$.

We summarize the above considerations for the backtransformation (9.11). First, compute once and for all $\mathbf{x}_0^\dagger$ as follows:

Algorithm 9.5. (Computation of $\mathbf{x}_0^\dagger$)

- Solve $(AN)^T(AN)\mathbf{c}^\dagger = (AN)^T\mathbf{b}$ for $\mathbf{c}^\dagger$, i.e., $\mathbf{c}^\dagger = (AN)^\dagger \mathbf{b}$,
- compute $\mathbf{x}_0^\dagger = N\mathbf{c}^\dagger$.

Then, for any regularized approximation $\mathbf{z}$ of the standard form problem, compute $\mathbf{x}$ from (9.11) by means of

Algorithm 9.6. (Backtransformation)

- Determine the "basic solution" $\mathbf{x}_b = \begin{bmatrix} B_1^{-1}\mathbf{z} \\ 0 \end{bmatrix}$ of $B\mathbf{x} = \mathbf{z}$,
- solve $(AN)^T(AN)\mathbf{c} = -(A^T AN)^T \mathbf{x}_b$ for $\mathbf{c}$, i.e., $\mathbf{c} = -(AN)^\dagger A\mathbf{x}_b$,
- compute $\mathbf{x}_B = \mathbf{x}_b + N\mathbf{c}$, and put $\mathbf{x} = \mathbf{x}_0^\dagger + \mathbf{x}_B$.

Note that the backtransformation requires only about $(l + 2p + 1)n$ operations.

9.3. Implementation of Tikhonov Regularization

Due to the results of the foregoing section we can now restrict our attention to regularization problems in standard form. Most of the material of this section is taken from Eldén's paper [63], see also Voevodin [283].

Finite-dimensional Tikhonov regularization in standard form amounts to determining $\mathbf{x} \in \mathbb{R}^n$, for which

$$\|\mathbf{b} - A\mathbf{x}\|^2 + \alpha \|\mathbf{x}\|^2 \to \min, \qquad (9.16)$$

where, as before, $A \in \mathbb{R}^{m \times n}$ and $\mathbf{b} \in \mathbb{R}^m$. The norms in (9.16) and throughout this section always denote Euclidean norms. For simplicity, we restrict our discussion to the most relevant case where $m \geq n$ and $\mathcal{N}(A) = \{\mathbf{0}\}$; see [29] for the additional considerations that are necessary to treat the general case.

Given two orthogonal matrices $U \in \mathbb{R}^{m \times m}$ and $V \in \mathbb{R}^{n \times n}$, we observe that the left-hand side of (9.16) can be rewritten as

$$\|U(\mathbf{b} - AV^T V\mathbf{x})\|^2 + \alpha\|V\mathbf{x}\|^2 = \|\mathbf{c} - UAV^T\mathbf{v}\|^2 + \alpha\|\mathbf{v}\|^2, \qquad (9.17)$$
$$\mathbf{c} = U\mathbf{b}, \quad \mathbf{v} = V\mathbf{x}.$$

We may therefore choose U and V in such a way that UAV^T has a form which allows a particularly efficient computation of the minimizer $\mathbf{v}$ of the expression above. It

9.3. Implementation of Tikhonov Regularization

turns out convenient to reduce A to upper bidiagonal form, i.e., to choose U and V such that
$$UAV^T = \begin{bmatrix} J \\ 0 \end{bmatrix}, \qquad (9.18)$$
where $J \in \mathbb{R}^{n \times n}$ is upper bidiagonal.

We shall describe how this can be done by means of Householder transformations. Recall that a Householder transformation $H \in \mathbb{R}^{m \times m}$ is an orthogonal matrix defined by a vector $\mathbf{p} \in \mathbb{R}^m$, namely
$$H = I - \frac{2}{\|\mathbf{p}\|^2} \mathbf{p}\mathbf{p}^T.$$

H reflects $\mathbb{R}^m$ along the hyperplane which is orthogonal to $\mathbf{p}$, i.e.,
$$H\mathbf{x} = \begin{cases} \mathbf{x}, & \mathbf{x}^T\mathbf{p} = 0, \\ -\mathbf{x}, & \mathbf{x} = \gamma\mathbf{p}, \gamma \in \mathbb{R}. \end{cases}$$

Denoting by $\mathbf{a} = [a_{11}, \ldots, a_{m1}]^T$ the first column of A and by $\mathbf{e}_1$ the first unit vector in $\mathbb{R}^m$, the choice
$$\mathbf{p} = \mathbf{a} + \operatorname{sign}(a_{11})\|\mathbf{a}\|\mathbf{e}_1, \qquad \|\mathbf{p}\|^2 = 2\|\mathbf{a}\|(\|\mathbf{a}\| + |a_{11}|),$$
gives a Householder transformation H that maps $\mathbf{a}$ onto $-\operatorname{sign}(a_{11})\|\mathbf{a}\|\mathbf{e}_1$, and hence,
$$HA = \begin{bmatrix} \rho_1 & \tilde{a}_{11} & \cdots & \tilde{a}_{1n-1} \\ 0 & * & \cdots & * \\ \vdots & \vdots & & \vdots \\ 0 & * & \cdots & * \end{bmatrix}, \qquad \rho_1 = -\operatorname{sign}(a_{11})\|\mathbf{a}\|.$$

Defining accordingly the Householder transformation $\tilde{H} \in \mathbb{R}^{(n-1) \times (n-1)}$ which maps the new entries $\tilde{\mathbf{a}} = [\tilde{a}_{11}, \ldots, \tilde{a}_{1n-1}]^T$ onto the first unit vector in $\mathbb{R}^{n-1}$, we have
$$HA \begin{bmatrix} 1 & 0 \\ 0 & \tilde{H} \end{bmatrix}^T = \begin{bmatrix} \rho_1 & \tilde{\rho}_1 & 0 & \cdots & 0 \\ 0 & * & * & \cdots & * \\ \vdots & \vdots & \vdots & & \vdots \\ 0 & * & * & \cdots & * \end{bmatrix}.$$

Proceeding in a similar fashion, we subsequently reduce the $(m-1) \times (n-1)$ submatrix indicated by stars on the right-hand side. After n such transformations and about $2(mn^2 - n^3/3)$ operations we eventually arrive at the desired decomposition (9.18) with U and V being the accumulated products of the two sequences of Householder transformations. Note that the above algorithm – which constitutes the cost dominating part of the entire implementation of Tikhonov regularization – is completely independent of the particular value of the regularization parameter α. Thus, doing this transformation once and for all, one can easily compute solutions v_α for a series of regularization parameters without essential change of the total cost. It is only this latter property which makes Tikhonov regularization a competitive algorithm as compared, e.g., to iterative methods.

At this point it becomes clear why it is important to do the above computations only for regularization problems in standard form. For a general form problem the term $\|v\|$ in (9.17) has to be replaced by $\|BV^T v\|$; the matrix BV^T, however, is in general a full matrix with no obvious structure, and any algorithm for solving the minimization problem will take another $O(n^3)$ operations.

As is apparent from (9.17), the Householder matrices which make up U need not be stored, provided the new right-hand side vector c is updated as soon as a new factor H has been determined. The matrix V, however, is required to recover x from v at the end of the algorithm. Nevertheless, V should not be computed explicitly; rather, it is recommended to store the vectors p which define the transformations, and which can share storage with eliminated parts of A. Applying these Householder transformations to v one after another, x is computed in about n^2 operations. Usually, this backsubstitution needs only to be done once at the very end, since most parameter choice rules for α can be evaluated as soon as v is known, cf. Section 9.4. This is so because v and x are unitarily equivalent.

We mention that the reduction to bidiagonal form coincides with the first part of the algorithm by Golub and Kahan for computing the singular value decomposition of A. Instead of Householder transformations one could also use Givens rotations. We refer to the books by Golub and van Loan [96] or Björck [29] for this and many further details.

Decomposing next the vector $c = [c_1, c_2]^T$ of (9.17) according to the matrix decomposition on the right-hand side of (9.18), the quadratic form (9.16), (9.17) to be minimized assumes the convenient form

$$\|c_1 - Jv\|^2 + \|c_2\|^2 + \alpha \|v\|^2 = \left\| \begin{bmatrix} c_1 \\ 0 \end{bmatrix} - \begin{bmatrix} J \\ \sqrt{\alpha} I \end{bmatrix} v \right\|^2 + \|c_2\|^2.$$

Since this minimization problem is independent of the actual value of c_2 we face a standard least-squares problem

$$\begin{bmatrix} J \\ \sqrt{\alpha} I \end{bmatrix} v = \begin{bmatrix} c_1 \\ 0 \end{bmatrix} \tag{9.19}$$

with a unique solution and a particularly structured matrix on the left-hand side. This structure enables efficient solution of the minimization problem in only $O(n)$ operations by means of Givens rotations as we will see next.

A Givens rotation (or plane rotation) R is an orthogonal matrix which looks like an identity matrix except for four differing elements at positions (i,i), (i,j), (j,i) and (j,j), where R has entries $c = \cos\theta$, $s = \sin\theta$, $-s$, and c, respectively; here θ is a given angle. It follows that multiplication of R to some matrix $A = [a_{ij}]$ from the left only affects those columns of A which have nonzero entries in the ith or jth row. If, say, the kth column of A is affected, the corresponding entries are replaced in the following way:

$$a_{ik} \leftarrow ca_{ik} + sa_{jk}, \qquad a_{jk} \leftarrow -sa_{ik} + ca_{jk}.$$

For example, if $a_{jk} \neq 0$ and if we choose c and s to be

$$c = \frac{a_{ik}}{\sigma}, \quad s = \frac{a_{jk}}{\sigma}, \qquad \text{with } \sigma = \sqrt{a_{ik}^2 + a_{jk}^2}, \tag{9.20}$$

then a_{jk} is replaced by zero, and a_{ik} becomes σ.

In this way we will annihilate the entries of $\sqrt{\alpha}\,I$ in (9.19). Note that orthogonal transformations do not alter the solution of the least-squares problem. We first take i and j to be the top rows of the two submatrices in (9.19), i.e., $i = 1$, $j = n+1$, and apply the corresponding Givens rotation to annihilate the top entry of the lower submatrix $\sqrt{\alpha}\,I$. Recall that this rotation affects *all* columns with nonzero entries in the top rows of these two submatrices. Thus, the second column is also affected, and in general this will produce a new nonzero entry at position $(n+1, 2)$. No further column changes, but the right-hand side does. After this transformation all but the second column have zero entries in rows $n+1$ and $n+2$. Therefore, choosing $i = n+2$ and $j = n+1$, a second Givens rotation can be applied to annihilate the newly created nonzero entry at $(n+1, 2)$ without introducing any further nonzero entries in the matrix. We sketch this idea with a figure for the case $n = 3$; plus signs denote nonzero entries, stars indicate modified entries which in general will be nonzero. The right-hand side is separated by a vertical line, the horizontal line indicates the matrix decomposition as shown in (9.19).

$$\left[\begin{array}{cccc|c} ++ & & & & + \\ ++ & & & & + \\ & ++ & & & + \\ & & + & & + \\ \hline + & & & & \\ & + & & & \\ & & + & & \\ & & & + & \end{array}\right] \rightarrow \left[\begin{array}{cccc|c} ** & & & & * \\ ++ & & & & + \\ & ++ & & & + \\ & & + & & + \\ \hline * & & & & * \\ & + & & & \\ & & + & & \\ & & & + & \end{array}\right] \rightarrow \left[\begin{array}{cccc|c} ++ & & & & + \\ ++ & & & & + \\ & ++ & & & + \\ & & + & & + \\ \hline & & & & * \\ * & & & & * \\ & + & & & \\ & & & + & \end{array}\right]$$

Note that in our particular application σ in (9.20) is simply $(\alpha + a_{jk}^2)^{1/2}$, hence the computations for (9.20) take only three operations and one square root evaluation. The annihilation process does not require any additional computations; all that remains to be computed is the effect of the rotation onto the second column and the right-hand side, which takes six operations if we account for the fact that the entry at position $(n+1, 2)$ has been zero. (At this first step this has also been the case for the corresponding right-hand side entry, but this is no longer true in subsequent steps). The determination of the second Givens rotation takes another three operations and one square root evaluation and, using the zero structure of the matrix, it takes only two more operations to determine the new right-hand side. This amounts to 14 operations and two square root evaluations. Note that the new entry at position $(n+2, 2)$ is given by σ from (9.20), and hence its square is available in later stages of the algorithm without computations.

Since the submatrix consisting of rows 2 through $2n$ now has the same structure as the entire matrix had before, we can proceed in the same way until we eventually have reduced the matrix to upper bidiagonal form. At this stage we have obtained a QR-decomposition of the matrix in (9.19): denoting by U_α the product of all Givens rotations, there is an upper bidiagonal matrix $J_\alpha \in \mathbb{R}^{n \times n}$ with

$$U_\alpha \begin{bmatrix} J \\ \sqrt{\alpha} I \end{bmatrix} = \begin{bmatrix} J_\alpha \\ 0 \end{bmatrix}, \qquad (9.21)$$

and the solution $\mathbf{v}_\alpha$ of (9.19) can be computed from

$$\mathbf{v}_\alpha = J_\alpha^{-1} \mathbf{d}_1 \quad \text{with} \quad \begin{bmatrix} \mathbf{d}_1 \\ \mathbf{d}_2 \end{bmatrix} = U_\alpha \begin{bmatrix} \mathbf{c}_1 \\ 0 \end{bmatrix} \qquad (9.22)$$

with another $2n$ operations. In the way we have described the algorithm above, the matrix U_α need not be stored. However, for most parameter choice rules, and in particular for an implementation of iterated Tikhonov regularization, it is advantageous to have the sines and cosines of the Givens rotations ready to do another solve of (9.19) with a different right-hand side, cf. Section 9.4 below.

The entire implementation of Tikhonov regularization is summarized in

Algorithm 9.7. (Tikhonov regularization)

- If necessary, transform the problem into standard form, cf. Algorithm 9.4, compute $\mathbf{x}_0^\dagger$ by Algorithm 9.5, and evaluate the right-hand side of (9.12);

- use Householder transformations to reduce A to upper bidiagonal form, cf. (9.18),

- select α and do a QR-decomposition of (9.19) by means of Givens rotations,

- solve for $\mathbf{v}_\alpha$ as shown in (9.22),

- if necessary, modify α and continue in Step 3,

- (backsubstitution) apply the Householder transformations that make up V^T in (9.18) to find $\mathbf{x}_\alpha = V^T \mathbf{v}_\alpha$,

- if necessary, use $\mathbf{x}_\alpha$ as input $\mathbf{z}$ for Algorithm 9.6 to recover the solution of the general form problem. (Note the different use of the symbol $\mathbf{x}$ here and in Section 9.2.)

We briefly comment on the implementation of iterated Tikhonov type methods, cf. Section 5.1. As shown in (5.37), the iterated Tikhonov approximations are obtained from classical Tikhonov approximations $\mathbf{x}_\alpha = V^T \mathbf{v}_\alpha$ by solving further minimization problems with modified right-hand sides, such as $\mathbf{b} - A\mathbf{x}_\alpha$. It is readily

Table 9.1: Operation counts for Tikhonov regularization

transformation to standard form	$(l + 2p + 1)mn$
bidiagonalization	$2(mn^2 - \frac{1}{3}n^3)$
solution of (9.19)	$16n$ plus $2n$ roots
parameter choice rule	$O(n)$
backsubstitution	n^2

seen that the least-squares problem (9.19) can again be used for the subsequent computations, when changing c_1 on the right-hand side to $c_1 - J v_\alpha$ (note that in this way x_α is not required, and therefore, no expensive backsubstitution is necessary). From (9.19) one can even proceed with any other regularization parameter. Each iteration only takes about $18n$ operations.

In spite of the many attractive features of this particular way of implementing Tikhonov regularization, other algorithms may be preferred if the matrices A and B in the general form problem have special structure. The algorithm presented above cannot take advantage of any such structure. The overall amount of work will always be $O(n^3)$ operations. Especially for Toeplitz-type matrices this does not seem to be efficient. A considerable amount of research has been devoted to finding better algorithms with only $O(n^2)$ operations. A description of these algorithms, however, is beyond the scope of this book. Instead, we refer to [65, 146, 201] and references therein.

One option that remains for solving the standard form problem (9.6) is the use of iteration methods. Since the coefficient matrix $A^T A + \alpha B^T B$ is positive definite the conjugate gradient method is a natural candidate to use. If there are measurement errors in the data it is not necessary to do a lot of iterations (the convergence is slow, anyway). In fact, the iteration error will not depend much on α in the early stage of the iteration, and it might be easiest to just put $\alpha = 0$, i.e., not to regularize the problem at all. Rather, one could use the inherent regularizing property of conjugate gradients. For the implementation of iterative schemes see Section 9.5.

9.4. Updating the Regularization Parameter

A number of parameter choice rules have been proposed in Chapters 4 and 5: some, like the discrepancy principle, apply to general regularization methods and others, like the quasioptimality criterion, only make sense for (iterated) Tikhonov regularization. Here, we will elaborate on how to implement parameter choice rules for Tikhonov regularization in the context of Algorithm 9.7.

Two of the parameter choice rules that we have considered, i.e., the discrepancy principle and the order optimal rule (5.28) from Section 5.1, use a-priori knowledge concerning the noise level δ in the data. Both rules determine the regularization

234   9. Numerical Realization

parameter as the solution of a nonlinear equation, namely

$$\phi_q(\alpha) := \alpha^q \mathbf{b}^T (AA^T + \alpha I)^{-q} \mathbf{b} = \tau \delta^2, \qquad (9.23)$$

with $q = 2$ and $q = 3$ for the discrepancy principle and the rule (5.28), respectively, and some number $\tau > 1$. To find the solution of (9.23) we propose the following iteration:

$$\alpha_{k+1} = \frac{\alpha_k \phi'_q(\alpha_k)}{\alpha_k \phi'_q(\alpha_k) + \phi_q(\alpha_k) - \tau \delta^2} \alpha_k \qquad k \geq 0. \qquad (9.24)$$

Here, α_0 is an initial guess and ϕ'_q is the derivative of ϕ_q, i.e.,

$$\phi'_q(\alpha) = q\alpha^{q-1}(A^T \mathbf{b})^T (A^T A + \alpha I)^{-q-1} A^T \mathbf{b}. \qquad (9.25)$$

The following result establishes global quadratic convergence for this scheme provided the initial guess α_0 is *greater* than the solution of (9.23).

Proposition 9.8. *Let Q be the orthoprojector onto $\mathcal{R}(A)$, and assume that $\|\mathbf{b}\| > \tau \delta$ and $\|(I - Q)\mathbf{b}\| \leq \delta$. Denote the positive solution of (9.23) by α_*. Then the iteration (9.24) converges monotonically to α_* from above provided $\alpha_0 \geq \alpha_*$, and the convergence is locally quadratic.*

Proof. Recall from Secion 5.1 that (9.23) has a unique positive solution α_* under the stipulated assumptions. Introduce $\beta = \alpha^{-1}$, $\beta_* = \alpha_*^{-1}$, and define

$$\psi_q(\beta) := \phi_q(\frac{1}{\beta}) = \mathbf{b}^T(\beta AA^T + I)^{-q}\mathbf{b}.$$

The Newton iteration applied to $\psi_q(\beta) = \tau \delta^2$ with $\beta_0 = \alpha_0^{-1}$ yields the iteration

$$\beta_{k+1} = \beta_k + \frac{\tau \delta^2 - \psi_q(\beta_k)}{\psi'_q(\beta_k)}, \qquad k \geq 0. \qquad (9.26)$$

The first and second derivative of ψ_q are given by

$$\psi'_q(\beta) = -q(A^T\mathbf{b})^T (\beta A^T A + I)^{-q-1} A^T \mathbf{b},$$
$$\psi''_q(\beta) = q(q+1)(AA^T\mathbf{b})^T (\beta AA^T + I)^{-q-2} AA^T \mathbf{b},$$

and hence, ψ is a convex, decreasing function of β. Clearly, we have $\psi_q(\beta_*) = \tau\delta^2$. Furthermore, $\psi'_q(\beta_*) = 0$ can only occur when $A^T\mathbf{b} = \mathbf{0}$. The latter, however, implies $(I - Q)\mathbf{b} = \mathbf{b}$ and hence, $\tau\delta^2 = \psi_q(\beta_*) = \|\mathbf{b}\|^2$, in contradiction to the assumption on $\|\mathbf{b}\|$. Consequently, by standard properties of the Newton iteration (cf., e.g., [261]), β_k converges monotonically to β_* from below, and the convergence is locally quadratic. It follows that $\alpha_k = \beta_k^{-1} \to \alpha_*$, monotonically from above, and again convergence is locally quadratic. Rewriting (9.26) in terms of α_k yields (9.24), and the proof is complete. ∎

9.4. Updating the Regularization Parameter

The quantities ϕ_q and ϕ'_q which need to be evaluated for implementing (9.24) can be extracted or readily computed from intermediate quantities of Algorithm 9.7 as we show next.

Example 9.9. Recall from (9.17) that

$$(AA^T + \alpha I)^{-1}\mathbf{b} = \frac{1}{\alpha}(\mathbf{b} - A\mathbf{x}_\alpha) = \frac{1}{\alpha}U^T\left(\begin{bmatrix}\mathbf{c}_1\\\mathbf{c}_2\end{bmatrix} - \begin{bmatrix}J\\0\end{bmatrix}\mathbf{v}_\alpha\right). \tag{9.27}$$

Consequently, using (9.18),

$$(A^TA + \alpha I)^{-1}A^T\mathbf{b} = \frac{1}{\alpha}A^TU^T\left(\begin{bmatrix}\mathbf{c}_1\\\mathbf{c}_2\end{bmatrix} - \begin{bmatrix}J\\0\end{bmatrix}\mathbf{v}_\alpha\right) = \frac{1}{\alpha}V^TJ^T(\mathbf{c}_1 - J\mathbf{v}_\alpha).$$

To compute $(AA^T + \alpha I)^{-2}\mathbf{b}$ the idea is to apply (9.27) with $\alpha(AA^T + \alpha I)^{-1}\mathbf{b} = \mathbf{b} - A\mathbf{x}_\alpha$ instead of $\mathbf{b}$, similar as in iterated Tikhonov regularization. As described in the foregoing section we replace $\mathbf{c}_1$ by $\mathbf{c}_1 - J\mathbf{v}_\alpha$ in (9.22), and compute

$$\mathbf{v}_\alpha^{II} = [J_\alpha^{-1}, 0]U_\alpha \begin{bmatrix}\mathbf{c}_1 - J\mathbf{v}_\alpha\\0\end{bmatrix}.$$

Then, some computations show that

$$(AA^T + \alpha I)^{-2}\mathbf{b} = \frac{1}{\alpha}(AA^T + \alpha I)^{-1}(\mathbf{b} - A\mathbf{x}_\alpha) = \frac{1}{\alpha^2}U^T\begin{bmatrix}\mathbf{c}_1 - J\mathbf{v}_\alpha - J\mathbf{v}_\alpha^{II}\\\mathbf{c}_2\end{bmatrix},$$

and using (9.18) once again,

$$(A^TA + \alpha I)^{-2}A^T\mathbf{b} = A^T(AA^T + \alpha I)^{-2}\mathbf{b} = \frac{1}{\alpha^2}V^TJ^T(\mathbf{c}_1 - J\mathbf{v}_\alpha - J\mathbf{v}_\alpha^{II}).$$

With these quantities the values of ϕ_2, ϕ'_2 and ϕ_3, ϕ'_3 can be implemented without expensive backsubstitution:

$$\phi_2(\alpha) = \|\mathbf{c}_1 - J\mathbf{v}_\alpha\|^2 + \|\mathbf{c}_2\|^2,$$

$$\phi'_2(\alpha) = \frac{2}{\alpha^2}(J^T(\mathbf{c}_1 - J\mathbf{v}_\alpha))^TJ^T(\mathbf{c}_1 - J\mathbf{v}_\alpha - J\mathbf{v}_\alpha^{II}),$$

$$\phi_3(\alpha) = (\mathbf{c}_1 - J\mathbf{v}_\alpha)^T(\mathbf{c}_1 - J\mathbf{v}_\alpha - J\mathbf{v}_\alpha^{II}) + \|\mathbf{c}_2\|^2,$$

$$\phi'_3(\alpha) = \frac{3}{\alpha^2}\|J^T(\mathbf{c}_1 - J\mathbf{v}_\alpha - J\mathbf{v}_\alpha^{II})\|^2.$$

Now we turn to the error-free parameter choice rules suggested in Sections 4.5 and 5.1, namely the L-curve criterion, generalized cross-validation, the quasioptimality criterion, and the rule from Theorem 4.35. All these rules determine their regularization parameter as the minimizer of a certain real function χ. Any scalar optimization technique may be applied for finding this parameter (cf., e.g., [53]), however, it must be emphasized that for any of these methods χ may have further non-global minima.

For the implementation of such optimization routines, it is typically necessary to evaluate χ and – if possible – derivatives of χ. In all the above parameter choice methods the function χ is somehow related to ϕ_2 or ϕ_3 from (9.23). For instance, as follows from (5.42) and (9.25), the quasioptimality criterion chooses the minimizer of ϕ'_3, whereas the function ϕ_3/α is minimized in the rule of Theorem 4.35. (Note that for Tikhonov regularization (4.121) coincides with ϕ_3.) To compute the curvature (4.130) for the L-curve criterion, the first two derivatives of ϕ_2 are required (recall that ϕ_2 is the squared residual norm). While ϕ'_2 is given by (9.25), another differentiation readily yields

$$\phi_2'' = \frac{1}{\alpha}\phi'_2 - \frac{2}{\alpha}\phi'_3.$$

For the L-curve criterion, $\|\mathbf{x}_\alpha\|^2$ and its first two derivatives are required as well. Recall that $\|\mathbf{x}_\alpha\| = \|\mathbf{v}_\alpha\|$ because of (9.17). The derivatives of $\|\mathbf{x}_\alpha\|^2$, on the other hand, turn out to be related to those of ϕ_2, since

$$\frac{d}{d\alpha}\|\mathbf{x}_\alpha\|^2 = -2(A^T\mathbf{b})^T(A^TA + \alpha I)^{-3}A^T\mathbf{b} = -\frac{1}{\alpha}\phi'_2(\alpha).$$

More straightforward is the variant of the L-curve criterion proposed by Regińska [236]. To evaluate the function ψ of (4.133) we only need the residual norm, i.e., ϕ_2, and the norm of the approximation, i.e., $\|\mathbf{x}_\alpha\| = \|\mathbf{v}_\alpha\|$. In other words, all quantities required for these first three parameter choice rules can be computed as in Example 9.9 without any backtransformations.

In generalized cross-validation, cf. (4.125), the function to be minimized is

$$\chi(\alpha) = \frac{\phi_2(\alpha)}{(\frac{\alpha}{m}\operatorname{trace}[(AA^T + \alpha I)^{-1}])^2}.$$

Eldén [66] has shown how to evaluate the trace term in this function from knowledge of the upper bidiagonal matrix J_α of (9.21). The clue is the identity

$$J_\alpha^T J_\alpha = J^T J + \alpha I = V(A^TA + \alpha I)V^T,$$

which follows readily from forming the cross product matrices of (9.21) and (9.18). The right-hand side is a similarity transformation of $A^TA + \alpha I$ which does not affect the trace, i.e.,

$$\operatorname{trace}[(A^TA + \alpha I)^{-1}] = \operatorname{trace}[(J_\alpha^T J_\alpha)^{-1}].$$

Now, recall that the trace of a matrix equals the sum of its eigenvalues, and that the $n \times n$ and $m \times m$ matrices $A^TA + \alpha I$ and $AA^T + \alpha I$ have the same eigenvalues except for $m - n$ additional eigenvalues α of $AA^T + \alpha I$. This gives the final representation of the cross-validation function χ:

$$\chi(\alpha) = \frac{m^2 \phi_2(\alpha)}{(m - n + \alpha \operatorname{trace}[(J_\alpha^T J_\alpha)^{-1}])^2}. \tag{9.28}$$

We now show how to determine the trace of this latter matrix in $O(n)$ operations. Let

$$J_\alpha = \begin{pmatrix} \gamma_{1,1} & \gamma_{1,2} & & \\ & \gamma_{2,2} & \gamma_{2,3} & \\ & & \ddots & \ddots & \\ & & & \gamma_{n-1,n-1} & \gamma_{n-1,n} \\ & & & & \gamma_{n,n} \end{pmatrix}, \quad J_\alpha^{-T} = [\mathbf{c}_1, \ldots, \mathbf{c}_n],$$

and denote by $\mathbf{e}_i$, $i = 1, \ldots, n$, the ith Cartesian vector of $\mathbb{R}^n$. Since the trace of $(J_\alpha^T J_\alpha)^{-1} = J_\alpha^{-1} J_\alpha^{-T}$ equals the sum of squares of all elements of J_α^{-T}, we have to compute

$$\text{trace}[J_\alpha^{-1} J_\alpha^{-T}] = \|\mathbf{c}_1\|^2 + \ldots + \|\mathbf{c}_n\|^2. \tag{9.29}$$

Comparing the columns of the matrix equation $I = J_\alpha^{-T} J_\alpha^T$, we obtain the following relations between the vectors $\mathbf{c}_i$:

$$\mathbf{e}_i = \gamma_{i,i} \mathbf{c}_i + \gamma_{i,i+1} \mathbf{c}_{i+1}, \quad i = 1, \ldots, n-1, \quad \mathbf{e}_n = \gamma_{n,n} \mathbf{c}_n.$$

Since J_α is upper triangular, J_α^{-T} is lower triangular, and hence, $\mathbf{e}_i$ is orthogonal to $\mathbf{c}_{i+1}$ for all $i = 1, \ldots, n-1$. We can therefore use the Pythagorean theorem to conclude

$$\gamma_{i,i}^2 \|\mathbf{c}_i\|^2 = \|\mathbf{e}_i - \gamma_{i,i+1} \mathbf{c}_{i+1}\|^2 = 1 + \gamma_{i,i+1}^2 \|\mathbf{c}_{i+1}\|^2, \tag{9.30}$$

$i = 1, \ldots, n-1$. Thus, the function χ to be minimized in generalized cross-validation can be evaluated as follows:

Algorithm 9.10. (Generalized cross-validation)

- With $\|\mathbf{c}_n\|^2 = 1/\gamma_{n,n}^2$, compute $\|\mathbf{c}_i\|^2$ for $i = n-1$ down to $i = 1$ from (9.30),
- determine the trace of $(J_\alpha^T J_\alpha)^{-1}$ from (9.29), and insert its value into (9.28).

9.5. Implementation of Iterative Methods

While there is a clear need for some sophistication when implementing Tikhonov regularization, the implementation of iterative regularization methods is much more straightforward. For standard form regularization (where $B = I$ in Section 9.2) one simply substitutes the matrix A for the continuous operator T, and the right-hand side $\mathbf{b}$ for y in one of the iterative schemes suggested in Chapters 6 or 7. It follows that the major amount of work per iteration are matrix vector products with A and A^T. This reveals one of the big advantages of iterative regularization techniques: several different structure patterns of A can be used with success for fast matrix vector multiplications.

For example, convolution type integral equations typically lead to matrices A, which are either Toeplitz or close to Toeplitz. Matrix vector products with Toeplitz

9. Numerical Realization

matrices can efficiently be computed by FFT techniques in $O(n \log n)$ operations by embedding the Toeplitz matrix into a double size circulant matrix, cf., e.g., [280]. Besides, recent developments like the multipole method (cf. [237]) or wavelet based decompositions (cf. [8, 25]) have lead to fast implementations of matrix vector multiplications with dense matrices A arising from the discretization of weakly singular integral equations without convolution structure. Other applications like systems of integral equations over hyperplanes, as they arise, e.g., in Radon transform inversion (cf. Section 1.2), can be discretized so as to obtain comparatively sparse matrices A. As a consequence, the matrix vector product is again significantly cheaper than $O(n^2)$. On vector and parallel computers, finally, matrix vector products with dense and unstructured matrices A can be implemented far more efficiently than, for instance, direct solvers as required in Section 9.3.

If regularization is subject to a different norm in the coefficient space (i.e., $B \neq I$ in Sections 9.1 and 9.2) then, in view of the substitution (9.11), the iterative algorithm should be applied to the associated standard form problem (9.12). In other words, the kth iterate $\mathbf{x}_k$ for the general form problem is given as

$$\mathbf{x}_k = \mathbf{x}_0^\dagger + B_A^\dagger \mathbf{z}_k, \qquad \mathbf{z}_k = g_k((B_A^\dagger)^T A^T A B_A^\dagger) (B_A^\dagger)^T A^T \overline{\mathbf{b}},$$

which leads to iteration schemes similar to Examples 8.12 and 8.13 in Section 8.3. In principle, this could be implemented by first computing $\overline{A} = A B_A^\dagger$ explicitly as in Section 9.2. However, this would take $O(n^2)$ operations and, above all, any potential structure of A would be lost. Rather, one should keep the above factored form and perform products with $(B_A^\dagger)^T$ and $B_A^\dagger$ whenever necessary. These products can be implemented as in Algorithm 9.4 (computation of $(B_A^\dagger)^T \mathbf{a}_k$) and Algorithm 9.6 (evaluation of $B_A^\dagger \mathbf{z}$). These implementations only take $O(n)$ operations when B is a banded matrix of the form (9.13), which is the least we have to pay per iteration anyway.

We exemplify this approach for the Landweber method as a prototype iteration. Here the recursion for the sequence $\{\mathbf{z}_k\}$ is given by, cf. Example 8.12,

$$\mathbf{z}_k = \mathbf{z}_{k-1} + (B_A^\dagger)^T A^T (\overline{\mathbf{b}} - A B_A^\dagger \mathbf{z}_{k-1}),$$

where $k \in \mathbb{N}$ and $\mathbf{z}_0 = 0$. After backsubstitution we obtain

$$\mathbf{x}_k = \mathbf{x}_{k-1} + B_A^\dagger (B_A^\dagger)^T A^T (\mathbf{b} - A\mathbf{x}_{k-1}), \qquad k \in \mathbb{N}, \qquad (9.31)$$

with $\mathbf{x}_0 = \mathbf{x}_0^\dagger$. In other words, regularization in standard form amounts to a premultiplication of the normal equation system by suitable generalized inverses of B and of B^T.

Other iterative regularization methods have to be modified accordingly to deal with general form regularization. For the ν-methods we have to replace the command $x_k^\delta = x^* + T^* z_k^\delta$ in Algorithm 6.13 by

$$\mathbf{x}_k = \mathbf{x}_0^\dagger + B_A^\dagger (B_A^\dagger)^T A^T \mathbf{z}_k.$$

9.5. Implementation of Iterative Methods

For the conjugate gradient method the resulting algorithm coincides with the version of Algorithm 8.14, with T and $L_T^\dagger$ replaced by A and $B_A^\dagger$, respectively, and y replaced by $\mathbf{b}$.

Before we turn to a description of stopping rules, we comment on the need for a proper scaling of the equation $A\mathbf{x} = \mathbf{b}$ before iteration. The Landweber iteration and the ν-methods (but not CGNE) require that the matrix $\overline{A} = AB_A^\dagger$ of the standard form problem has norm less or equal to one. In practice this is achieved by rescaling A and $\mathbf{b}$ with an appropriate factor, namely with the reciprocal of $\|AB_A^\dagger\|$ (or simply $\|A\|$ in case of regularization in standard form). This number can be approximated with a few steps of the power method:

$$\mathbf{w}_{k+1} = AB_A^\dagger (B_A^\dagger)^T A^T \mathbf{w}_k, \qquad \mathbf{w}_0 = \overline{\mathbf{b}}.$$

After few steps ($k = 2$ or 3 typically suffices),

$$\left(\frac{\|\mathbf{w}_{k+1}\|}{\|\mathbf{w}_k\|}\right)^{\frac{1}{2}} \approx \|AB_A^\dagger\|$$

typically holds up to a small relative error. To be on the safe side, one may add about 10% to this approximation before rescaling the problem. Somewhat better is an estimation of $\|AB_A^\dagger\|$ with the Lanczos process; we refer to [117] for the details.

The implementation of stopping rules for iterative regularization methods is completely different from the implementation of parameter choice rules for Tikhonov regularization. Since iterative regularization methods are completely sequential, the only way of finding a reasonable stopping parameter k_* is by trying all smaller iteration indices $k \leq k_*$. This may of course be a disadvantage if k_* is large.

Concerning the discrepancy principle, the stopping index $k(\delta, y^\delta)$ is easily found: the iteration is stopped as soon as the residual norm $\|\mathbf{b} - A\mathbf{x}_k\|$ drops for the first time below the given tolerance. Since the residual is usually available without extra computations (e.g., d_k in Algorithm 8.14), all that is required are m operations to compute its norm.

A sequential implementation of the order-optimal stopping rule for the ν-method has been derived in Section 6.3, cf. Algorithm 6.17. This algorithm carries over to regularization problems in general form after the obvious replacements.

An implementation of the type of minimization problems that arise in most error-free stopping rules is somewhat more delicate. The detection of a non-global minimum must be based on further considerations or on computational experience. In general, all that can be done when a local minimum has been found is a small number of additional iterations, with the possibility of backing up the iterate corresponding to the observed minimum.

The method of generalized cross-validation is in general not implementable since no suitable factorization of AA^T is available. A solution to this problem may be the use of Monte Carlo cross-validation, cf. Section 4.5. As far as CGNE is concerned, we emphasize that the same recursion coefficients α_k and β_k must be used for computing the CGNE iterates *and* the Monte Carlo estimate of the trace term denominator;

otherwise the sequences $\{r_k\}$ of the corresponding residual polynomials would be different which, in principle, could lead to a failure of the cross-validation approach.

Concerning the L-curve, it appears as if the corner definition (4.132) from [236] is advantageous in the context of iteration methods. Its implementation (4.133) is straightforward. For applications where the corner definition via the curvature of the graph is preferred, Hansen and O'Leary [124] suggest fitting a 2D cubic spline to the discrete points on the L-curve (perhaps after local smoothing of the points), and then locating the point on the L-curve closest to the corner of the spline. Easier to implement would be an approximation of the curvature function based on central differences for the first and second derivatives: besides being significantly more basic (no spline subroutines are required), the amount of work for this implementation does not increase with k. Another advantage of this latter alternative would be that the resulting corner would not depend on subsamplings of the given points (like throwing away regularization parameters which are clearly outside the critical range). On the other hand, central differences may be more sensitive to noisy L-curve points than a cubic spline.

We conclude by warning once again that the L-curve is *not* suited for the ν-methods: it is typically full of wiggles because of the non-monotonicity of the corresponding residual polynomials.

10. Tikhonov Regularization of Nonlinear Problems

10.1. Introduction

We have seen in the chapters above that the theory for linear ill-posed problems is very well developed and can be considered as rather complete. A first step towards nonlinear problems was the combination of linear equations with convex constraints (see Section 5.4). Now we want to deal with the fully nonlinear case, where the theory is by far not so well developed as in the linear one, i.e., we want to solve

$$F(x) = y, \qquad (10.1)$$

where $F : \mathcal{D}(F) \subset \mathcal{X} \to \mathcal{Y}$ is a nonlinear operator between Hilbert spaces $\mathcal{X}$ and $\mathcal{Y}$. Under ill-posedness of the nonlinear problem we will always mean that the solutions do not depend continuously on the data. Throughout this chapter we assume that

(i) F is continuous and

(ii) F is weakly (sequentially) closed, i.e., for any sequence $\{x_n\} \subset \mathcal{D}(F)$, weak convergence of x_n to x in $\mathcal{X}$ and weak convergence of $F(x_n)$ to y in $\mathcal{Y}$ imply that $x \in \mathcal{D}(F)$ and $F(x) = y$.

If $\mathcal{D}(F)$ is weakly closed, which is e.g. the case if it is closed and convex, and if F is weakly continuous, then F is weakly closed.

More general than in the linear case, where we looked for the minimum-norm solution, for the nonlinear problem (10.1), we choose the concept of an x^*-minimum-norm solution $x^\dagger$, i.e.:

$$F(x^\dagger) = y$$

and

$$\|x^\dagger - x^*\| = \min\{\|x - x^*\| \mid F(x) = y\}.$$

The reason for treating the general x^* case is that for nonlinear problems $x^* = 0$ plays no special role. In the contrary, the choice of x^* will be very crucial, especially for the local results about convergence rates in the next section. Of course, available a-priori information about the solutions of $F(x) = y$ has to enter into the selection of x^*. In the case of multiple solutions x^* plays the role of a selection criterion. As for linear problems, an x^*-minimum-norm solution need not exist, and even if it does, it need not be unique, since F is nonlinear. In the following we assume the existence of an x^*-minimum-norm solution $x^\dagger$ for the data $y \in \mathcal{Y}$, which, by the weak closedness of F, follows from the attainability assumption that equation (10.1) has an exact solution. The case that (10.1) is only solvable in the least-squares sense is more complicated (see [27]).

10. Tikhonov Regularization of Nonlinear Problems

As for linear problems, we want to find a criterion to decide whether a nonlinear problem is ill-posed or not. If the nonlinear operator F is compact, one can give a sufficient condition for ill-posedness of (10.1) being similar to its compact linear counterpart. There injectivity of the linear operator is such a sufficient condition, provided that $\mathcal{X}$ is infinite-dimensional (see Proposition 2.7). This condition carries over to the nonlinear case in the sense that local injectivity around $x^\dagger$ of the nonlinear compact operator F implies the ill-posedness of problem (10.1) provided that $\mathcal{D}(F)$ is infinite-dimensional around $x^\dagger$ (see (10.2)).

Proposition 10.1. *Let F be a (nonlinear) compact and weakly closed operator, and let $\mathcal{D}(F)$ be weakly closed. Moreover, assume that $F(x^\dagger) = y$ and that there exists an $\varepsilon > 0$ such that $F(x) = \bar{y}$ has a unique solution for all $\bar{y} \in \mathcal{R}(F) \cap \mathcal{B}_\varepsilon(y)$. If there exists a sequence $\{x_n\} \subset \mathcal{D}(F)$ satisfying*

$$x_n \rightharpoonup x^\dagger \quad \text{but} \quad x_n \not\to x^\dagger, \tag{10.2}$$

then F^{-1} (defined on $\mathcal{R}(F) \cap \mathcal{B}_\varepsilon(y)$) is not continuous in y.

Proof. Let $\{x_n\} \subset \mathcal{D}(F)$ be a sequence satisfying (10.2). Since the compactness and weak closedness of F imply that F maps weakly into strongly convergent sequences, $y_n := F(x_n) \to F(x^\dagger)$. Since $y_n \in \mathcal{R}(F) \cap \mathcal{B}_\varepsilon(y)$ for n sufficiently large, x_n is the unique solution of $F(x) = y_n$, i.e., $x_n = F^{-1}(y_n)$. Now $y_n \to y$ and $F^{-1}(y_n) = x_n \not\to x^\dagger = F^{-1}(y)$ imply that F^{-1} is not continuous in y. ∎

Since there is an easy way for linear problems to characterize their stability via the closedness of the range of the linear operator (cf. Proposition 2.4), it were helpful if one could characterize the stability of a nonlinear problem through conditions on its linearization (if F is Fréchet-differentiable). However, besides the well known Inverse Function Theorem, the connection between the stability or ill-posedness of a nonlinear problem and its linearization is not as strong as one might think. Although F' is compact if F is compact (cf., e.g., [290, Proposition 7.33]), it might happen that $\mathcal{R}(F'(x^\dagger))$ is finite-dimensional so that the generalized inverse of the linearization is bounded. It is even possible to construct a nonlinear operator such that the corresponding nonlinear problem is everywhere ill-posed, whereas the linearized operator has a finite-dimensional range at a dense set of points, i.e., the linearization is well-posed almost everywhere (see [74, Example A.1]). On the other hand, well-posed nonlinear problems may have ill-posed linearizations (see [74, Example A.2]).

If the problem of solving (10.1) is ill-posed, we have to regularize it. We will treat Tikhonov regularization, since this method is widely used, especially in connection with the output least-squares formulation of parameter estimation problems (see [40, 41]). The role of Tikhonov regularization to stabilize parameter estimation problems has been studied in [40, 41, 163, 194]. Stability and convergence for general nonlinear problems have been treated in Seidman and Vogel [253]. Convergence rates for general nonlinear problems have been presented more recently in [74, 217].

10.2. Convergence Analysis

We will summarize some of these results in the next section. Parameter selection methods are treated in Section 10.3. Finally, we deal with Tikhonov regularization of nonlinear problems in Hilbert scales in Section 10.4.

10.2. Convergence Analysis

As in the linear case, we replace problem (10.1) by the minimization problem

$$\|F(x) - y^\delta\|^2 + \alpha \|x - x^*\|^2 \to \min, \quad x \in \mathcal{D}(F), \tag{10.3}$$

where $\alpha > 0$, $y^\delta \in \mathcal{Y}$ is an approximation of the exact right-hand side y of problem (10.1) and $x^* \in \mathcal{X}$. Under the given assumptions on F problem (10.3) admits a solution. Since F is nonlinear, the solution will not be unique, in general. As in the linear case, any solution to (10.3) will be denoted by x_α^δ. Before we address the question of convergence, we will first prove that the problem of solving (10.3) is stable in the sense of continuous dependence of the solutions on the data y^δ.

Theorem 10.2. *Let $\alpha > 0$ and let $\{y_k\}$ and $\{x_k\}$ be sequences where $y_k \to y^\delta$ and x_k is a minimizer of (10.3) with y^δ replaced by y_k. Then there exists a convergent subsequence of $\{x_k\}$ and the limit of every convergent subsequence is a minimizer of (10.3).*

Proof. By definition of x_k we have

$$\|F(x_k) - y_k\|^2 + \alpha \|x_k - x^*\|^2 \leq \|F(x) - y_k\|^2 + \alpha \|x - x^*\|^2 \tag{10.4}$$

for any $x \in \mathcal{D}(F)$. Hence, $\{\|x_k\|\}$ and $\{\|F(x_k)\|\}$ are bounded. Therefore, a subsequence $\{x_m\}$ of $\{x_k\}$ and $\bar{x}$ exist such that

$$x_m \rightharpoonup \bar{x} \quad \text{and} \quad F(x_m) \rightharpoonup F(\bar{x}).$$

By weak lower semicontinuity of the norm we have

$$\|\bar{x} - x^*\| \leq \liminf \|x_m - x^*\|$$

and

$$\|F(\bar{x}) - y^\delta\| \leq \liminf \|F(x_m) - y_m\|. \tag{10.5}$$

Moreover, (10.4) implies

$$\begin{aligned}
\|F(\bar{x}) - y^\delta\|^2 + \alpha \|\bar{x} - x^*\|^2 &\leq \liminf \left(\|F(x_m) - y_m\|^2 + \alpha \|x_m - x^*\|^2 \right) \\
&\leq \limsup \left(\|F(x_m) - y_m\|^2 + \alpha \|x_m - x^*\|^2 \right) \\
&\leq \lim_{m \to \infty} \left(\|F(x) - y_m\|^2 + \alpha \|x - x^*\|^2 \right) \\
&= \|F(x) - y^\delta\|^2 + \alpha \|x - x^*\|^2
\end{aligned}$$

244   10. Tikhonov Regularization of Nonlinear Problems

for all $x \in \mathcal{D}(F)$. This implies that $\bar{x}$ is a minimizer of (10.3) and that

$$\lim_{m \to \infty} (\|F(x_m) - y_m\|^2 + \alpha \|x_m - x^*\|^2) = \|F(\bar{x}) - y^\delta\|^2 + \alpha \|\bar{x} - x^*\|^2. \qquad (10.6)$$

Now, assume that $x_m \not\to \bar{x}$. Then $c := \limsup \|x_m - x^*\| > \|\bar{x} - x^*\|$ and there exists a subsequence $\{x_n\}$ of $\{x_m\}$ such that $x_n \rightharpoonup \bar{x}$, $F(x_n) \to F(\bar{x})$ and $\|x_n - x^*\| \to c$. As a consequence of (10.6), we obtain

$$\lim_{n \to \infty} \|F(x_n) - y_n\|^2 = \|F(\bar{x}) - y^\delta\|^2 + \alpha(\|\bar{x} - x^*\|^2 - c^2) < \|F(\bar{x}) - y^\delta\|^2$$

in contradiction to (10.5). This argument shows that $x_m \to \bar{x}$. ∎

In the next theorem we show that under the same conditions on $\alpha(\delta)$ as in the linear case solutions of (10.3) converge towards an x^*-minimum-norm solution of (10.1).

Theorem 10.3. *Let $y^\delta \in \mathcal{Y}$ with $\|y - y^\delta\| \leq \delta$ and let $\alpha(\delta)$ be such that $\alpha(\delta) \to 0$ and $\delta^2/\alpha(\delta) \to 0$ as $\delta \to 0$. Then every sequence $\{x_{\alpha_k}^{\delta_k}\}$, where $\delta_k \to 0$, $\alpha_k := \alpha(\delta_k)$ and $x_{\alpha_k}^{\delta_k}$ is a solution of (10.3), has a convergent subsequence. The limit of every convergent subsequence is an x^*-minimum-norm solution. If in addition, the x^*-minimum-norm solution $x^\dagger$ is unique, then*

$$\lim_{\delta \to 0} x_{\alpha(\delta)}^\delta = x^\dagger.$$

Proof. Let α_k and $x_{\alpha_k}^{\delta_k}$ be as above and let $x^\dagger$ be an x^*-minimum-norm solution. Then by the definition of $x_{\alpha_k}^{\delta_k}$

$$\|F(x_{\alpha_k}^{\delta_k}) - y^{\delta_k}\|^2 + \alpha_k \|x_{\alpha_k}^{\delta_k} - x^*\|^2 \leq \delta_k^2 + \alpha_k \|x^\dagger - x^*\|^2$$

and hence

$$\lim_{k \to \infty} F(x_{\alpha_k}^{\delta_k}) = y \qquad (10.7)$$

and

$$\limsup_{k \to \infty} \|x_{\alpha_k}^{\delta_k} - x^*\| \leq \|x^\dagger - x^*\|. \qquad (10.8)$$

Therefore, $\{x_{\alpha_k}^{\delta_k}\}$ is bounded. Hence, there exist an element $x \in \mathcal{X}$ and a subsequence again denoted by $\{x_{\alpha_k}^{\delta_k}\}$ such that

$$x_{\alpha_k}^{\delta_k} \rightharpoonup x \quad \text{as} \quad k \to \infty. \qquad (10.9)$$

This together with (10.7) implies that $x \in \mathcal{D}(F)$ and that $F(x) = y$. Now by the weak lower semicontinuity of the norm, (10.8), (10.9), and the definition of $x^\dagger$,

$$\|x - x^*\| \leq \limsup_{k \to \infty} \|x_{\alpha_k}^{\delta_k} - x^*\| \leq \|x^\dagger - x^*\| \leq \|x - x^*\|.$$

10.2. Convergence Analysis

This shows that $\|x - x^*\| = \|x^\dagger - x^*\|$, i.e., x is also an x^*-minimum-norm solution. This together with (10.8), (10.9) and the fact that

$$\|x_{\alpha_k}^{\delta_k} - x\|^2 = \|x_{\alpha_k}^{\delta_k} - x^*\|^2 + \|x^* - x\|^2 + 2\langle x_{\alpha_k}^{\delta_k} - x^*, x^* - x \rangle$$

yields that

$$\limsup_{k \to \infty} \|x_{\alpha_k}^{\delta_k} - x\|^2 \leq 2\|x^* - x\|^2 + 2 \lim_{k \to \infty} \langle x_{\alpha_k}^{\delta_k} - x^*, x^* - x \rangle = 0$$

and hence that

$$\lim_{k \to \infty} x_{\alpha_k}^{\delta_k} = x.$$

If $x^\dagger$ is unique, the assertion about the convergence of $x_{\alpha(\delta)}^\delta$ follows from the fact that every sequence has a subsequence converging towards $x^\dagger$. ∎

We now turn to the convergence rate analysis. In the next theorem, we will give sufficient conditions for the rate $\|x_\alpha^\delta - x^\dagger\| = O(\sqrt{\delta})$. These conditions will also imply the rate $\|F(x_\alpha^\delta) - y\| = O(\delta)$ for the residuals. In the linear case, these conditions are equivalent to the condition that $x^\dagger - x^* \in \mathcal{R}(F^*)$, which is there sufficient for the rates $O(\sqrt{\delta})$ and $O(\delta)$, respectively (see Section 5.1). There this result has been proved using spectral-theoretic techniques. Since F is now nonlinear, these techniques are no longer available. Hence, the proof of the next theorem differs from the corresponding linear situation.

Theorem 10.4. *Let $\mathcal{D}(F)$ be convex, let $y^\delta \in \mathcal{Y}$ with $\|y - y^\delta\| \leq \delta$ and let $x^\dagger$ be an x^*-minimum-norm solution. Moreover, let the following conditions hold:*

(i) F is Fréchet-differentiable,

(ii) there exists $\gamma \geq 0$ such that $\|F'(x^\dagger) - F'(x)\| \leq \gamma \|x^\dagger - x\|$ for all $x \in \mathcal{D}(F)$ in a sufficiently large ball around $x^\dagger$,

(iii) there exists $w \in \mathcal{Y}$ satisfying $x^\dagger - x^* = F'(x^\dagger)^* w$ and

(iv) $\gamma \|w\| < 1$.

Then for the choice $\alpha \sim \delta$, we obtain

$$\|x_\alpha^\delta - x^\dagger\| = O(\sqrt{\delta}) \quad \text{and} \quad \|F(x_\alpha^\delta) - y^\delta\| = O(\delta).$$

Proof. Since x_α^δ is a minimizer of (10.3), $F(x^\dagger) = y$ and $\|y - y^\delta\| \leq \delta$ imply that

$$\|F(x_\alpha^\delta) - y^\delta\|^2 + \alpha \|x_\alpha^\delta - x^*\|^2 \leq \delta^2 + \alpha \|x^\dagger - x^*\|^2$$

and hence that

$$\begin{aligned}
\|F(x_\alpha^\delta) - y^\delta\|^2 + \alpha \|x_\alpha^\delta - x^\dagger\|^2 &\leq \delta^2 + \alpha(\|x^\dagger - x^*\|^2 + \|x_\alpha^\delta - x^\dagger\|^2 \\
&\qquad - \|x_\alpha^\delta - x^*\|^2) \qquad (10.10) \\
&= \delta^2 + 2\alpha \langle x^\dagger - x^*, x^\dagger - x_\alpha^\delta \rangle.
\end{aligned}$$

246   10. Tikhonov Regularization of Nonlinear Problems

Let now $\alpha \sim \delta$. Then (10.10) implies that $x_\alpha^\delta \in \mathcal{B}_\rho(x^\dagger)$ for any fixed $\rho > 2\|x^\dagger - x^*\|$, provided δ is sufficiently small, which we assume to hold in the following. Therefore, condition (ii) is applicable assuming that it holds for all $x \in \mathcal{D}(F) \cap \mathcal{B}_\rho(x^\dagger)$. Note that conditions (i) and (ii) imply that

$$F(x_\alpha^\delta) = F(x^\dagger) + F'(x^\dagger)(x_\alpha^\delta - x^\dagger) + r_\alpha^\delta \tag{10.11}$$

holds with
$$\|r_\alpha^\delta\| \leq \frac{\gamma}{2}\|x_\alpha^\delta - x^\dagger\|^2. \tag{10.12}$$

By condition (iii), (10.10) implies that

$$\|F(x_\alpha^\delta) - y^\delta\|^2 + \alpha\|x_\alpha^\delta - x^\dagger\|^2 \leq \delta^2 + 2\alpha\langle w, F'(x^\dagger)(x^\dagger - x_\alpha^\delta)\rangle. \tag{10.13}$$

Combining (10.11) – (10.13) leads to

$$\begin{aligned}\|F(x_\alpha^\delta) - y^\delta\|^2 + \alpha\|x_\alpha^\delta - x^\dagger\|^2 &\leq \delta^2 + 2\alpha\langle w, (y - y^\delta) + (y^\delta - F(x_\alpha^\delta)) + r_\alpha^\delta\rangle \\ &\leq \delta^2 + 2\alpha\delta\|w\| + 2\alpha\|w\|\,\|F(x_\alpha^\delta) - y^\delta\| \\ &\quad + \alpha\gamma\|w\|\,\|x_\alpha^\delta - x^\dagger\|^2\end{aligned}$$

and hence to

$$(\|F(x_\alpha^\delta) - y^\delta\| - \alpha\|w\|)^2 + \alpha(1 - \gamma\|w\|)\|x_\alpha^\delta - x^\dagger\|^2 \leq (\delta + \alpha\|w\|)^2.$$

This together with (iv) implies that

$$\|F(x_\alpha^\delta) - y^\delta\| \leq \delta + 2\alpha\|w\|$$

and that

$$\|x_\alpha^\delta - x^\dagger\| \leq \frac{\delta + \alpha\|w\|}{\sqrt{\alpha}(1 - \gamma\|w\|)^{\frac{1}{2}}}.$$

The assertion now follows with $\alpha \sim \delta$. ∎

Remark 10.5. Even if the x^*-minimum-norm solution is not unique, it is a consequence of Theorem 10.4 that only one x^*-minimum-norm solution $x^\dagger$ can fulfill conditions (ii) – (iv). We have seen from the proof of Theorem 10.4 that the radius ρ of the sufficiently large ball around $x^\dagger$ in condition (ii) has to be larger than $2\|x^\dagger - x^*\|$. If $x^\dagger$ is unique, then the convergence of x_α^δ towards $x^\dagger$ follows already from Theorem 10.3. Therefore, $\rho > 2\|x^\dagger - x^*\|$ can then be relaxed to any $\rho > 0$.

Note that the weak closedness of F is not needed for the convergence rate result; it was used only to obtain existence, stability, and convergence of the regularized solutions.

If F is twice Fréchet-differentiable, conditions (ii) and (iv) may be replaced by the weaker condition

(ii)' $2\langle w, \int_0^1 (1-t)F''[x^\dagger + t(x_\alpha^\delta - x^\dagger)](x_\alpha^\delta - x^\dagger)^2\,dt\rangle \leq \gamma\|x_\alpha^\delta - x^\dagger\|^2$ with $\gamma < 1$.

10.2. Convergence Analysis

To see this, note that the left-hand side of (ii)' equals $2\langle w, r_\alpha^\delta \rangle$ with r_α^δ as in (10.11).

Remark 10.6. In the linear case, the results about convergence and convergence rates for attainable right-hand sides y also hold under the same conditions for the case of non-attainability, where problem (10.1) is solved in the least-squares sense. For nonlinear problems the case that (10.1) is solvable only in the least-squares sense was handled in [27, 194]. It turns out that the condition that $\delta^2/\alpha \to 0$ has to be replaced by the stronger requirement that δ/α is bounded to guarantee the convergence towards a least-squares solution and that it has to be replaced by the condition that $\delta/\alpha \to 0$ or the condition that the metric projection of y onto $\mathcal{R}(F)$ is unique to guarantee convergence towards an x^*-minimum-norm-least-squares solution. The convergence rate $O(\sqrt{\delta})$ can be obtained if in addition to the conditions of Theorem 10.4 the metric projection of y onto $\mathcal{R}(F)$ is unique and if the boundary of $\mathcal{R}(F)$ is convex locally around $F(x^\dagger)$.

In the linear case, the best possible convergence rate for $\|x_\alpha^\delta - x^\dagger\|$ is $O(\delta^{\frac{2}{3}})$, and this rate is obtained under the condition $x^\dagger - x^* \in \mathcal{R}(F^*F)$ (see Section 5.1). We show next that the same is true in the nonlinear case if we replace $\mathcal{R}(F^*F)$ by $\mathcal{R}(F'(x^\dagger)^*F'(x^\dagger))$. With some $\mu \in [1/2, 1]$, it is even possible to prove results for $x^\dagger - x^* \in \mathcal{R}((F'(x^\dagger)^*F'(x^\dagger))^\mu)$. The proof we give differs from the original one in [217], the conditions there can be even slightly relaxed. It rather uses an idea from [157, 268], namely to compare x_α^δ with an element different from $x^\dagger$ as opposed to the proof of Theorem 10.4.

Theorem 10.7. *Under the assumptions of Theorem 10.4, if $x^\dagger$ is an element in the interior of $\mathcal{D}(F)$ and if $x^\dagger - x^* = (F'(x^\dagger)^*F'(x^\dagger))^\mu v$ for some $\mu \in [1/2, 1]$, then, for the choice $\alpha \sim \delta^{\frac{2}{2\mu+1}}$, we obtain*

$$\|x_\alpha^\delta - x^\dagger\| = O(\delta^{\frac{2\mu}{2\mu+1}}).$$

Proof. The main idea of the proof is to compare x_α^δ with

$$z_\alpha := x^\dagger - \alpha(F'(x^\dagger)^*F'(x^\dagger) + \alpha I)^{-1}F'(x^\dagger)^*w \tag{10.14}$$

with w as in condition (iii) of Theorem 10.4. Since $\|z_\alpha - x^\dagger\| \leq \sqrt{\alpha}\|w\|$ and $x^\dagger$ is an element in the interior of $\mathcal{D}(F)$, we have that $z_\alpha \in \mathcal{D}(F)$ for α sufficiently small, which we assume in the following. Conditions (i) and (ii) of Theorem 10.4 imply similarly to (10.11) and (10.12) that

$$F(z_\alpha) = F(x^\dagger) + F'(x^\dagger)(z_\alpha - x^\dagger) + s_\alpha \tag{10.15}$$

holds with

$$\|s_\alpha\| \leq \frac{\gamma}{2}\|z_\alpha - x^\dagger\|^2. \tag{10.16}$$

From the minimization property (10.3) it follows that

$$\|F(x_\alpha^\delta) - y^\delta\|^2 + \alpha\|x_\alpha^\delta - x^*\|^2 \le \|F(z_\alpha) - y^\delta\|^2 + \alpha\|z_\alpha - x^*\|^2.$$

Together with (10.11), (10.15) and condition (iii) of Theorem 10.4 we now obtain the estimate

$$\begin{aligned}\|x_\alpha^\delta - x^\dagger\|^2 &\le \tfrac{1}{\alpha}(\|F(z_\alpha) - y^\delta\|^2 - \|F(x_\alpha^\delta) - y^\delta\|^2) + \|z_\alpha - x^\dagger\|^2 \\ &\quad + 2\langle z_\alpha - x^\dagger, x^\dagger - x^*\rangle - 2\langle x_\alpha^\delta - x^\dagger, x^\dagger - x^*\rangle \\ &= 2\langle r_\alpha^\delta, w\rangle - \tfrac{1}{\alpha}\|F(x_\alpha^\delta) - y^\delta + \alpha w\|^2 \\ &\quad + \|z_\alpha - x^\dagger\|^2 + \tfrac{1}{\alpha}\|s_\alpha + y - y^\delta\|^2 \\ &\quad + \langle 2F'(x^\dagger)(z_\alpha - x^\dagger) + \alpha w, w\rangle + \tfrac{1}{\alpha}\|F'(x^\dagger)(z_\alpha - x^\dagger)\|^2 \\ &\quad + 2\langle y - y^\delta, w\rangle + \tfrac{2}{\alpha}\langle s_\alpha + y - y^\delta, F'(x^\dagger)(z_\alpha - x^\dagger)\rangle.\end{aligned} \quad (10.17)$$

Using (10.12), (10.16) and (10.14), we obtain

$$2\langle r_\alpha^\delta, w\rangle \le \gamma\|w\|\,\|x_\alpha^\delta - x^\dagger\|^2,$$

$$\tfrac{1}{\alpha}\|s_\alpha + y - y^\delta\|^2 \le \tfrac{2}{\alpha}(\tfrac{\gamma^2}{4}\|z_\alpha - x^\dagger\|^4 + \delta^2),$$

$$\langle 2F'(x^\dagger)(z_\alpha - x^\dagger) + \alpha w, w\rangle + \tfrac{1}{\alpha}\|F'(x^\dagger)(z_\alpha - x^\dagger)\|^2 = \alpha^3\|(F'(x^\dagger)F'(x^\dagger)^* + \alpha I)^{-1}w\|^2,$$

and

$$2\langle y - y^\delta, w\rangle + \tfrac{2}{\alpha}\langle s_\alpha + y - y^\delta, F'(x^\dagger)(z_\alpha - x^\dagger)\rangle$$
$$\le \gamma\|w\|\,\|z_\alpha - x^\dagger\|^2 + 2\delta\alpha\|(F'(x^\dagger)F'(x^\dagger)^* + \alpha I)^{-1}w\|.$$

Combining these estimates with (10.17) and condition (iv) of Theorem 10.4 yields

$$\|x_\alpha^\delta - x^\dagger\|^2 = O(\|z_\alpha - x^\dagger\|^2 + \tfrac{1}{\alpha}\|z_\alpha - x^\dagger\|^4 + \alpha^3\|(F'(x^\dagger)F'(x^\dagger)^* + \alpha I)^{-1}w\|^2 + \tfrac{\delta^2}{\alpha}).$$

This together with $F'(x^\dagger)^*w = (F'(x^\dagger)^*F'(x^\dagger))^\mu v$ and spectral theory implies that

$$\|x_\alpha^\delta - x^\dagger\|^2 = O(\alpha^{2\mu}e_\alpha + \tfrac{\delta^2}{\alpha})$$

with

$$e_\alpha := \int_0^\infty \frac{\lambda^{2\mu}\alpha^{2-2\mu} + \lambda^{2\mu-1}\alpha^{3-2\mu}}{(\lambda+\alpha)^2}\,d\|E_\lambda v\|^2,$$

where $\{E_\lambda\}$ is the spectral family of $F'(x^\dagger)^*F'(x^\dagger)$. Since $e_\alpha = o(1)$ for $1/2 \le \mu < 1$ and $O(1)$ for $\mu = 1$, the assertion follows for the choice $\alpha \sim \delta^{\frac{2}{2\mu+1}}$. ∎

Remark 10.8. It can be seen from the proof above that for an appropriate parameter choice, one can even obtain the rate $o(\delta^{\frac{2\mu}{2\mu+1}})$ for $\mu < 1$.

In [219], a result for a parameter estimation problem was presented, where the rate $O(\delta^{\frac{2}{3}})$ was obtained without the assumption that the exact solution $x^\dagger$ has to be in the interior of $\mathcal{D}(F)$.

It was shown in [39] that, as for constrained linear problems (cf. Theorem 5.19), the result of Theorem 10.4 remains valid if condition (iii) is replaced by

$$x^\dagger = P_{\mathcal{D}(F)}(x^* + F'(x^\dagger)^*w),$$

where $P_{\mathcal{D}(F)}$ is the metric projector onto $\mathcal{D}(F)$.

We have already mentioned that in the linear case $O(\delta^{\frac{2}{3}})$ is the best possible rate if $\alpha \sim \delta^{\frac{2}{3}}$, except for the trivial case that $x^\dagger = x^*$. It is also known that the condition $x^\dagger - x^* \in \mathcal{R}(F^*F)$ is necessary and sufficient for the rate $O(\delta^{\frac{2}{3}})$. Analogous assertions also hold in the nonlinear case (see [217, Theorem 2.7]).

Finally, we want to mention that in computational reality it is, in general, not possible to solve the nonlinear minimization problem (10.3) exactly. So instead of solving (10.3) one is really looking for a solution $x_\alpha^{\delta,\eta}$ of the problem

$$\|F(x_\alpha^{\delta,\eta}) - y^\delta\|^2 + \alpha\|x_\alpha^{\delta,\eta} - x^*\|^2 \leq \|F(x) - y^\delta\|^2 + \alpha\|x - x^*\|^2 + \eta$$

for all $x \in \mathcal{D}(F)$, where $\eta \geq 0$ is a small parameter. All results concerning convergence and convergence rates remain valid for this approach [74, 217, 219, 253], as long as it is guaranteed that $\eta/\alpha \to 0$. Of course, for nonlinear minimization problems it is in general not easy to find global minimizers. This question was treated in [38].

The combination of Tikhonov regularization with finite-dimensional approximation of the space $\mathcal{X}$ and of the operator F was treated in [169, 219].

10.3. A-posteriori Parameter Choice Rules

As in the linear case, one wants to choose the regularization parameter α such that the error $\|x_\alpha^\delta - x^\dagger\|$ is as small as possible. Unfortunately, in the a-priori parameter selection strategies suggested in the theorems of the last section, the regularization parameter depends on the smoothness information about the exact solution $x^\dagger$. Since, in general, this information is not available, these strategies are of little use in practice. It was shown in Sections 4.3 and 4.4 that for linear problems there are a-posteriori parameter selection methods which always yield the optimal rates without needing any information about the exact solution.

It follows from the proof of Theorem 10.4 that Morozov's discrepancy principle, where $\alpha(\delta)$ is determined as the solution of

$$\|F(x_\alpha^\delta) - y^\delta\| = \delta, \tag{10.18}$$

yields the rate $O(\sqrt{\delta})$ under the assumptions of that theorem *if this $\alpha(\delta)$ exists*. However, for nonlinear problems, (10.18) has only a solution under very restrictive assumptions on the regularized solutions (cf. [163]). Even then, one would have to solve two nonlinear problems simultaneously.

One can show that under certain conditions the problem

$$\|x - x^*\| \to \min, \quad x \in \mathcal{M}^\delta := \{x \in \mathcal{D}(F) \mid \|F(x) - y^\delta\| \leq \delta\} \tag{10.19}$$

has a stable solution x^δ and that, under the conditions of Theorem 10.4, we obtain the rate $\|x^\delta - x^\dagger\| = O(\sqrt{\delta})$. However, this problem is still complicated to solve. Therefore, we consider the Lagrange multiplier formulation of the constrained optimization problem (10.19) which leads to the following modified Tikhonov regularization ($1/\alpha$ plays the role of the Lagrange multiplier):

$$(\|F(x) - y^\delta\| - \delta)^2 + \alpha \|x - x^*\|^2 \to \min, \quad x \in \mathcal{D}(F). \tag{10.20}$$

One can show as in the section above that this problem has a stable solution, again denoted by x_α^δ for all $\alpha > 0$, and that these regularized solutions converge towards an x^*-minimum-norm solution in the sense of Theorem 10.3 if $\alpha(\delta) = O(\delta^2)$. We will show in the next proposition that with this a-priori parameter choice the rate $O(\sqrt{\delta})$ is obtained if the conditions of Theorem 10.4 are satisfied. Note that for this parameter choice no information about the exact solution $x^\dagger$ is needed and that only one nonlinear problem has to be solved. For the proof of the proposition we need the following lemma.

Lemma 10.9. *Let $\mathcal{D}(F)$ be convex, let $y^\delta \in \mathcal{Y}$ with $\|y - y^\delta\| \leq \delta$ and assume that $x^* \in \mathcal{D}(F)$ with $F(x^*) \neq y$. Then problem (10.19) has a solution x^δ with*

$$\|F(x^\delta) - y^\delta\| = \delta$$

$\delta > 0$ *sufficiently small.*

Proof. Since $x^\dagger \in \mathcal{M}^\delta$, $\mathcal{M}^\delta$ is not empty. The existence of a minimizing element x^δ now follows from the weak closedness of F and the weak lower semicontinuity of the norm.

Let us now assume that $\delta < \|F(x^*) - y\|$. Then $x^* \notin \mathcal{M}^\delta$. We will prove that $\|F(x^\delta) - y^\delta\| = \delta$ by assuming the contrary, namely that $\|F(x^\delta) - y^\delta\| < \delta$. Since $\mathcal{D}(F)$ is convex, $x(t) := tx^* + (1-t)x^\delta \in \mathcal{D}(F)$ for all $t \in [0,1]$. By the continuity of F, this implies that there is a $t \in (0,1]$ with $x(t) \in \mathcal{M}^\delta$. Moreover,

$$\|x(t) - x^*\| = (1-t)\|x^\delta - x^*\| < \|x^\delta - x^*\|,$$

which is a contradiction to x^δ being a minimizer of $\|x - x^*\|$ on $\mathcal{M}^\delta$. ■

The assumption that $x^* \in \mathcal{D}(F)$ is quite natural, since x^* is an initial guess for $x^\dagger \in \mathcal{D}(F)$; $F(x^*) \neq y$ just means that x^* is not a solution.

Proposition 10.10. *Let all conditions of Theorem 10.4 hold and assume that $x^* \in \mathcal{D}(F)$ with $F(x^*) \neq y$. If x_α^δ is a solution of problem (10.20), then for the*

10.3. A-posteriori Parameter Choice Rules

choice $\alpha = \alpha(\delta) = O(\delta^2)$, we obtain

$$\|x_\alpha^\delta - x^\dagger\| = O(\sqrt{\delta}) \quad \text{and} \quad \|F(x_\alpha^\delta) - y^\delta\| = O(\delta).$$

Proof. Let x^δ be as in Lemma 10.9 with $\delta > 0$ so small that $\|F(x^\delta) - y^\delta\| = \delta$. Since x_α^δ is a minimizer of (10.20), we obtain that

$$(\|F(x_\alpha^\delta) - y^\delta\| - \delta)^2 + \alpha\|x_\alpha^\delta - x^*\|^2 \leq \alpha\|x^\delta - x^*\|^2 \leq \alpha\|x^\dagger - x^*\|^2.$$

This together with $\alpha = \alpha(\delta) = O(\delta^2)$ and $\|y - y^\delta\| \leq \delta$ implies that

$$\|F(x_\alpha^\delta) - y\| = O(\delta). \tag{10.21}$$

Moreover, we obtain similarly to the proof of Theorem 10.4 that

$$(\|F(x_\alpha^\delta) - y^\delta\| - \delta)^2 + \alpha\|x_\alpha^\delta - x^\dagger\|^2 \leq \alpha\|w\|(2\delta + 2\|F(x_\alpha^\delta) - y^\delta\| + \gamma\|x_\alpha^\delta - x^\dagger\|^2)$$

and hence that

$$(1 - \gamma\|w\|)\|x_\alpha^\delta - x^\dagger\|^2 \leq 2\|w\|(\delta + \|F(x_\alpha^\delta) - y^\delta\|),$$

which together with (10.21) implies the assertion. ∎

It is remarkable that no lower bound on the parameter α is necessary as for ordinary Tikhonov regularization.

Like Morozov's discrepancy principle the approach above will not yield the optimal rate $O(\delta^{\frac{2}{3}})$ if the assumptions of Theorem 10.7 are satisfied.

In [246], a generalization of the parameter selection method (5.28) to nonlinear problems was presented that yields the optimal rate $O(\delta^{\frac{2}{3}})$ under the assumptions of Theorem 10.7 if the nonlinear operator F satisfies some additional conditions.

In the linear case, the parameter selection criterion (5.28) is based on minimizing the functional

$$\frac{1}{2}\|x_\alpha - x^\dagger\|^2 + c\frac{\delta^2}{\alpha}, \tag{10.22}$$

where c is a fixed positive constant, since one can show that

$$\|x_\alpha^\delta - x^\dagger\|^2 = O\left(\frac{1}{2}\|x_\alpha - x^\dagger\|^2 + c\frac{\delta^2}{\alpha}\right).$$

For nonlinear problems, this estimate is only valid if F satisfies certain conditions (see below). We will now show in a more heuristic way how the parameter choice (5.28) may be generalized to the nonlinear case. We will assume that F is twice continuously Fréchet-differentiable.

The first order necessary condition for the minimum of the functional in (10.22) is

$$\alpha^2 \langle \frac{dx_\alpha}{d\alpha}, x_\alpha - x^\dagger \rangle = c\delta^2. \tag{10.23}$$

10. Tikhonov Regularization of Nonlinear Problems

Since x_α is a minimizer of (10.3) (with y^δ replaced by y), it satisfies the first order necessary condition

$$F'(x_\alpha)^*(F(x_\alpha) - y) + \alpha(x_\alpha - x^*) = 0. \tag{10.24}$$

A formal differentiation of this equation with respect to α yields

$$(F'(x_\alpha)^*)'(F(x_\alpha) - y, \frac{dx_\alpha}{d\alpha}) + F'(x_\alpha)^* F'(x_\alpha) \frac{dx_\alpha}{d\alpha} + \alpha \frac{dx_\alpha}{d\alpha} = x^* - x_\alpha.$$

If we neglect the second derivative term in this equation, we obtain the approximation

$$\frac{dx_\alpha}{d\alpha} \approx (F'(x_\alpha)^* F'(x_\alpha) + \alpha I)^{-1}(x^* - x_\alpha). \tag{10.25}$$

Combining (10.23) – (10.25) and using the approximation

$$F(x_\alpha) - y \approx F'(x_\alpha)(x_\alpha - x^\dagger)$$

yields

$$\begin{aligned}
c\delta^2 &\approx \alpha^2 \langle (F'(x_\alpha)^* F'(x_\alpha) + \alpha I)^{-1}(x^* - x_\alpha), x_\alpha - x^\dagger \rangle \\
&= \alpha \langle (F'(x_\alpha) F'(x_\alpha)^* + \alpha I)^{-1}(F(x_\alpha) - y), F'(x_\alpha)(x_\alpha - x^\dagger) \rangle \\
&\approx \alpha \langle (F'(x_\alpha) F'(x_\alpha)^* + \alpha I)^{-1}(F(x_\alpha) - y), F(x_\alpha) - y \rangle.
\end{aligned}$$

If we now replace y by y^δ and x_α by x_α^δ, we arrive at the following implementable strategy

$$\alpha \langle (F'(x_\alpha^\delta) F'(x_\alpha^\delta)^* + \alpha I)^{-1}(F(x_\alpha^\delta) - y^\delta), F(x_\alpha^\delta) - y^\delta \rangle = c\delta^2. \tag{10.26}$$

Note that for a linear operator F and $x^* = 0$ this strategy coincides with (5.28).

It was shown in [246] that (10.26) always has a solution provided that c is chosen large enough, that $\|x^\dagger - x^*\|$ is small enough, and that F is not *too nonlinear*. More precisely, F has to satisfy the following conditions: F is twice Fréchet-differentiable and there exist constants $c_1, c_2, c_3 > 0$ such that for all $w, x, z \in \mathcal{X}$ and $y \in \mathcal{Y}$ there are elements $l(x, y, z) \in \mathcal{Y}$ and $m(x, z, F'(x)w), n(x, z, F'(x)w) \in \mathcal{X}$ satisfying

$$\begin{aligned}
(F'(z)^* - F'(x)^*)y &= F'(x)^* l(x, z, y) \\
(F'(z) - F'(x))w &= F'(x) m(x, z, F'(x)w) \\
(F'(x)^* - F'(z)^*)F'(x)w &= F'(x)^* F'(x) n(x, z, F'(x)w)
\end{aligned}$$

with

$$\begin{aligned}
\|l(x, z, y)\| &\le c_1 \|x - z\| \|y\| \\
\|m(x, z, F'(x)w)\| &\le c_2 \|x - z\| \|F'(x)w\| \\
\|n(x, z, F'(x)w)\| &\le c_3 \|x - z\| \|F'(x)w\|.
\end{aligned}$$

Moreover, these conditions guarantee convergence properties similar to (5.31), especially the parameter selection method (10.26) yields the optimal rate $O(\delta^{\frac{2}{3}})$ if the conditions in Theorem 10.7 are satisfied for $\mu = 1$.

10.4. Regularization in Hilbert Scales

We have seen in Section 8.5 for linear ill-posed problems that one can obtain better rates than $O(\delta^{\frac{2}{3}})$ with Tikhonov regularization if the exact solution is smooth enough and if the regularizing norm in $\mathcal{X}$ is replaced by a stronger one. We shall show in this section that this also holds for Tikhonov regularization of nonlinear problems. A big advantage of this approach is that rates are obtained by merely requiring smoothness conditions for the exact solution as in the linear case and not smoothness and closeness conditions that are necessary for ordinary Tikhonov regularization of nonlinear problems (cf. Theorems 10.4 and 10.7).

Let $(\mathcal{X}_s)_{s \in \mathbb{R}}$ be a Hilbert scale induced by a densely defined unbounded selfadjoint strictly positive operator L on $\mathcal{X}$, i.e., $\mathcal{X}_s$ is the completion of $\bigcap_{k=0}^{\infty} \mathcal{D}(L^k)$ with respect to the Hilbert space norm $\|x\|_s := \|L^s x\|$ (see Section 8.4). Moreover, we assume that the operator F and the x^*-minimum-norm solution $x^\dagger$ satisfy the following assumptions.

Assumption 10.11.

(i) $\mathcal{D}(F)$ is convex, and F is continuous and Fréchet-differentiable on $\mathcal{X}$ and weakly (sequentially) closed on $\mathcal{X}_s$ for some $s \geq 0$ (not necessarily on $\mathcal{X}$!),

(ii) a unique x^*-minimum-norm solution $x^\dagger$ exists in $\mathcal{X}_s$, i.e., $x^\dagger \in \mathcal{D}(F) \cap \mathcal{X}_s$, $F(x^\dagger) = y$ and $\|x^\dagger - x^*\|_s = \min\{\|x - x^*\|_s \mid F(x) = y\}$,

(iii) there exist $\varepsilon > 0$ and $\gamma > 0$ such that $\|F'(x^\dagger) - F'(x)\| \leq \gamma \|x^\dagger - x\|_{\mathcal{X}}$ for all $x \in \mathcal{D}(F) \cap \mathcal{B}_\varepsilon(x^\dagger)$,

(iv) $\|F'(x^\dagger) x\| \sim \|x\|_{-a}$ for all $x \in \mathcal{X}$ for some $a > 0$.

As in the linear case, the number a in Assumption 10.11 (iv) can be interpreted as a *degree of ill-posedness* of the linearized problem in $x^\dagger$. Instead of regularizing problem (10.1) via the solution of (10.3), we now approximate $x^\dagger$ by a solution x_α^δ of the regularized problem

$$\|F(x) - y^\delta\|^2 + \alpha \|x - x^*\|_s^2 \to \min, \quad x \in \mathcal{D}(F) \cap \mathcal{X}_s. \qquad (10.27)$$

One can show that stability and convergence in $\mathcal{X}_s$ hold under similar conditions as in Theorems 10.2 and 10.3, respectively.

We will show in the next theorem that convergence rates are obtained in $\mathcal{X}$ under the smoothness condition

$$x^\dagger - x^* \in \mathcal{X}_u \quad \text{for some } u \in [s, a+2s], \ s \geq a. \qquad (10.28)$$

Theorem 10.12. *Let $y^\delta \in \mathcal{Y}$ with $\|y - y^\delta\| \leq \delta$, and let Assumption 10.11 and (10.28) hold. Then for the choice $\alpha \sim \delta^{\frac{2(a+s)}{a+u}}$ we obtain*

$$\|x_\alpha^\delta - x^\dagger\| = O(\delta^{\frac{u}{a+u}}).$$

254   10. Tikhonov Regularization of Nonlinear Problems

Proof. Let us assume that $\alpha \sim \delta^{\frac{2(a+s)}{a+u}}$. Since for this choice $\delta^2/\alpha \to 0$ as $\delta \to 0$, it follows as in Theorem 10.3 that

$$x_\alpha^\delta \to x^\dagger \quad \text{in } \mathcal{X}_s \quad \text{as} \quad \delta \to 0. \tag{10.29}$$

Similarly to the proof of Theorem 10.4 we obtain that

$$\|F(x_\alpha^\delta) - y^\delta\|^2 + \alpha \|x_\alpha^\delta - x^\dagger\|_s^2 = O(\delta^2 + \alpha \langle x^\dagger - x^*, x^\dagger - x_\alpha^\delta \rangle_s)$$

and that

$$F(x_\alpha^\delta) = F(x^\dagger) + F'(x^\dagger)(x_\alpha^\delta - x^\dagger) + r_\alpha^\delta$$

with

$$\|r_\alpha^\delta\| \le \frac{\gamma}{2} \|x_\alpha^\delta - x^\dagger\|^2 \quad \text{for } \delta > 0 \text{ sufficiently small}.$$

This together with Assumption 10.11 (ii), (iv), (10.28), and

$$\|F(x_\alpha^\delta) - y^\delta\|^2 = \|y - y^\delta + r_\alpha^\delta\|^2 + \|F'(x^\dagger)(x_\alpha^\delta - x^\dagger)\|^2$$
$$+ 2\langle F'(x^\dagger)(x_\alpha^\delta - x^\dagger), y - y^\delta + r_\alpha^\delta \rangle$$

implies that

$$\|x_\alpha^\delta - x^\dagger\|_{-a}^2 + \alpha \|x_\alpha^\delta - x^\dagger\|_s^2 \tag{10.30}$$
$$= O(\delta^2 + \alpha \|x_\alpha^\delta - x^\dagger\|_{2s-u} + \|x_\alpha^\delta - x^\dagger\|_{-a}(\delta + \|x_\alpha^\delta - x^\dagger\|^2)).$$

Since $\|\cdot\| = \|\cdot\|_0$, an application of the interpolation inequality (8.23) (with $q = -a$ and $r = 0$) together with (10.29) and $s \ge a$ shows that

$$\|x_\alpha^\delta - x^\dagger\|_{-a} \|x_\alpha^\delta - x^\dagger\|^2 \le \|x_\alpha^\delta - x^\dagger\|_{-a}^{\frac{a+3s}{a+s}} \|x_\alpha^\delta - x^\dagger\|_s^{\frac{2a}{a+s}} = o(\|x_\alpha^\delta - x^\dagger\|_{-a}^2).$$

This together with (10.30) and (8.23) (with $q = -a$ and $r = 2s - u$) implies that

$$\|x_\alpha^\delta - x^\dagger\|_{-a}^2 + \alpha \|x_\alpha^\delta - x^\dagger\|_s^2 \tag{10.31}$$
$$= O(\delta^2 + \delta \|x_\alpha^\delta - x^\dagger\|_{-a} + \alpha \|x_\alpha^\delta - x^\dagger\|_{-a}^{\frac{u-s}{a+s}} \|x_\alpha^\delta - x^\dagger\|_s^{\frac{a+2s-u}{a+s}}).$$

For the following analysis we use the estimate

$$c^p \le e + dc^q \Rightarrow c^p = O(e + d^{\frac{p}{p-q}}), \tag{10.32}$$

which holds for all $c, d, e \ge 0$ and $p > q \ge 0$.

Using (10.32) with

$$c = \|x_\alpha^\delta - x^\dagger\|_{-a}, \qquad p = 2,$$
$$d = \alpha \|x_\alpha^\delta - x^\dagger\|_s^{\frac{a+2s-u}{a+s}}, \qquad q = \frac{u-s}{a+s},$$
$$e = \delta^2 + \delta \|x_\alpha^\delta - x^\dagger\|_{-a},$$

and (10.31) (with the second term on its left-hand side omitted), we obtain that

$$\|x_\alpha^\delta - x^\dagger\|_{-a}^2 = O(\delta^2 + \delta \|x_\alpha^\delta - x^\dagger\|_{-a} + \alpha^{\frac{2(a+s)}{2a+3s-u}} \|x_\alpha^\delta - x^\dagger\|_s^{\frac{2(a+2s-u)}{2a+3s-u}}).$$

A further application of (10.32) with

$$c = \|x_\alpha^\delta - x^\dagger\|_{-a}, \qquad p = 2,$$
$$d = \delta, \qquad q = 1,$$
$$e = \delta^2 + \alpha^{\frac{2(a+s)}{2a+3s-u}} \|x_\alpha^\delta - x^\dagger\|_s^{\frac{2(a+2s-u)}{2a+3s-u}},$$

and the fact that $\sqrt{f+g} \leq \sqrt{f} + \sqrt{g}$ for all $f, g \geq 0$, yields

$$\|x_\alpha^\delta - x^\dagger\|_{-a} = O(\delta + \alpha^{\frac{a+s}{2a+3s-u}} \|x_\alpha^\delta - x^\dagger\|_s^{\frac{a+2s-u}{2a+3s-u}}). \qquad (10.33)$$

This together with (10.31) (with the first term on its left-hand side omitted) implies

$$\|x_\alpha^\delta - x^\dagger\|_s^2 = O(\delta^2 \alpha^{-1} + \delta \alpha^{\frac{u-a-2s}{2a+3s-u}} \|x_\alpha^\delta - x^\dagger\|_s^{\frac{a+2s-u}{2a+3s-u}}$$
$$+ \delta^{\frac{u-s}{a+s}} \|x_\alpha^\delta - x^\dagger\|_s^{\frac{a+2s-u}{a+s}} + \alpha^{\frac{u-s}{2a+3s-u}} \|x_\alpha^\delta - x^\dagger\|_s^{\frac{2a+4s-2u}{2a+3s-u}}).$$

We now apply (10.32) for three more times, where the role of the term dc^q in (10.32) is played by the second, third and fourth term in the estimate above, respectively. The appropriate settings of c, d, e, p, q are as follows: $c = \|x_\alpha^\delta - x^\dagger\|_s$, $p = 2$ and e equals always the rest of the terms in the estimate; d, q vary depending on the application, namely

$$d = \delta \alpha^{\frac{u-a-2s}{2a+3s-u}}, \qquad q = \frac{a+2s-u}{2a+3s-u}$$

if (10.32) is applied to the second term,

$$d = \delta^{\frac{u-s}{a+s}}, \qquad q = \frac{a+2s-u}{a+s}$$

if (10.32) is applied to the third term, and

$$d = \alpha^{\frac{u-s}{2a+3s-u}}, \qquad q = \frac{2a+4s-2u}{2a+3s-u}$$

if (10.32) is applied to the fourth term. So we finally arrive at

$$\|x_\alpha^\delta - x^\dagger\|_s = O(\delta \alpha^{-\frac{1}{2}} + \delta^{\frac{2a+3s-u}{3a+4s-u}} \alpha^{\frac{u-a-2s}{3a+4s-u}} + \delta^{\frac{u-s}{a+u}} + \alpha^{\frac{u-s}{2a+2s}}).$$

By the proposed choice $\alpha \sim \delta^{\frac{2a+2s}{a+u}}$ we obtain the rate

$$\|x_\alpha^\delta - x^\dagger\|_s = O(\delta^{\frac{u-s}{a+u}})$$

and hence with (10.33)

$$\|x_\alpha^\delta - x^\dagger\|_{-a} = O(\delta).$$

Finally, (8.23) (with $q = -a$ and $r = 0$) implies the assertion. ∎

Remark 10.13. Following the proof of Theorem 10.12, we see that, by the interpolation inequality (8.23), even the general convergence rates result

$$\|x_\alpha^\delta - x^\dagger\|_t = O(\delta^{\frac{u-t}{a+u}}) \quad \text{for all } -a \le t \le s$$

holds. In the linear case, we have shown that the result of Theorem 10.12 also holds for $u < s$ (cf. Theorem 8.23).

Of course, the parameter choice in Theorem 10.12 is not realizable, since usually the smoothness parameter u is not known. In Section 10.3 we proposed a variant of Tikhonov regularization, where no information about the exact solution was needed and where the rate $O(\sqrt{\delta})$ was obtained under the assumptions of Theorem 10.4. One can show that an analogous variant for Tikhonov regularization in Hilbert scales yields the rates of Theorem 10.12.

Theorem 10.14. *Let $y^\delta \in \mathcal{Y}$ with $\|y - y^\delta\| \le \delta$, and let Assumption 10.11 and (10.28) hold, and assume that $x^* \in \mathcal{D}(F)$ with $F(x^*) \ne y$. Moreover, let x_α^δ be a solution of the minimization problem*

$$(\|F(x) - y^\delta\| - \delta)^2 + \alpha\|x - x^*\|_s^2 \to \min, \quad x \in \mathcal{D}(F) \cap \mathcal{X}_s.$$

Then for the choice $\alpha = O(\delta^2)$ we obtain

$$\|x_\alpha^\delta - x^\dagger\| = O(\delta^{\frac{u}{a+u}}).$$

Proof. Similar to the proofs of Proposition 10.10 and Theorem 10.12. ∎

Remark 10.15. The a-priori parameter choice rule in Theorem 10.14 is independent of u and still yields order optimality for all $u \in [s, a+2s]$, $s \ge a$.

Note also that in this variant of Tikhonov regularization it is only necessary to solve just one nonlinear minimization problem and not a sequence of nonlinear problems, as it would be necessary if the regularization parameter α in (10.27) were chosen according to an a-posteriori parameter selection method (cf. Section 10.3).

10.5. Applications

In this section we present applications of Theorem 10.4 to obtain the rate $O(\sqrt{\delta})$ to some special nonlinear ill-posed problems. There will be two examples from the area of parameter estimation in two-point boundary value problems and one Hammerstein integral equation. The emphasis will be on the illustration of the abstract conditions (iii) and (ii)' of Theorem 10.4 and Remark 10.5, respectively. It will turn out that condition (iii) can be interpreted as a smoothness condition on the solution $x^\dagger$ and the initial guess x^*, and that condition (ii)' imposes a bound on the necessary accuracy of the a-priori estimation x^* of the exact solution $x^\dagger$. The illustrations of the stronger conditions in Theorem 10.7 to obtain the optimal rate $O(\delta^{\frac{2}{3}})$ can be found in [217]. For the first parameter estimation problem we will also show how the results on nonlinear Tikhonov regularization in Hilbert scales are applicable.

Example 10.16. We consider the identification of the parameter $c \in \mathcal{L}^2[0,1]$, $c \geq 0$ almost everywhere, in

$$- u_{ss} + cu = f, \qquad u(0) = u(1) = 0, \qquad (10.34)$$

where $f \in \mathcal{L}^2[0,1]$ is given, from noisy measurements $u^\delta \in \mathcal{L}^2[0,1]$. By the index ss we denote the second derivative with respect to s. As in our general theory we assume that the unperturbed observation u is attainable, i.e., there exists $c^\dagger \in \mathcal{L}^2[0,1]$, $c^\dagger \geq 0$ almost everywhere, with $u(c^\dagger) = u$. Here $u(c^\dagger) \in \mathcal{H}^2 \cap \mathcal{H}_0^1$ denotes the solution of (10.34) with $c = c^\dagger$; c plays now the role of x in the previous sections.

In the context of problem (10.1), the operator F is given by

$$F : \mathcal{D}(F) = \{c \in \mathcal{L}^2[0,1] \mid c \geq 0 \text{ almost everywhere}\} \to \mathcal{L}^2[0,1]$$
$$F(c) := u(c).$$

It can be shown (cf. [40]) that F is continuous and weakly closed. The problem of estimating c in (10.34) is ill-posed as can be seen from the following argument:

Let, for instance, f be the constant function $f := 16$. Then for the data

$$u(s) := 8s(1-s) \quad \text{and} \quad u_n := u + e_n, n \geq 2,$$

where

$$e_n(s) := \begin{cases} n^{-\frac{5}{4}}(2s)^{2n} - 4n^{-\frac{1}{4}}s, & s \leq \frac{1}{2}, \\ n^{-\frac{5}{4}}(2-2s)^{2n} - 4n^{-\frac{1}{4}}(1-s), & s > \frac{1}{2}, \end{cases}$$

the unique solutions in $\mathcal{D}(F)$ are given by

$$c = 0 \quad \text{and} \quad c_n = \frac{(e_n)_{ss}}{u + e_n}.$$

Observe that $u_n \to u$ in $\mathcal{L}^2$, but $\|c_n\|_{\mathcal{L}^2} \sim n^{\frac{1}{4}} \to \infty$, and hence $c_n \not\to c$ in $\mathcal{L}^2$. This means that the solution c does not depend continuously on u.

For $c \in \mathcal{D}(F)$ let $A(c) : \mathcal{H}^2 \cap \mathcal{H}_0^1 \to \mathcal{L}^2$ be defined by

$$A(c)\varphi = -\varphi_{ss} + c\varphi.$$

Then $F(c) = u(c) = A(c)^{-1} f$ and it can be shown that the first and second Fréchet-derivatives are given by (cf. [41])

$$F'(c)h = -A(c)^{-1}(hu(c)),$$
$$F''(c)(h,h) = 2A(c)^{-1}[hA(c)^{-1}(hu(c))].$$

Moreover, the adjoint $F'(c)^*$ is given by

$$F'(c)^*h = -u(c)A(c)^{-1}h.$$

Condition (iii) therefore takes the form

$$c^\dagger - c^* = F'(c^\dagger)^*w = -u(c^\dagger)A(c^\dagger)^{-1}w$$

for some $w \in \mathcal{L}^2$. Such an element w exists if the following smoothness condition

$$v := \frac{c^* - c^\dagger}{u(c^\dagger)} \in \mathcal{H}^2 \cap \mathcal{H}_0^1 \qquad (10.35)$$

holds, and is given by

$$w = A(c^\dagger)v. \qquad (10.36)$$

Detailed calculations in [74] show that condition (ii)' is satisfied for $\delta > 0$ sufficiently small, provided that

$$\left\| \frac{c^* - c^\dagger}{u(c^\dagger)} \right\|_{\mathcal{L}^2} < \frac{2\sqrt{3}}{\|A(c^\dagger)^{-1}\|_{\mathcal{L}^2, \mathcal{H}^2 \cap \mathcal{H}_0^1} \|u(c^\dagger)\|_{\mathcal{L}^\infty}}. \qquad (10.37)$$

An estimate of $u(c^\dagger)$ in (10.37) might be available. Note also that the operators $A(c)^{-1} : \mathcal{L}^2 \to \mathcal{H}^2 \cap \mathcal{H}_0^1$ are uniformly bounded for c in bounded sets of $\mathcal{L}^2$ (cf. [40]).

Condition (10.37) means that the difference between c^* and $c^\dagger$ has to be sufficiently small in two ways: *globally* and *locally*, where globally refers to the complete estimate and locally refers to the fact that the expression $u(c^\dagger)$ appears in the denominator, showing that the estimate c^* has to be better where $|u(c^\dagger)|$ is small.

Before we can draw the final conclusion, we observe that the c^*-minimum-norm solution $c^\dagger$ is unique. In fact, on the set $\{s \in [0,1] \mid u(c^\dagger)(s) \neq 0\}$ $c^\dagger$ is unique (cf. [168]) and on the complement of this set, $c^\dagger(s) = \max\{c^*(s), 0\}$. Now, by Remark 10.5 and Theorem 10.4, we obtain the rate

$$\|c_\alpha^\delta - c^\dagger\|_{\mathcal{L}^2} = O(\sqrt{\delta})$$

if (10.35) and (10.37) are valid and if $\alpha \sim \delta$.

Note that in this example, condition (10.37) is much weaker than the original condition (iv) of Theorem 10.4 which would require an estimate of $\|w\|_{\mathcal{L}^2}$. In view of (10.36) this would be equivalent to a bound for $\|v\|_{\mathcal{H}^2}$. This shows the clear advantage of (ii)' over (ii) and (iv).

We will now show that the results on Tikhonov regularization in Hilbert scales are applicable if we assume that the exact data $u = u(c^\dagger) > 0$. Then, it follows from the results above that $c^\dagger$ is unique and that $\|F'(c)h\|_{\mathcal{L}^2} \sim \|h\|_{-2}$, where $\mathcal{X}_{-2} = (\mathcal{H}^2 \cap \mathcal{H}_0^1)'$ is the anti-dual space of $\mathcal{H}^2 \cap \mathcal{H}_0^1$ (cf. [180]).

There are now different possibilities to regularize problem (10.34) in Hilbert scales. If $c^\dagger$ and c^* are elements of $\mathcal{H}^2 \cap \mathcal{H}_0^1$, we can approximate $c^\dagger$ by the regularized solution $\hat{c}^\delta$ solving

$$(\|F(c) - u^\delta\| - \delta)^2 + \delta^2 \|c - c^*\|_2^2 \to \min, \quad c \in \mathcal{D}(F) \cap \mathcal{X}_2, \qquad (10.38)$$

with $\mathcal{X}_2 = \mathcal{H}^2 \cap \mathcal{H}_0^1$; $\|\cdot\|_2$ is the usual Sobolev norm in $\mathcal{H}^2$ (cf. [180]). Since for this problem Assumption 10.11 is satisfied with $a = s = 2$, Theorem 10.14 implies, for instance, that, if

$$c^\dagger - c^* \in \mathcal{X}_6 = \{c \in \mathcal{H}^6[0,1] \mid c(0) = c(1) = c''(0) = c''(1) = c''''(0) = c''''(1) = 0\},$$

we shall obtain the rate

$$\|\hat{c}^\delta - c^\dagger\|_{\mathcal{L}^2} = O(\delta^{\frac{3}{4}}).$$

If $c^\dagger, c^* \in \mathcal{H}^2$, but $\notin \mathcal{H}_0^1$, the space $\mathcal{X}_2$ in (10.38) has to be replaced by $\widetilde{\mathcal{X}}_2 = \mathcal{H}^2$. Since by a result in [180], the interpolation space $[\mathcal{H}^2, (\mathcal{H}^2 \cap \mathcal{H}_0^1)']_{\frac{1}{2}} \sim \mathcal{L}^2$, this still fits into our Hilbert scales framework. Again we obtain the rate $O(\delta^{\frac{3}{4}})$ under the condition

$$c^\dagger - c^* \in \widetilde{\mathcal{X}}_6 = \{c \in \mathcal{H}^6[0,1] \mid c(0) = c'''(0), c(1) = c'''(1), c''(0) = c''(1) =$$
$$c''''(0) = c''''(1) = 0\}.$$

Suppose that $c^\dagger - c^*$ would satisfy the former condition for the rate $O(\delta^{\frac{3}{4}})$ and that $(c^\dagger)'''(0) - (c^*)'''(0) \neq 0$. Then $c^\dagger - c^* \notin \widetilde{\mathcal{X}}_6$ but $\in \widetilde{\mathcal{X}}_u$ for all $u < 7/2$. Thus, by Theorem 10.14 we can only guarantee the rate $O(\delta^\nu)$ for all $\nu < 7/11$. This shows that the choice of the regularizing space is crucial for obtaining good rates, since different spaces impose different boundary conditions on $c^\dagger - c^*$.

Example 10.17. We consider the identification of the parameter $a \in \mathcal{H}^1[0,1]$, $a \geq \underline{a}$, in

$$-(au_s)_s + cu = f, \qquad u(0) = u(1) = 0, \tag{10.39}$$

where $\underline{a} > 0$, $f \in \mathcal{L}^2[0,1]$ and $c \in \mathcal{L}^2[0,1]$, $c \geq 0$ almost everywhere, are given, from noisy measurements $u^\delta \in \mathcal{L}^2[0,1]$ for $u \in \mathcal{H}^2[0,1]$, i.e., $\|u^\delta - u\|_{\mathcal{L}^2} \leq \delta$. As in Example 10.16, we assume that u is attainable by an element $a^\dagger \in \mathcal{H}^1[0,1]$, $a^\dagger \geq \underline{a}$. Here, the operator F is given by

$$F : \mathcal{D}(F) = \{\mathcal{H}^1[0,1] \mid a \geq \underline{a} > 0\} \to \mathcal{L}^2[0,1]$$
$$F(a) := u(a),$$

where $u(a)$ is the solution of (10.39). The operator F is continuous and weakly closed; the problem $F(a) = u$ is ill-posed.

Let $A(a) : \mathcal{H}^2 \cap \mathcal{H}_0^1 \to \mathcal{L}^2$ be given by

$$A(a)\varphi = -(a\varphi_s)_s + c\varphi \tag{10.40}$$

for $a \in \mathcal{D}(F)$. Then $F(a) = u(a) = A(a)^{-1}f$ and it can be shown [41] that the first and second Fréchet-derivatives are given by

$$F'(a)h = A(a)^{-1}(hu_s(a))_s,$$
$$F''(a)(h,h) = 2A(a)^{-1}[h(A(a)^{-1}(hu_s(a))_s)_s]_s.$$

260   10. Tikhonov Regularization of Nonlinear Problems

Moreover, the adjoint $F'(a)^* : \mathcal{L}^2 \to \mathcal{H}^1$ is given by

$$F'(a)^* h - -B^{-1}[u_s(a)(A(a)^{-1}h)_s],$$

where $B : \mathcal{D}(B) := \{\varphi \in \mathcal{H}^2[0,1] \mid \varphi_s(0) = 0 = \varphi_s(1)\} \to \mathcal{L}^2[0,1]$ is defined by $B\varphi := -\varphi_{ss} + \varphi$. To give this characterization, we used the fact that B^{-1} is the adjoint of the embedding operator from $\mathcal{H}^1$ into $\mathcal{L}^2$, which is easy to show.

Condition (iii) takes the form

$$a^\dagger - a^* = F'(a^\dagger)^* w = -B^{-1}[u_s(a^\dagger)(A(a^\dagger)^{-1}w)_s]$$

for some $w \in \mathcal{L}^2$. Such an element w exists if the smoothness condition

$$a^\dagger - a^* \in \mathcal{H}^3 \cap \mathcal{D}(B) \quad \text{and} \quad B(a^* - a^\dagger) = u_s(a^\dagger)(A(a^\dagger)^{-1}w)_s$$

or equivalently if

$$v := \int_0^\bullet \frac{B(a^* - a^\dagger)}{u_s(a^\dagger)}(s)\,ds \in \mathcal{H}^2 \cap \mathcal{H}_0^1 \tag{10.41}$$

is satisfied, and it is given by

$$w = A(a^\dagger)v.$$

It was shown in [74] that condition (ii)' is satisfied for $\delta > 0$ sufficiently small, provided that

$$\left\| \int_0^\bullet \frac{B(a^* - a^\dagger)}{u_s(a^\dagger)}(s)\,ds \right\|_{\mathcal{L}^2} < \frac{1}{12.5 \|A(a^\dagger)^{-1}\|_{\mathcal{L}^2, \mathcal{H}^2 \cap \mathcal{H}_0^1} \|u(a^\dagger)\|_{\mathcal{H}^2}} \tag{10.42}$$

holds. As in Example 10.16, condition (10.42) means that the difference between a^* and $a^\dagger$ has to be sufficiently small *globally* and *locally*, where the local effect refers to the fact that the expression $u_s(a^\dagger)$ appears in the denominator, which shows that the estimate a^* has to be better where $|u_s(a^\dagger)|$ is small.

We will now show that $a^\dagger$ is unique. Assume that $u = u(a) = u(b)$, then due to (10.39), $((b-a)u_s)_s \equiv 0$ and hence $(b-a)u_s$ is a constant. Since $u(0) = u(1)$, there is a ξ such that $u_s(\xi) = 0$, hence $(b-a)u_s \equiv 0$. This shows that $a^\dagger$ is unique on the set $\mathcal{W} := \{s \in [0,1] \mid u_s(a^\dagger)(s) \neq 0\}$. The uniqueness of the a^*-minimum-norm solution now follows from the convexity of the set $\{a \in \mathcal{D}(F) \mid u(a) = u\} = \{a \in \mathcal{D}(F) \mid a = a^\dagger \text{ on } \mathcal{W}\}$. Thus, by Remark 10.5 and Theorem 10.4, we obtain the rate

$$\|a_\alpha^\delta - a^\dagger\|_{\mathcal{H}^1} = O(\sqrt{\delta})$$

if (10.41) and (10.42) are valid and if $\alpha \sim \delta$. The use of (ii)' and hence (10.42) offers an advantage over (ii) and (iv) of Theorem 10.4, which would require an estimate of $\|v\|_{\mathcal{H}^2}$.

Example 10.18. We consider the Hammerstein integral equation

$$F : \mathcal{D}(F) = \mathcal{H}^1[0,1] \to \mathcal{L}^2[0,1]$$

$$F(x)(s) := \int_0^s (s-t)x^3(t)\,dt.$$

Since F is continuous, weakly closed, compact and injective, Proposition 10.1 implies that the problem of solving $F(x) = y$ is ill-posed. In the sequel we assume that $x^\dagger$ exists with $F(x^\dagger) = y$. The first and second Fréchet-derivatives are given by

$$(F'(x)h)(s) = 3\int_0^s (s-t)x^2(t)h(t)\,dt,$$

$$(F''(x)(h,h))(s) = 6\int_0^s (s-t)x(t)h^2(t)\,dt.$$

Moreover, the adjoint $F'(x)^*$ is characterized by

$$(F'(x)^*h)(s) = B^{-1}\left(3x^2(s)\int_s^1 (t-s)h(t)\,dt\right)$$

with B as in Example 10.17.

An element w satisfying condition (iii) exists if $x^\dagger - x^* \in \mathcal{H}^3 \cap \mathcal{D}(B)$ and if

$$\frac{B(x^\dagger - x^*)}{3(x^\dagger)^2}(s) = \int_s^1 (t-s)w(t)\,dt.$$

This is equivalent to

$$\frac{B(x^\dagger - x^*)}{3(x^\dagger)^2} \in \{z \in \mathcal{H}^2[0,1] \mid z(1) = 0 = z'(1)\}, \tag{10.43}$$

and that w is given by

$$w = \left[\frac{B(x^\dagger - x^*)}{3(x^\dagger)^2}\right]_{ss}.$$

Provided that $x^\dagger \geq \gamma > 0$ and $x^* \in \mathcal{H}^4$, condition (10.43) is equivalent to the smoothness condition: $x^\dagger \in \mathcal{H}^4$, $x_s^\dagger(0) = x_s^*(0)$, $x_s^\dagger(1) = x_s^*(1)$, $x^\dagger(1) - x_{ss}^\dagger(1) = x^*(1) - x_{ss}^*(1)$, $x_{sss}^\dagger(1) = x_{sss}^*(1)$.

It was again shown in [74] that condition (ii)' holds for $\delta > 0$ sufficiently small if

$$\left\|B^{-1}\left[\frac{B(x^\dagger - x^*)}{x^\dagger}\right]\right\|_{\mathcal{H}^1} < \frac{1}{5}. \tag{10.44}$$

In case that $x^\dagger \geq \gamma > 0$, this condition may be replaced by

$$\|x^\dagger - x^*\|_{\mathcal{H}^1} \leq \frac{2}{25\|\frac{1}{x^\dagger}\|_{\mathcal{H}^1}}. \tag{10.45}$$

As in the previous examples we can now draw the conclusion that

$$\|x_\alpha^\delta - x^\dagger\|_{\mathcal{H}^1} = O(\sqrt{\delta})$$

for $\alpha \sim \delta$, provided that (10.43) and (10.44) (or (10.45)) hold.

For other applications of nonlinear Tikhonov regularization see [30, 82, 84, 85, 243, 245].

10.6. Convergence of Maximum Entropy Regularization

In Section 5.3, we gave an information-theoretic motivation for maximum entropy regularization, described several variants and gave an outline of convergence and stability results in the literature. In this section, we give a detailed stability and convergence analysis following [75]. We consider the following special form of maximum entropy regularization (denoting the prior distribution, which was denoted by x^* in (5.66), by m from now on for simplicity of notation, cf. also [142]):

$$\|Tx - y^\delta\|^2 + \alpha E(x) \to \min \tag{10.46}$$

with the (negative) entropy defined as

$$E(x) := \int_\Omega x(t) \log \frac{x(t)}{m(t)} \, dt. \tag{10.47}$$

Without a normalization for x and m, E need not be nonnegative.

We will give the precise requirements for the operator T, for x and m below and restrict, for technical simplicity, ourselves to the case where Ω is a bounded measurable subset of $\mathbb{R}$.

The basic idea for analyzing (10.46) is the following one: we define an operator H such that for all $v \in \mathcal{D}(H) \subseteq \mathcal{L}^2$,

$$E(Hv) = \|v\|_{\mathcal{L}^2}^2 + c \tag{10.48}$$

holds, where the constant c depends only on m. Then, by subsituting Hv for x in (10.46), we obtain the minimization problem

$$\|THv - y^\delta\|^2 + \alpha \|v\|^2 \to \min, \tag{10.49}$$

which is just Tikhonov regularization for the transformed problem

$$THv = y. \tag{10.50}$$

(In this section, we will use the notation "Hv" in spite of the fact that H is nonlinear). Since TH is nonlinear even if T is linear (which we assume here for simplicity), (10.49) has to be treated with the methods developed in Section 10.2. It turns out that (10.46) and (10.49) are equivalent, so that the convergence and stability results for (10.49) can be translated into corresponding results for (10.46). As opposed to the approach of [60] outlined in Section 5.3, this method works also if T is nonlinear (see [76]).

Now, we introduce the operator H which transforms (10.46) into (10.49). We start by specifying functions ϕ and $\tilde{\phi}$ which we will use to define this transformation:

Definition 10.19. *Let $\Omega \subset \mathbb{R}$ be a bounded measurable set and let $m \in \mathcal{L}^1(\Omega)$ be such that $m(t) > 0$ for all $t \in \Omega$. For any $\mu > 0$ we denote by ψ_μ the function*

$$\begin{aligned}\psi_\mu : \mathbb{R}_0^+ &\to [-\sqrt{\tfrac{\mu}{e}}, \infty) \\ a &\mapsto \operatorname{sign}(a - \tfrac{\mu}{e})\sqrt{a \log \tfrac{a}{\mu} + \tfrac{\mu}{e}}.\end{aligned} \tag{10.51}$$

10.6. Convergence of Maximum Entropy Regularization

We denote by ϕ_1, ϕ and $\tilde{\phi}$ the functions

$$\phi_1 : [-\tfrac{1}{\sqrt{e}}, \infty) \to \mathbb{R}_0^+ \qquad (10.52)$$
$$v \mapsto \psi_1^{-1}(v),$$

$$\phi : \{(t,v) \mid t \in \Omega \wedge v \in [-\sqrt{\tfrac{m(t)}{e}}, \infty)\} \to \mathbb{R}_0^+ \qquad (10.53)$$
$$(t,v) \mapsto \psi_{m(t)}^{-1}(v),$$

where $\cdot^{-1}$ denotes the inverse function,

$$\tilde{\phi} : \Omega \times \mathbb{R} \to \mathbb{R}_0^+$$
$$(t,v) \mapsto \begin{cases} \phi(t,v), & v \geq -\sqrt{\tfrac{m(t)}{e}}, \\ 0, & v < -\sqrt{\tfrac{m(t)}{e}}. \end{cases}$$

As will be shown in Lemma 10.22, ψ_μ is bijective for every $\mu > 0$ and thus, ϕ and $\tilde{\phi}$ are well-defined.

Definition 10.20. *Let $\Omega \subset \mathbb{R}$ be a bounded measurable set and $m \in \mathcal{L}^1(\Omega)$ such that $m(t) > 0$ for all $t \in \Omega$. We denote by $\Sigma(m)$ the following subset of $\mathcal{L}^2(\Omega)$:*

$$\Sigma(m) := \left\{ v \in \mathcal{L}^2(\Omega) \mid \forall t \in \Omega : v(t) \geq -\sqrt{\tfrac{m(t)}{e}} \right\}. \qquad (10.54)$$

Similarly, we denote by $\mathcal{E}_m(\Omega)$ the following subset of $\mathcal{L}^1(\Omega)$:

$$\mathcal{E}_m(\Omega) := \left\{ x \in \mathcal{L}^1(\Omega) \mid \forall t \in \Omega : x(t) \geq 0 \wedge \int_\Omega x(t) \log \tfrac{x(t)}{m(t)} \, dt < \infty \right\}. \qquad (10.55)$$

$\mathcal{E}_m(\Omega)$ *is the set of all positive $\mathcal{L}^1$-functions with finite entropy. Note that the sets defined in Definition 10.20 are convex.*

Definition 10.21. *Let $\Omega \subset \mathbb{R}$ be a bounded measurable set and m a continuous bounded function with $m(t) \geq m_0 > 0$ for all $t \subset \Omega$. We denote by H and $\tilde{H}$ the superposition operators (Nemytskii operators, see, e.g., [162, Chapter 17]) $H : \Sigma(m) \to \mathcal{L}^1(\Omega)$ defined by*

$$Hv(t) := \phi(t, v(t)) \qquad (10.56)$$

and $\tilde{H} : \mathcal{L}^2(\Omega) \to \mathcal{L}^1(\Omega)$ defined by

$$\tilde{H}v(t) := \tilde{\phi}(t, v(t)) \qquad (10.57)$$

respectively.

264   10. Tikhonov Regularization of Nonlinear Problems

We will show in Lemma 10.23 that the operators H and $\tilde{H}$ actually act into $\mathcal{L}^1(\Omega)$. It will turn out that (10.48) holds, so that H actually transforms (10.46) into (10.49).

First, we investigate the basic properties of the superposition operators H and $\tilde{H}$. We give only short proofs for the following lemmata and propositions. The details can be found in [172].

Lemma 10.22. *For each $\mu > 0$, ψ_μ is bijective, strictly increasing and concave. For fixed t, the function ϕ is bijective, convex, strictly increasing with v, and (partially) differentiable with respect to variable v. For fixed t, the first derivative $\dfrac{\partial \phi}{\partial v}$ is continuous for $v \geq -\sqrt{(m(t)/e)}$, and the following inequality holds for all $t \in \Omega$, $v \in [-\sqrt{(m(t)/e)}, \infty)$:*

$$\left| \frac{\partial \phi}{\partial v}(t, v) \right| \leq 2\sqrt{\frac{m(t)}{e}} + 2|v| \tag{10.58}$$

Moreover, for $v > -\sqrt{(m(t)/e)}$, ϕ is twice (partially) differentiable with respect to variable v and for every fixed t, $\dfrac{\partial^2 \phi}{\partial v^2}$ is continuous.

Proof. For fixed $\mu > 0$, the function ψ_μ has continuous first and second derivatives for all $a > 0$ where $\psi_\mu(a) \neq 0$. It turns out that ψ_μ has only one zero-crossing, namely at $a = \mu/e$. This gives existence and continuity of the derivatives of ψ_μ on the intervals $(0, \mu/e)$ and $(\mu/e, \infty)$, and it can be shown (after some computation) that $\psi_\mu'(a) > 0$ and $\psi_\mu''(a) < 0$ there. Bijectivity of function ψ_μ then follows from the fact that $\psi_\mu(0) = -\sqrt{\mu/e}$. Now applying de l'Hospital's rule one shows that $\lim_{a \to (\mu/e)} \psi_\mu'(a)$ and $\lim_{a \to (\mu/e)} \psi_\mu''(a)$ exist, giving existence and continuity of these derivatives for $a \in (0, \infty)$ by the Mean Value Theorem. This together with the fact that

$$\lim_{v \to \sqrt{\frac{m(t)}{e}}} \frac{\partial \phi}{\partial v}(t, v) = \lim_{a \to 0} \frac{1}{\psi_{m(t)}'(a)} = 0$$

gives the differentiability results. From $\dfrac{\partial \phi}{\partial v}(t, v) = 1/\psi_{m(t)}'(\phi(t, v))$, one obtains

$$\frac{\partial \phi}{\partial v}(t, v) = \frac{2v}{1 + \log \frac{\phi(t,v)}{m(t)}}. \tag{10.59}$$

As ϕ is increasing with v and $\phi(t, \sqrt{(m(t)/e)}) = m(t)$, $0 \leq \dfrac{\partial \phi}{\partial v}(t, v) \leq 2v$ for $v \geq +\sqrt{(m(t)/e)}$, and from the convexity of ϕ for fixed t we get

$$0 \leq \frac{\partial \phi}{\partial v}(t, v) \leq \frac{\partial \phi}{\partial v}\left(t, \sqrt{\frac{m(t)}{e}}\right) = 2\sqrt{\frac{m(t)}{e}}$$

for $v < +\sqrt{(m(t)/e)}$. From these inequalities (10.58) follows. ∎

10.6. Convergence of Maximum Entropy Regularization

This lemma also implies that the function ϕ defined in (10.53) is well-defined.

Lemma 10.23. *For all $v \in \mathcal{L}^2(\Omega)$, $\tilde{H}v \in \mathcal{L}^1(\Omega)$.*

Proof. From the definition of $\tilde{\phi}$ (see Definition 10.19) we have $\tilde{\phi}(t, v) \geq 0$; (10.58) together with boundedness of m implies the existence of $c_1, c_2 > 0$ such that

$$\tilde{\phi}(t, v) \leq c_1 + c_2 v^2 \qquad \text{for } v \geq -\sqrt{\frac{m(t)}{e}}.$$

But then,

$$\int_\Omega \left|(\tilde{H}v)(t)\right| dt = \int_\Omega \left|\tilde{\phi}(t, v(t))\right| dt \leq c_1 \text{ meas}(\Omega) + c_2 \int_\Omega v^2(t)\, dt,$$

so that for $v \in \mathcal{L}^2(\Omega)$, we obtain that $\tilde{H}v \in \mathcal{L}^1(\Omega)$. ∎

Proposition 10.24. *The operator H defined in Definition 10.21 is continuous on $\Sigma(m)$, $\tilde{H}$ is continuous on $\mathcal{L}^2(\Omega)$. Both operators are bounded.*

Proof. This is a consequence of Lemma 10.22 and Theorems 17.1 and 17.2 in [162] as soon as we have shown that $\tilde{\phi}$ satisfies the *Caratheodory condition* (see, e.g., [162, p. 349]), that is, for fixed t, $\tilde{\phi}(t, v)$ is continuous and for fixed v, it is measurable. Continuity for fixed t is again a consequence of Lemma 10.22 together with the fact that $\phi(t, -\sqrt{(m(t)/e)}) = 0$. Measurability for fixed v can be seen from the fact that

$$\phi(t, v) = m(t)\phi_1\left(\frac{v}{\sqrt{m(t)}}\right) \tag{10.60}$$

which follows from the definitions of ϕ, ϕ_1 and ψ, since m is a positive continuous function bounded away from 0, together with the measurability of ϕ_1. ∎

Proposition 10.25. *The operator $H : \Sigma(m) \to \mathcal{L}^1(\Omega)$ is injective.*

Proof. If $v \neq w$ on a set of positive measure, then $Hv \neq Hw$ on the same set by the definition of H and Lemma 10.22. ∎

We now return to our problem: the aim is to solve the (operator) equation

$$Tx = y, \tag{10.61}$$

where we now assume that T is a bounded linear operator acting from $\mathcal{L}^1(\Omega)$ into $\mathcal{L}^2(\Omega^*)$, where $\Omega \subset \mathbb{R}$ is bounded and measurable and $\Omega^* \subset \mathbb{R}$ is measurable. Its solution is approximated by the solution of

Problem 10.26. *Let $m : \Omega \to \mathbb{R}$ be a given function (the "prior distribution") which satisfies the inequalities $m_0 \leq m(t) \leq M_0$ with given constants $m_0, M_0 > 0$, y^δ*

266   10. Tikhonov Regularization of Nonlinear Problems

be an approximation of the right-hand side of (10.61) with $\|y - y^\delta\|_{\mathcal{L}^2} \leq \delta$. Minimize

$$\|Tx - y^\delta\|_{\mathcal{L}^2}^2 + \alpha E(x), \tag{10.62}$$

where E is defined by (10.47), subject to $x \in \mathcal{E}_m(\Omega)$ (defined by (10.55)).

Note that due to the strict convexity of the functional in (10.62) and the convexity of $\mathcal{E}_m(\Omega)$, Problem 10.26 has at most one solution.

It can be easily seen from Definitions 10.19 to 10.21 that, if $x = Hv$, then, for all $t \in \Omega$,

$$v^2(t) = \psi_{m(t)}^2(x(t)) = x(t)\log\frac{x(t)}{m(t)} + \frac{m(t)}{e}.$$

Therefore, we have for the (negative) entropy

$$E(Hv) = \int_\Omega \left(\psi_{m(t)}^2(\phi(t, v(t))) - \frac{m(t)}{e}\right) dt = \int_\Omega v^2(t)\, dt + c$$

(note that Ω as well as m are bounded). Thus, substituting Hv for x in (10.62) gives

Problem 10.27. With m and y^δ as in Problem 10.26, determine $x = Hv$, where v minimizes

$$\|THv - y^\delta\|_{\mathcal{L}^2(\Omega^*)}^2 + \alpha\|v\|_{\mathcal{L}^2(\Omega)}^2 \tag{10.63}$$

subject to $v \in \Sigma(m)$, defined by (10.54).

Proposition 10.28. *Problems 10.26 and 10.27 are equivalent.*

Proof. As shown above, the operator H maps any $v \in \Sigma(m)$ into the set $\mathcal{E}_m(\Omega)$, since $E(Hv) < \infty$. By Proposition 10.25, H is injective. On the other hand, for any function $x \in \mathcal{E}_m(\Omega)$, $v = H^{-1}x$ is given by $v(t) = \psi_{m(t)}(x(t))$. But then

$$\int_\Omega v^2(t)\, dt = \int_\Omega \left(x(t)\log\frac{x(t)}{m(t)} + \frac{m(t)}{e}\right) dt < \infty,$$

hence $v \in \mathcal{L}_2(\Omega)$. Moreover, due to (10.51), $v \in \Sigma(m)$. Thus, H is bijective as an operator acting from $\Sigma(m)$ into $\mathcal{E}_m(\Omega)$. This proves the assertion. ∎

For the solvability of these problems see Remark 10.39. We finally note that solving Problem 10.27 subject to $v \in \Sigma(m)$ gives the same result as minimizing (10.63) with H replaced by $\tilde{H}$ subject to $v \in \mathcal{L}^2(\Omega)$. Indeed, if $v \notin \Sigma(m)$, then there is a set $\Omega_1 \subset \Omega$ with positive measure such that $v(t) < -\sqrt{m(t)/e}$ for $t \in \Omega_1$. If we define

$$v_1(t) := \max(v(t), -\sqrt{\frac{m(t)}{e}}),$$

10.6. Convergence of Maximum Entropy Regularization

then $\tilde{H}v = \tilde{H}v_1 = Hv_1$, so that

$$\|T\tilde{H}v - y^\delta\|_{\mathcal{L}^2(\Omega^*)} = \|THv_1 - y^\delta\|_{\mathcal{L}^2(\Omega^*)},$$

but $\|v_1\|_{\mathcal{L}^2(\Omega)} < \|v\|_{\mathcal{L}^2(\Omega)}$. This argument shows that a solution to the unconstrained modification of Problem 10.27 (H replaced by $\tilde{H}$ and minimization subject to $v \in \mathcal{L}^2(\Omega)$) must give a result in $\Sigma(m)$, where $\tilde{H}$ coincides with H. Thus, the constrained Problem 10.27 and its unconstrained modification are equivalent.

We conclude these technical preparations by proving some auxiliary results, which we will need for the main theorems of this section. We start with the differentiability of the operators H and $\tilde{H}$:

Proposition 10.29. *The operator $\tilde{H}$ defined in Definition 10.21 is Fréchet-differentiable at each point of $\mathcal{L}^2(\Omega)$,*

$$[\tilde{H}'(v)h](s) = \frac{\partial\tilde{\phi}}{\partial v}(s, v(s))h(s)$$

for $v, h \in \mathcal{L}^2(\Omega)$, $s \in \Omega$.

Proof. To apply Theorem 20.2 of [162], with Lemma 10.22 the only thing left to show is that $\frac{\partial\tilde{\phi}}{\partial v}$ satisfies the Caratheodory condition. This is done in the same way as in the proof of Proposition 10.24: continuity for fixed t is shown by Lemma 10.22 together with the fact that $\frac{\partial\tilde{\phi}}{\partial v}(t, -\sqrt{(m(t)/e)}) = 0$; measurability for fixed v can be seen from

$$\frac{\partial\phi}{\partial v}(t, v) = \sqrt{m(t)}\phi_1'\left(\frac{v}{\sqrt{m(t)}}\right)$$

which follows from (10.60), since m is a positive continuous function bounded away from 0, and ϕ_1 has a measurable derivative ∎

Remark 10.30. Note that $\Sigma(m)$, the domain of H, has no interior point, so that the notion of Fréchet-differentiability is not defined there. But H can be viewed as the restriction of $\tilde{H}$ to $\Sigma(m)$, so that we may define the derivative of H as

$$H'(v)\,h := \tilde{H}'(v)\,h \tag{10.64}$$

for $v \in \Sigma(m)$, $h \in \mathcal{L}^2(\Omega)$. This makes sense, as for $v \in \Sigma(m)$ and h such that $v + h \in \Sigma(m)$, (10.64) implies that

$$H(v+h) = H(v) + H'(v)\,h + r_v(h)$$

where (for fixed v and for $h \to 0$) we have $r_v(h) = o(\|h\|)$. Due to the assumptions we will make in the formulation of Theorem 10.41 and due to convexity of $\Sigma(m)$, we will need this estimate only for such v and h anyway. In this sense, Proposition 10.29 then also holds for H instead of $\tilde{H}$.

Lemma 10.31. *For fixed t, $\dfrac{\partial^2 \phi}{\partial v^2}$ is decreasing in v.*

Proof. Differentiating (10.60) twice gives

$$\frac{\partial^2 \phi}{\partial v^2}(t,v) = \phi_1''\left(\frac{v}{\sqrt{m(t)}}\right). \tag{10.65}$$

Thus, it is sufficient to show that ϕ_1'' is decreasing. By Lemma 10.22 we have existence and continuity of the second derivative and with the same arguments as used there we obtain existence of $\phi_1'''(v)$ for $v \in (-1/\sqrt{e}, 1/\sqrt{e})$ and $v \in (1/\sqrt{e}, \infty)$. It can be shown after some computation that in these intervals, ϕ_1 satisfies the differential equation

$$\phi_1'''(v) = 2\phi_1'(v)\frac{2(\phi_1(v)\log\phi_1(v) + \frac{1}{e})(4 + \log\phi_1(v)) - 3\phi_1(v)(1 + \log\phi_1(v))^2}{\phi_1^2(v)(1 + \log\phi_1(v))^4}.$$

Due to Lemma 10.22, $\phi_1'(v) > 0$. The denominator on the right-hand side is also nonnegative for all v. With

$$\chi(a) = 2\left(a\log a + \frac{1}{e}\right)(4 + \log a) - 3a(1 + \log a)^2$$

we have

$$(a\,\chi'(a))' = -(1 + \log a)^2 \leq 0$$

as well as $\chi(1/e) = \chi'(1/e) = 0$. This implies, that $\chi(a) \leq 0$ for all $a > 0$. Since

$$\phi_1'''(v) = 2\phi_1'(v)\frac{\chi(\phi_1(v))}{\phi_1^2(v)(1 + \log\phi_1(v))^4},$$

this implies that $\phi_1'''(v) \leq 0$ for all $v \neq 1/\sqrt{e}$, so that $\phi_1''(v)$ is decreasing. ∎

Lemma 10.33 below will be crucial for obtaining a convergence rate. For sake of clarity we divide the proof into two steps and first consider the special case $m = 1$:

Lemma 10.32. *Let ϕ_1 be as defined by (10.52). For $a_* > 0$, let $v_* := \psi_1(a_*)$ (i.e., $a_* = \phi_1(v_*)$). Then, for all $v \geq \underline{v} := -1/\sqrt{e}$, the inequality*

$$(1 + \log a_*)\int_0^1 (1-\tau)\phi_1''(v_* + \tau(v - v_*))\,d\tau \leq \overline{\chi}_1(v_*) := 1 - \frac{\phi_1(v_*)}{(v_* - \underline{v})^2} < 1$$

holds. The function $\overline{\chi}_1$ is strictly increasing.

10.6. Convergence of Maximum Entropy Regularization

Proof. As shown in Lemma 10.31, the second derivative of ϕ_1 is decreasing. As $v \geq \underline{v}$, we can conclude that, since $\phi_1(\underline{v}) = 0$,

$$(1 + \log a_*) \int_0^1 (1 - \tau)\phi_1''(v_* + \tau(v - v_*))\, d\tau$$

$$\leq (1 + \log a_*) \int_0^1 (1 - \tau)\phi_1''(v_* + \tau(\underline{v} - v_*))\, d\tau$$

From (10.59) (note (10.52)), $a_* = \phi_1(v_*)$ and $\phi_1(v_*) \log \phi_1(v_*) = v_*^2 - \underline{v}^2$ (see Definition 10.19), we get

$$(1 + \log a_*) \int_0^1 (1 - \tau)\phi_1''(v_* + \tau(\underline{v} - v_*))\, d\tau$$

$$= \frac{1 + \log a_*}{(v_* - \underline{v})^2}(\phi_1'(v_*)(v_* - \underline{v}) - \phi_1(v_*))$$

$$= \frac{1}{(v_* - \underline{v})^2}(2v_*(v_* - \underline{v}) - \phi_1(v_*)(1 + \log \phi_1(v_*)))$$

$$= \frac{2v_*(v_* - \underline{v}) - (v_*^2 - \underline{v}^2) - \phi_1(v_*)}{(v_* - \underline{v})^2} = 1 - \frac{\phi_1(v_*)}{(v_* - \underline{v})^2} = \overline{\chi}_1(v_*) < 1.$$

We now show that $\overline{\chi}_1'(v_*) > 0$ almost everywhere. Calculating the derivative and using (10.59), we get

$$\overline{\chi}_1'(v_*) = \frac{2\phi_1(v_*) - \phi_1'(v_*)(v_* - \underline{v})}{(v_* - \underline{v})^3} = \frac{2\phi_1(v_*)(1 + \log \phi_1(v_*)) - 2v_*(v_* - \underline{v})}{(v_* - \underline{v})^3(1 + \log \phi_1(v_*))}.$$

By the convexity of ϕ_1 we have $\phi_1(v_*) \geq \phi_1(0) + \phi_1'(0)v_* = (1/e) + \sqrt{2}v_*/\sqrt{e}$. Again using the fact that $\phi_1(v_*)\log\phi_1(v_*) = v_*^2 - \underline{v}^2$ we get for $v_* > 0$ (i.e., for $1 + \log \phi_1(v_*) > 0$) that

$$\overline{\chi}_1'(v_*) = \frac{2(\phi_1(v_*) + v_*\underline{v} - \underline{v}^2)}{(v_* - \underline{v})^3(1 + \log\phi_1(v_*))} \geq \frac{2\left(\frac{1}{e} + \sqrt{\frac{2}{e}}v_* - \frac{1}{\sqrt{e}}\underline{v} - \frac{1}{e}\right)}{(v_* - \underline{v})^3(1 + \log \phi_1(v_*))}$$

$$= \frac{2(\sqrt{2} - 1)v_*}{\sqrt{e}(v_* - \underline{v})^3(1 + \log \phi_1(v_*))} > 0.$$

For $v_* < 0$ (i.e., for $1 + \log \phi_1(v_*) < 0$) the convexity of ϕ_1 implies that $\phi_1(v_*) \leq ((\underline{v} - v_*)\phi_1(0) + v_*\phi_1(\underline{v}))/\underline{v} = (1 + v_*\sqrt{e})/e$. Thus, we have

$$\overline{\chi}_1'(v_*) \geq \frac{2\left(\frac{1}{e} + \frac{v_*}{\sqrt{e}} + v_*\underline{v} - \underline{v}^2\right)}{(v_* - \underline{v})^3(1 + \log \phi_1(v_*))} = 0. \quad\blacksquare$$

Lemma 10.33. *For any $x_0 \in \mathcal{E}_m(\Omega)$ and v_0 such that $x_0 = Hv_0$ we have for all $t \in \Omega$ and $v \in \Sigma(m)$*

$$\left(1 + \log \frac{x_0(t)}{m(t)}\right) \int_0^1 (1 - \tau)\frac{\partial^2 \phi}{\partial v^2}(t, v_0(t) + \tau(v(t) - v_0(t)))\, d\tau$$

$$\leq \overline{\chi}(t, x_0(t)) := 1 - \frac{\phi(t, v_0(t))}{\left(v_0(t) + \sqrt{\frac{m(t)}{e}}\right)^2} = 1 - \frac{x_0(t)}{\left(\psi_{m(t)}(x_0(t)) + \sqrt{\frac{m(t)}{e}}\right)^2}.$$

10. Tikhonov Regularization of Nonlinear Problems

Moreover, if $x_0(t)/m(t)$ is bounded on Ω, then

$$\chi(x_0) := \sup_{t \in \Omega} \overline{\chi}(t, x_0(t)) < 1.$$

Proof. This follows from Lemma 10.32 with $a_* = x_0(t)/m(t)$ together with

$$\psi_{m(t)}(x(t)) = \sqrt{m(t)}\psi_1\Big(\frac{x(t)}{m(t)}\Big), \qquad (10.66)$$

(which follows from Definition 10.19 and implies that $v_* = \psi_1(a_*) = \psi_{m(t)}(x_0(t))/\sqrt{m(t)} = v_0(t)/\sqrt{m(t)}$), (10.60), and (10.65). These equations also imply that $\overline{\chi}(t, v_0(t)) = \overline{\chi}_1(v_0(t)/\sqrt{m(t)})$, from which the second assertion follows immediately, since $\overline{\chi}_1$ is increasing. ∎

In order to apply the results of Section 10.2, we must investigate the properties of the operator $T \circ H$ in (10.63). Especially, we have to check if (or, under which conditions) $T \circ H$ is continuous and weakly (sequentially) closed. Of course, due to Proposition 10.24, $T \circ H$ is continuous as soon as T is.

Recall that an operator is called *weakly compact* if it maps each closed bounded set into a weakly (sequentially) compact set. We will show that H is weakly compact; one can use this fact to show (along the lines of the proof of Proposition 10.35 below) that $T \circ H$ is weakly closed as soon as H is. But since a nonlinear superposition operator defined on $\mathcal{L}^2(\Omega)$ need not be weakly closed, we impose an additional a-priori assumption on the solution, thereby restricting the domain of H in order to make H weakly closed: We will assume that the solution of (10.50) which we want to approximate lies in a set $\mathcal{U} \subset \mathcal{L}^2(\Omega)$ which is compact in measure; this set will then also be used as a restriction for the regularized solutions in Problem 10.27. These assumptions have to be translated into assumptions about the original equation (10.61) and about Problem 10.26.

Recall that a sequence $\{v_n\}$ of measurable functions acting from Ω into $\mathbb{R}$ converges to v in measure (we will write this as $v_n \xrightarrow[\text{meas}]{} v$) if for every $\varepsilon > 0$

$$\lim_{n \to \infty} \text{meas}(\{t \mid |v_n(t) - v(t)| > \varepsilon\}) = 0$$

holds. This is convergence in the metric space of measurable functions with

$$\text{dist}(u,v) := \inf_{\varepsilon > 0}\{\varepsilon + \text{meas}(\{t \mid |u(t) - v(t)| \geq \varepsilon\})\}$$

as the distance of u and v (see, e.g., [58, Lemma III.2.7]). Any set of such functions is called *compact in measure* if it is compact with respect to this metric.

Lemma 10.34. *The operator H defined in Definition 10.21 is weakly compact.*

Proof. This lemma is a consequence of Theorem 17.5 and Lemma 1.3 in [162]. These results state that a superposition operator acting from $\mathcal{L}^p(\Omega)$ into $\mathcal{L}^1(\Omega)$ is

10.6. Convergence of Maximum Entropy Regularization

weakly compact if and only if there exists a continuous even function M satisfying

$$\lim_{u \to +\infty} \frac{M(u)}{u} = +\infty, \tag{10.67}$$

where the inequality

$$M(\phi(t, u)) \leq \alpha(t) + \beta|u|^p \tag{10.68}$$

holds for all $t \in \Omega$ and all $u \in \mathbb{R}$ with $\alpha \in \mathcal{L}^1(\Omega)$ and $\beta \in \mathbb{R}$. For our operator H, such a function M can be taken as

$$M(u) := |u| \log \frac{|u|}{m_0} + \frac{m_0}{e} = \psi^2_{m_0}(|u|),$$

where m_0 is a lower bound of the function m (see Definition 10.21 and (10.51)). This M satisfies (10.67); (10.68) (with $p = 2$) can be shown by standard techniques taking into account the concavity of ψ. For a detailed proof see [172]. ∎

Proposition 10.35. *Let H be defined as in Definition 10.21, $\Sigma(m)$ as in (10.54), $T : \mathcal{L}^1(\Omega) \to \mathcal{L}^2(\Omega^*)$ be a bounded linear operator. Then $T \circ H$ is weakly (sequentially) closed on every subset $\mathcal{U}$ of $\Sigma(m)$ that is compact in measure.*

Proof. Take any sequence $\{v_n\}$ of functions in $\mathcal{U}$ such, that

$$v_n \rightharpoonup v \quad , \quad \mathcal{K} \circ H v_n \rightharpoonup g.$$

As $\{v_n \mid n \in \mathbb{N}\}$ is bounded, Lemma 10.34 implies that $\{Hv_n \mid n \in \mathbb{N}\}$ is weakly compact, so that there is a weakly convergent subsequence

$$Hv_{n_k} \rightharpoonup x.$$

Since T is weakly (sequentially) closed,

$$Tx = g.$$

On the other hand, $\{v_n \mid n \in \mathbb{N}\}$ is compact in measure. Now, if a sequence converges weakly and is compact in measure, then it converges in measure and the limits are the same (see, e.g., [162, p. 7]). Thus, we have

$$v_{n_k} \xrightarrow[\text{meas}]{} v.$$

Lemma 17.5 in [162] (together with its proof) implies that the continuous Nemytskii operator H maps a sequence converging to v in measure into a sequence converging to Hv in measure. Thus

$$Hv_{n_k} \xrightarrow[\text{meas}]{} Hv.$$

Since, if a sequence of functions converges weakly and converges in measure, the limits are the same almost everywhere, we obtain that

$$Hv = x \quad \text{a.e.}$$

272   10. Tikhonov Regularization of Nonlinear Problems

and thus
$$g = Tx = T \circ H\, v \quad \text{a.e.}$$

Since $v \in \mathcal{U}$ due to the compactness of $\mathcal{U}$ in measure, this implies that $T \circ H$ is weakly sequentially closed on $\mathcal{U}$. ∎

Since we are concerned with maximum entropy regularization, i.e., with Problem 10.26 rather than with Problem 10.27 (which fulfills only an auxiliary role), it is desirable to formulate all assumptions on the range space of H; i.e., an assumption about $\mathcal{U}$ like that in Proposition 10.35 should be reformulated in terms of $H(\mathcal{U})$:

Lemma 10.36. $\mathcal{U} \subset \Sigma(m)$ *is compact in measure if and only if $H(\mathcal{U})$ is compact in measure.*

Proof. To prove this, we first define the function $\overline{\psi} : \Omega \times \mathbb{R} \to \mathbb{R}$ as
$$\overline{\psi}(t, x) = \psi_{m(t)}(|x|).$$

Let G be the superposition operator defined by $\overline{\psi}$, i.e., $Gx(t) := \overline{\psi}(t, x(t))$. One sees from the definitions of ϕ and ψ (see Definition 10.19) that for any $v \in \Sigma(m)$ the equation $v = G(H(v))$ holds, so that $\mathcal{U} = G(H(\mathcal{U}))$. Thus, by Lemma 17.5 in [162], it is sufficient to show that both ϕ and $\overline{\psi}$ satisfy the Caratheodory-condition. It has already been shown in Proposition 10.24 that ϕ satisfies this condition. Continuity of $\overline{\psi}$ for fixed t is equivalent to continuity of ψ_μ for all μ, which has been shown in Lemma 10.22. Measurability can be seen from (10.66) taking into account that m is measurable and bounded away from 0. ∎

An important example for a set whose closure is compact in measure is any set of $\mathcal{H}^1$-functions bounded in the $\mathcal{H}^1$-norm, since $\mathcal{H}^1$ is compactly embedded into $\mathcal{L}^2$. For this case, the assumption that v lies in such a set becomes an a-priori smoothness assumption together with an a-priori bound for v and its (weak) derivative. Thus, the assumptions we need for verifying the weak closedness of H are not as severe as claimed in [60].

We are now in the position to state the main results of this section:

Theorem 10.37. *Let $\mathcal{E}_m(\Omega)$ be as in Definition 10.20. Let $\mathcal{U}$ be a convex subset of $\mathcal{E}_m(\Omega)$ that is compact in measure. Let $H : \overline{U} \to \mathcal{L}^1(\Omega)$ be the operator defined in Definition 10.21 (restricted to $\overline{U} := H^{-1}(\mathcal{U})$), $T : \mathcal{L}^1(\Omega) \to \mathcal{L}^2(\Omega^*)$ be a continuous linear operator. Let $\alpha > 0$, $\{y_n\}, \{x_n\}$ and $\{v_n\}$ be sequences such that $y_n \to y^\delta$, v_n is a minimizer of (10.63) subject to $v_n \in H^{-1}(\mathcal{U})$ with y^δ replaced by y_n, and $x_n = Hv_n$ (i.e., x_n is a minimizer of (10.62) subject to $x_n \in U$ with y^δ replaced by y_n).*

Then, $\{v_n\}$ converges (in the $\mathcal{L}^2$-norm) to the unique minimizer of (10.63) in $H^{-1}(\mathcal{U})$, while $\{x_n\}$ converges in the $\mathcal{L}^1$-norm to the unique minimizer of (10.62) in $\mathcal{U}$.

10.6. Convergence of Maximum Entropy Regularization

Proof. Concerning Problem 10.27 (i.e., the minimization of (10.63)), this follows from Theorem 10.2 with $F = T \circ H$, taking into account that Problem 10.26 (and hence also the equivalent Problem 10.27) has a unique solution (see Remark 10.39). As $\mathcal{U} \subset \mathcal{E}_m(\Omega)$, H is bijective as a function acting from $\overline{\mathcal{U}}$ into $\mathcal{U}$ (see Proposition 10.28 and its proof). Proposition 10.24, Proposition 10.35 and Lemma 10.36 then show that the assumptions needed for Theorem 10.2 are fulfilled. By continuity of H with respect to the appropriate norms (cf. Proposition 10.24), $\mathcal{L}^2$-convergence of $\{v_n\}$ implies $\mathcal{L}^1$-convergence of $\{x_n\}$. The equivalence of Problems 10.26 and 10.27 (see Proposition 10.28) then implies that each x_n is a minimizer (in $\mathcal{U}$) of (10.62) with y^δ replaced by y_n, and that $\{x_n\}$ converges to the minimizer of (10.62). ∎

Thus, maximum entropy regularization is stable (in the $\mathcal{L}^1$-norm) with respect to $\mathcal{L}^2$-perturbations in the data for fixed $\alpha > 0$. Here (and in analogous contexts below), the convexity of $\mathcal{U}$ is only needed to guarantee the uniqueness of the maximum entropy solution and of its regularizations. Without the convexity of $\mathcal{U}$, the results still hold, but convergence has to be understood in a set-valued sense as in Theorem 10.2.

In analogy to the best-approximate solution (with respect to the norm, cf. Definition 2.1), and with the concept of a minimum-norm solution (i.e., an x^*-minimum-norm solution in the sense of Section 10.2 with $x^* = 0$) we now introduce the *maximum entropy solution* of an equation:

Definition 10.38. *For a set $\mathcal{U}$ of nonnegative functions, we call a function $x_E^\dagger \in \mathcal{U}$ a $\mathcal{U}$-maximum-entropy solution of (10.61) if $T x_E^\dagger = y$ and*

$$E(x_E^\dagger) = \min\{E(x) \mid x \in \mathcal{U},\ T x = y\}$$

holds.

Remark 10.39. Note that, if $\mathcal{U}$ is convex, then there exists at most one $\mathcal{U}$-maximum-entropy solution due to the strict convexity of E. In general, the existence of a solution of (10.61) in $\mathcal{U}$ need not imply that a $\mathcal{U}$-maximum-entropy solution exists. We note, however, that, if v is a minimum-norm solution of (10.50) and $x = Hv$, then, by definition of H, x is a $\mathcal{U}$-maximum-entropy solution of (10.61). On the other hand, if a $\mathcal{U}$-maximum-entropy solution x of (10.61) exists, then a $v \in H^{-1}(\mathcal{U})$ satisfying $x = Hv$ exists (as H is surjective) and is a minimum-norm solution of (10.50) due to (10.48). Hence, as soon as (10.50) has a minimum-norm solution, (10.61) has a $\mathcal{U}$-maximum-entropy solution; this, in turn, is certainly the case if $T \circ H$ is weakly closed (see Section 10.1), hence especially, if $\mathcal{U}$ is compact in measure. Under the same conditions, Problems 10.26 and 10.27 are solvable (cf. Section 10.2 and Proposition 10.28).

Theorem 10.40. *Let $\mathcal{U}$ be as in Theorem 10.37. Let (10.61) have a solution in $\mathcal{U}$. Let y^δ satisfy $\|y - y^\delta\|_{\mathcal{L}^2} \leq \delta$. Let H and T be as in Theorem 10.37. Let*

274   10. Tikhonov Regularization of Nonlinear Problems

$\alpha = \alpha(\delta) > 0$ be such that

$$\alpha(\delta) \to 0, \quad \frac{\delta^2}{\alpha(\delta)} \to 0$$

as $\delta \to 0$. For any $\delta > 0$, let x_α^δ solve Problem 10.26 (constrained to $\mathcal{U}$). Then $\{x_\alpha^\delta\}$ converges in the $\mathcal{L}^1$-norm to the unique $\mathcal{U}$-maximum-entropy solution of (10.61) as $\delta \to 0$.

Proof. Taking into account the equivalence of Problems 10.26 and 10.27 (see Proposition 10.28), the result follows from Theorem 10.3 with $F = T \circ H$ (taking into account the uniqueness of a $\mathcal{U}$-maximum-entropy solution). As above, Proposition 10.24, Proposition 10.35 and Lemma 10.36 show that the assumptions needed are fulfilled. ∎

We have seen in previous sections, that for obtaining convergence rates for regularization methods, additional conditions have to be imposed on the exact solution. These conditions have the form of "source conditions" (cf., e.g., (3.30) or (iii) in Theorem 10.4). A similar condition is also needed for obtaining convergence rates for maximum entropy regularization. As in the conditions just quoted, it involves the adjoint of T. However, since here T is considered as an operator from $\mathcal{L}^1(\Omega)$ into $\mathcal{L}^2(\Omega^*)$, T^* is the (Banach-space) adjoint mapping $\mathcal{L}^2(\Omega^*)$ into $\mathcal{L}^\infty(\Omega)$ in the next theorem:

Theorem 10.41. *Let $\mathcal{U}$, H and T be as in Theorem 10.37, and let (10.61) have a solution in $\mathcal{U}$. By $x_E^\dagger$ we denote the $\mathcal{U}$-maximum-entropy solution of (10.61), $v^\dagger$ be such that $x_E^\dagger = Hv^\dagger$. Assume that*

$$1 + \log \frac{x_E^\dagger}{m} \in \mathcal{R}(T^*) \tag{10.69}$$

and that $x_E^\dagger/m$ is bounded. Let $y^\delta \in \mathcal{L}^2(\Omega^*)$ satisfy $\|y - y^\delta\|_{\mathcal{L}^2} \leq \delta$. Let, for $\alpha, \delta > 0$, x_α^δ solve Problem 10.26 (constrained to $\mathcal{U}$), and v_α^δ be such that $x_\alpha^\delta = Hv_\alpha^\delta$. Then, for $\alpha \sim \delta$, we have the convergence rates

$$\|v_\alpha^\delta - v^\dagger\|_{\mathcal{L}^2} = O(\sqrt{\delta}), \tag{10.70}$$

and

$$\|x_\alpha^\delta - x_E^\dagger\|_{\mathcal{L}^1} = O(\sqrt{\delta}). \tag{10.71}$$

Proof. As above, Proposition 10.24, Proposition 10.35 and Lemma 10.36 show continuity and weak closedness of the operator $F = T \circ H$. Existence of a minimum-norm solution of (10.50) follows from the arguments given in Remark 10.39. Since T is a continuous linear operator, Proposition 10.29 ensures Fréchet-differentiability of $T \circ H$. Thus, condition (i) of Theorem 10.4 is satisfied. Condition (iii) of that Theorem demands existence of a $w \in \mathcal{L}^2(\Omega^*)$ such that

$$v^\dagger = (T \circ H)'(v^\dagger)^* w, \tag{10.72}$$

10.6. Convergence of Maximum Entropy Regularization

where * denotes the Hilbert space adjoint. Observing that (cf. Proposition 10.29 and Remark 10.30)

$$(H'(v)z)(t) = (H'(v)^*z)(t) = \frac{\partial \phi}{\partial v}(t, v(t))z(t)$$

(now, * again denotes the Banach space adjoint, i.e., $H'(v)^* : \mathcal{L}^\infty(\Omega) \to \mathcal{L}^2(\Omega)$), (10.59) implies that (10.72) is equivalent to

$$T^*w(t) = \frac{v^\dagger(t)}{\frac{\partial \phi}{\partial v}(t, v^\dagger(t))} = \frac{1}{2}\left(1 + \log \frac{\phi(t, v^\dagger(t))}{m(t)}\right). \tag{10.73}$$

This, together with $x_E^\dagger = Hv^\dagger$, shows that (10.73), i.e., condition (iii) of Theorem 10.4 is in our situation equivalent to (10.69).

We could now proceed to check conditions (ii) and (iv) of Theorem 10.4. For this, we would need additional assumptions. Thus, we proceed along the lines of Remark 10.5, replacing (ii) and (iv) of Theorem 10.4 by a weaker condition. However, since H is not twice Fréchet-differentiable, we have to modify this argument as follows: For r_α^δ as in (10.11), i.e.,

$$r_\alpha^\delta = (T \circ H)v_\alpha^\delta - (T \circ H)v^\dagger - (T \circ H)'(v^\dagger)(v_\alpha^\delta - v^\dagger)$$

we have

$$r_\alpha^\delta = T\left(Hv_\alpha^\delta - Hv^\dagger - H'(v^\dagger)(v_\alpha^\delta - v^\dagger)\right) = T\sigma_\alpha^\delta,$$

where $\sigma_\alpha^\delta \in \mathcal{L}^1(\Omega)$ is given by

$$\sigma_\alpha^\delta(t) = \int_0^1 (1-\tau) \frac{\partial^2 \phi}{\partial v^2}\left(t, v^\dagger(t) + \tau(v_\alpha^\delta(t) - v^\dagger(t))\right)(v_\alpha^\delta(t) - v^\dagger(t))^2 \, d\tau.$$

Now, (10.73) and the fact that $x_E^\dagger(t) = \phi(t, v^\dagger(t))$ (for $t \in \Omega$) give

$$2\langle w, r_\alpha^\delta \rangle = 2\langle w, T\sigma_\alpha^\delta \rangle = 2\langle T^*w, \sigma_\alpha^\delta \rangle = \langle 1 + \log \frac{x_E^\dagger}{m}, \sigma_\alpha^\delta \rangle.$$

As $x_E^\dagger/m$ is bounded, Lemma 10.33 now implies that

$$2\langle w, r_\alpha^\delta \rangle \leq \rho \|v_\alpha^\delta - v^\dagger\|^2$$

for some $\rho < 1$. Inserting this into the proof of Theorem 10.4 we obtain the convergence rate (10.70).

In order to get (10.71), we make use of Lemmata 10.22 and 10.31. The latter implies that for fixed t, $\frac{\partial \phi}{\partial v}$ is concave, so that we can estimate

$$0 \leq \frac{\partial \phi}{\partial v}(t, v) \leq \frac{\partial \phi}{\partial v}(t, 0) + v \frac{\partial^2 \phi}{\partial v^2}(t, 0) = \sqrt{\frac{2m(t)}{e}} + \frac{2v}{3},$$

where the last equality follows (with de l'Hospital's rule) from (10.59). We now estimate $|x_\alpha^\delta(t) - x_E^\dagger(t)|$:

$$|x_\alpha^\delta(t) - x_E^\dagger(t)| = |\phi(t, v_\alpha^\delta(t)) - \phi(t, v^\dagger(t))|$$
$$= \int_{\min(v_\alpha^\delta(t), v^\dagger(t))}^{\max(v_\alpha^\delta(t), v^\dagger(t))} \frac{\partial \phi}{\partial v}(t, u)\, du \leq \sqrt{\frac{2m(t)}{e}}\, |v_\alpha^\delta(t) - v^\dagger(t)| + \frac{|v_\alpha^\delta(t)^2 - v^\dagger(t)^2|}{3}.$$

With $m(t) \leq M_0$ for all $t \in \Omega$ (see Definition 10.21), this implies that

$$\|x_\alpha^\delta - x_E^\dagger\|_{\mathcal{L}^1} = O(\|v_\alpha^\delta - v^\dagger\|_{\mathcal{L}^2}),$$

which, together with (10.70), implies (10.71). ∎

The source condition (10.69) is very similar to (5.71) needed for the analogous convergence rates result (10.71). Since these results are proven by completely different methods, this shows that these conditions (and also convergence in $\mathcal{L}^1$) are quite natural. A convergence analysis for finite-dimensional approximations of (10.46) and a numerical example for a problem from physical chemistry (cf. Section 1.3) can be found in [75].

Since in Theorem 10.4 the weak closedness is not needed for the result about convergence rates as soon as the existence of (regularized) solutions is guaranteed, a variant of Theorem 10.41 where $\mathcal{U}$ need not be compact in measure can be proven: there, one has either to *assume* that Problem 10.26 is solvable or to relax that problem by taking as regularized solution $x_\alpha^{\delta,\eta}$ any element in $\mathcal{U}$ such that

$$\|Tx_\alpha^{\delta,\eta} - y^\delta\|^2 + \alpha E(x_\alpha^{\delta,\eta}) \leq \|Tx - y^\delta\|^2 + \alpha E(x) + \eta$$

for all $x \in \mathcal{U}$, which is what one does in practice anyway (cf. the end of Section 10.2). The tolerance $\eta > 0$ has to be linked to the noise level by requiring $\eta = O(\delta^2)$ (cf. [75]).

The results of this section have been carried over to the case that T is nonlinear in [76]. In [173], analogous results are proven for maximum entropy type regularization methods of the form

$$\|Tx - y^\delta\|^2 + \alpha E(Lx) \to \min,$$

where L is a differential operator.

11. Iterative Methods for Nonlinear Problems

11.1. The Nonlinear Landweber Iteration

We now turn to an extension of the Landweber iteration (6.2) to nonlinear ill-posed problems. We shall assume throughout this section that F is a map between Hilbert spaces $\mathcal{X}$ and $\mathcal{Y}$, and that F has a Fréchet-derivative $F'(\cdot)$ which is continuous in the open convex set $\mathcal{D}(F) \subset \mathcal{X}$.

It is easy to see that the update $T^*(y^\delta - Tx_{k-1}^\delta)$ in the Landweber iteration for linear problems $Tx = y^\delta$ is the direction of the negative gradient at $x = x_{k-1}^\delta$ of the quadratic functional
$$\|y^\delta - Tx\|^2.$$
Replacing Tx in this functional by $F(x)$ we obtain
$$F'(x)^*(y^\delta - F(x))$$
as the negative gradient of the corresponding functional for nonlinear problems. With this in mind, we define the *nonlinear Landweber iteration* by updating x_{k-1}^δ along this direction via

$$x_k^\delta = x_{k-1}^\delta + F'(x_{k-1}^\delta)^*(y^\delta - F(x_{k-1}^\delta)), \quad k \in \mathbb{N}. \tag{11.1}$$

As always, $x_0 = x^*$ is an initial guess which may incorporate a-priori knowledge of an exact solution $x^\dagger$. Obviously, (11.1) reduces to the familiar Landweber iteration (6.2) when $F(x) = Tx$ is linear.

For nonlinear problems, iteration methods like (11.1) will not have a global convergence property, as opposed to, e.g., Tikhonov regularization, cf. Theorem 10.2. Moreover, since the argument x_{k-1}^δ of the Fréchet-derivative in (11.1) changes in each iteration, the sequence $\{x_k^\delta\}$ will in general not remain within a certain invariant subspace $\mathcal{X}_\mu$ like in the linear case. As a consequence, we will have to impose strong additional conditions on F to guarantee convergence rates.

We illustrate this latter remark with an example. In Section 10.2 we have derived conditions which imply the convergence rate $O(\sqrt{\delta})$ for Tikhonov regularization of nonlinear problems with appropriate regularization parameters, cf. Theorem 10.4. The source condition

$$x^\dagger - x^* = F'(x^\dagger)^* w_0, \quad w_0 \in \mathcal{Y}, \tag{11.2}$$

is a natural one, since it reduces to a sufficient and almost necessary condition for the same rate of convergence in the linear case, whether we use Tikhonov regularization (cf. Section 5.1) or Landweber iteration (cf. Section 6.1). Assuming this condition

to hold for $x_0 = x^*$ in the nonlinear Landweber iteration, we observe from (11.1) that
$$x^\dagger - x_1^\delta = x^\dagger - x^* - F'(x^*)^*(y^\delta - F(x^*)),$$
and hence, (11.2) with x^* replaced by x_1^δ, i.e.,
$$x^\dagger - x_1^\delta = F'(x^\dagger)^* w_1, \qquad w_1 \in \mathcal{Y}, \tag{11.3}$$
can only hold if $F'(x^*)^*(y^\delta - F(x^*)) \in \mathcal{R}(F'(x^\dagger)^*)$. Since we do not want to impose any restrictions on y^δ besides $\|y^\delta - y\| \le \delta$ we conclude that a necessary condition for (11.3) to hold for all perturbations y^δ of y is that
$$\mathcal{R}(F'(x^*)^*) \subset \mathcal{R}(F'(x^\dagger)^*).$$
In other words, we can find a linear operator Q_{x^*} (depending on x^* and of course on $x^\dagger$) with
$$F'(x)^* = F'(x^\dagger)^* Q_x \tag{11.4}$$
for $x = x^*$. With (11.4) we obtain
$$x^\dagger - x_1^\delta = F'(x^\dagger)^* \bigl(w_0 - Q_{x^*}(y - F(x^*)) + Q_{x^*}(y - y^\delta) \bigr),$$
and it follows that Q_{x^*} must be bounded if we further want to maintain a "smallness property" of the preimage w_1 like assumption (iv) on w_0 in Theorem 10.4. So far, we have only considered x_1^δ, but it is obvious that in order to maintain these properties *throughout* the iteration we have to impose factorizations (11.4) for *all* arguments $x = x_k^\delta$ instead of just x^*. This is of course a very severe restriction, showing that we cannot expect a general convergence rate result like Theorem 10.4 for the nonlinear Landweber iteration.

We mention, however, that, if (11.4) is satisfied in a neighbourhood of $x^\dagger$ and if, in addition, Q_x is Lipschitz continuous in x near $x^\dagger$, i.e.,
$$\|Q_x - I\| \le c\|x - x^\dagger\|,$$
then the convergence rate $O(\sqrt{\delta})$ can be established for the Landweber iteration with an appropriate stopping rule (the same discrepancy principle that we shall analyze in Theorem 11.5 below); these results have been obtained in [120], essentially by comparing the nonlinear Landweber iteration with its linear counterpart (6.2), where $T = F'(x^\dagger)$.

Here we will only investigate the convergence of the sequences $\{x_k\}$ and $\{x_k^\delta\}$ under a considerably weaker condition, cf. (11.6) below. Although still restrictive, this condition is quite natural, as we are going to explain now. If $F'(\cdot)$ is Lipschitz continuous and $x, \tilde{x} \in \mathcal{D}(F)$, then the error bound
$$\|F(\tilde{x}) - F(x) - F'(x)(\tilde{x} - x)\| \le c\|\tilde{x} - x\|^2 \tag{11.5}$$
for the Taylor approximation of F holds true. This upper bound has proved useful in the analysis of Tikhonov regularization, cf., e.g., (10.12), but it turns out that

11.1. The Nonlinear Landweber Iteration

(11.5) carries too little information about the local behaviour of F around $x^\dagger$ to draw conclusions about convergence of the nonlinear Landweber iteration from the iteration history. The crucial point is that the left-hand side of (11.5) can be much smaller than the right-hand side for certain pairs of points $\tilde{x}$ and x, whatever close to each other they are. For example, fix $x \in \mathcal{D}(F)$, and assume that F is continuous and compact. Then $F'(x)$ is compact, and hence, for every sequence $\{\tilde{x}_n\}$ with $\|\tilde{x}_n - x\| = \varepsilon$ for all $n \in \mathbb{N}$, the left-hand side of (11.5) goes to zero as $n \to \infty$ whereas the right-hand side remains $c\varepsilon^2$ for all n. In order to remedy this situation we impose a different assumption on the remainder term of the Taylor approximation, namely

$$\|F(\tilde{x}) - F(x) - F'(x)(\tilde{x} - x)\| \leq \eta \|F(\tilde{x}) - F(x)\|, \qquad \eta < \frac{1}{2}, \qquad (11.6)$$

to be valid for $\tilde{x} - x$ sufficiently small. From (11.6) follows immediately with the triangle inequality that

$$\frac{1}{1+\eta} \|F'(x)(\tilde{x} - x)\| \leq \|F(\tilde{x}) - F(x)\| \leq \frac{1}{1-\eta} \|F'(x)(\tilde{x} - x)\|. \qquad (11.7)$$

In particular, if some $F'(x)$ has a nontrivial nullspace, then F must be constant along any affine subspace through x and parallel to $\mathcal{N}(F'(x))$, and vice versa. It is therefore clear that this restriction is still a severe one.

We remark that (11.6) is stronger than (11.5) in the sense that the upper bound on the right-hand side of (11.6) involves the gap between the images rather than the distance between the arguments, whereas on the other hand (11.5) is stronger than (11.6) in the sense that the Taylor estimate comes with the second power of the distance between x and $\tilde{x}$ on the right-hand side. This second power may sometimes be useful, though, to establish $\eta < 1/2$ in (11.6), provided $\tilde{x}$ and x are sufficiently close.

Example 11.1. We consider once more Example 10.17, i.e.,

$$-(au_s)_s + cu = f, \qquad u(0) = g_0, \quad u(1) = g_1, \qquad (11.8)$$

where, $c, f \in \mathcal{L}^2[0,1]$ with $c \geq 0$ almost everywhere, and g_0, g_1 are given. We are concerned with the parameter-to-solution mapping $F : a \mapsto u$, and want to determine a from knowledge of u, assuming that $a \geq \underline{a} > 0$. We use the same notation as in Example 10.17, except that we simply write u instead of $u(a)$ for the solution $F(a)$ of (11.8); in particular, we make use of the operator $A(a)$ of (10.40). Let g be the linear interpolant which satisfies the boundary conditions of (11.8): then $F(a) - g \in \mathcal{D}(A(a))$, and

$$A(a)(F(a) - g) = f + \gamma a_s - cg,$$

where $\gamma = g_1 - g_0 \in \mathbb{R}$ is the slope of g.

11. Iterative Methods for Nonlinear Problems

To verify (11.6) we need to estimate the error of the Taylor approximation

$$F(\tilde{a}) - F(a) - F'(a)(\tilde{a} - a)$$
$$= A(a)^{-1}\Big(A(a)(F(\tilde{a}) - F(a)) - A(a)F'(a)(\tilde{a} - a)\Big)$$
$$= A(a)^{-1}\Big(A(a)(F(\tilde{a}) - g) + A(a)(g - F(a)) - [(\tilde{a} - a)u_s]_s\Big)$$
$$= A(a)^{-1}\Big((A(a) - A(\tilde{a}))(F(\tilde{a}) - g)$$
$$\quad + A(\tilde{a})(F(\tilde{a}) - g) - A(a)(F(a) - g) - [(\tilde{a} - a)u_s]_s\Big)$$
$$= A(a)^{-1}\Big((A(a) - A(\tilde{a}))(F(\tilde{a}) - g) + \gamma\tilde{a}_s - \gamma a_s - [\tilde{a}u_s]_s + [au_s]_s\Big)$$
$$= A(a)^{-1}\Big((A(a) - A(\tilde{a}))(F(\tilde{a}) - g) + A(\tilde{a})(u - g) - A(a)(u - g)\Big).$$

Thus, we have established the functional equation

$$F(\tilde{a}) - F(a) - F'(a)(\tilde{a} - a) = A(a)^{-1}(A(a) - A(\tilde{a}))(F(\tilde{a}) - F(a)).$$

Under the given assumptions the operator $A(a) - A(\tilde{a})$ – as a mapping from $\mathcal{L}^2$ into the anti-dual $\mathcal{X}_{-2} = (\mathcal{H}^2 \cap \mathcal{H}_0^1)'$ of $\mathcal{H}^2 \cap \mathcal{H}_0^1$ is bounded by $\|a - \tilde{a}\|_{\mathcal{H}^1}$, and $A(a)^{-1}$ is bounded as operator from $\mathcal{X}_{-2}$ back into $\mathcal{L}^2$ (cf., e.g., [40, Lemma 2.6]). Therefore, if $\tilde{a}$ is sufficiently close to a in $\mathcal{H}^1$, then we have

$$\|A(a)^{-1}(A(a) - A(\tilde{a}))\|_{\mathcal{L}^2, \mathcal{L}^2} \leq \frac{1}{2},$$

and hence, (11.6) holds.

Note that for this example we have $F(a) = F(\tilde{a})$ if and only if $[(\tilde{a} - a)u_s]_s = 0$ for $u = u(a)$, i.e., if and only if

$$\tilde{a} - a \in \text{span}\{\frac{1}{u_s}\}. \tag{11.9}$$

On the other hand, as we have established before for any F satisfying (11.6), the right-hand side of (11.9) is the nullspace of $F'(a)$ and also the nullspace of $F'(\tilde{a})$ if and only if $u = F(a) = F(\tilde{a})$. We mention, however, that the range of $F'(a)^*$ is not invariant under a, and therefore, the theory developed in [120] concerning convergence rates of the nonlinear Landweber iteration does not apply to this example. Nevertheless, this theory can be applied to identify the parameter function c in (11.8), cf. Example 10.16; details are provided in [120].

Before we turn to a convergence analysis of the nonlinear Landweber iteration we want to emphasize – in view of Theorem 10.2 concerning Tikhonov regularization – that for fixed iteration index k the iterate x_k^δ depends continuously on the data y^δ, since x_k^δ is the result of a combination of continuous operations. This will be an important point in our argument below, and quite similar to a general theory developed by Alifanov and Rumjancev [6] (see also Proposition 3.4).

To begin with, we formulate the following extension of the monotonicity property (Proposition 6.3) of the Landweber iteration to nonlinear problems which satisfy Assumption (11.6).

11.1. The Nonlinear Landweber Iteration

Proposition 11.2. *Let F be Fréchet-differentiable with $\|F'(\cdot)\| \leq 1$, and assume that (11.6) holds in $\mathcal{D}(F)$. Then, for any solution $x \in \mathcal{D}(F)$ of $F(x) = y$, a sufficient condition for x_{k+1}^δ to be a better approximation of x than x_k^δ is that*

$$\|y^\delta - F(x_k^\delta)\| > 2 \frac{1+\eta}{1-2\eta} \delta. \tag{11.10}$$

Proof. From definition (11.1) we have

$$\|x - x_{k+1}^\delta\|^2 - \|x - x_k^\delta\|^2$$
$$= 2\langle x_k^\delta - x, x_{k+1}^\delta - x_k^\delta \rangle + \|x_{k+1}^\delta - x_k^\delta\|^2$$
$$= 2\langle F'(x_k^\delta)(x_k^\delta - x), y^\delta - F(x_k^\delta) \rangle$$
$$\quad + \langle y^\delta - F(x_k^\delta), F'(x_k^\delta)F'(x_k^\delta)^*(y^\delta - F(x_k^\delta)) \rangle$$
$$= 2\langle y^\delta - F(x_k^\delta), y^\delta - F(x_k^\delta) - F'(x_k^\delta)(x - x_k^\delta) \rangle$$
$$\quad - \langle y^\delta - F(x_k^\delta), (I - F'(x_k^\delta)F'(x_k^\delta)^*)(y^\delta - F(x_k^\delta)) \rangle - \|y^\delta - F(x_k^\delta)\|^2$$
$$\leq 2\langle y^\delta - F(x_k^\delta), y^\delta - F(x_k^\delta) - F'(x_k^\delta)(x - x_k^\delta) \rangle - \|y^\delta - F(x_k^\delta)\|^2,$$

since $I - F'(x_k^\delta)F'(x_k^\delta)^*$ is positive semidefinite by assumption. Using the Cauchy-Schwarz inequality and (11.6) we obtain from this the inequality

$$\|x - x_{k+1}^\delta\|^2 - \|x - x_k^\delta\|^2$$
$$\leq \|y^\delta - F(x_k^\delta)\| \Big(2\eta \|y - F(x_k^\delta)\| + 2\delta - \|y^\delta - F(x_k^\delta)\|\Big)$$
$$\leq \|y^\delta - F(x_k^\delta)\| \Big((2\eta - 1)\|y^\delta - F(x_k^\delta)\| + 2(1+\eta)\delta\Big). \tag{11.11}$$

The assertion now follows since the right-hand side is negative if (11.10) holds. ∎

We mention that the factor in front of δ on the right-hand side of (11.10) is always greater than 2, which is the corresponding number in Proposition 6.3. From the proof of Proposition 11.2 we easily extract an inequality which is important in its own right:

Corollary 11.3. *Under the assumptions of Proposition 11.2, if $x_k^\delta \in \mathcal{D}(F)$ and $\|y^\delta - F(x_k^\delta)\| \geq \tau\delta$ for all $0 \leq k < k_*$ with some $\tau > 2(1+\eta)/(1-2\eta)$, then*

$$k_*(\tau\delta)^2 \leq \sum_{k=0}^{k_*-1} \|y^\delta - F(x_k^\delta)\|^2 \leq \frac{\tau}{(1-2\eta)\tau - 2(1+\eta)} \|x - x^*\|^2. \tag{11.12}$$

In particular, if $y^\delta = y$ (i.e., if $\delta = 0$) and all iterates x_k remain in $\mathcal{D}(F)$ then

$$\sum_{k=0}^{\infty} \|y - F(x_k)\|^2 < \infty. \tag{11.13}$$

11. Iterative Methods for Nonlinear Problems

Proof. From (11.11) we conclude that for $0 \le k < k_*$,

$$\|x - x_{k+1}^\delta\|^2 - \|x - x_k^\delta\|^2 \le \|y^\delta - F(x_k^\delta)\|^2 \left(2\eta - 1 + \frac{2}{\tau}(1+\eta)\right).$$

Adding up these inequalities for k from 0 through $k_* - 1$ we obtain

$$\|x - x^*\|^2 - \|x - x_{k_*}^\delta\|^2 \ge \left(1 - 2\eta - \frac{2}{\tau}(1+\eta)\right) \sum_{k=0}^{k_*-1} \|y^\delta - F(x_k^\delta)\|^2,$$

which yields (11.12). Obviously, if $\delta = 0$ then k_* may be any positive integer in (11.12), which proves (11.13). ∎

The consequences of this corollary are two-fold. First, if the nonlinear Landweber iteration is run with precise data $y = y^\delta$, then (11.13) shows that the residual norms of the iterates tend to zero as $k \to \infty$. That is, if the iteration converges, then the limit is necessarily a solution of $F(x) = y$. Second, if $y^\delta \ne y$, then (11.12) implies that there must be a unique iteration index k_* such that $\|y^\delta - F(x_k^\delta)\| \ge \tau\delta$ for all $k < k_*$ but is violated at $k = k_*$. This argument is the justification for the discrepancy principle that will be employed as a stopping rule below. But first we shall consider the behaviour of the iteration given precise data.

Theorem 11.4. *Let F have a continuous Fréchet-derivative, with $\|F'(\cdot)\| \le 1$ in a ball $\mathcal{B}_{2\rho}(x^*) \subset \mathcal{D}(F)$ around x^*. If $F(x) = y$ is solvable in $\mathcal{B}_\rho(x^*)$ and (11.6) holds in $\mathcal{B}_{2\rho}(x^*)$, then the nonlinear Landweber iteration applied to y converges to a solution of $F(x) = y$.*

Proof. We restrict $\mathcal{D}(F)$ to $\mathcal{B}_{2\rho}(x^*)$ in order to apply Proposition 11.2. Let x be any solution of $F(x) = y$ in $\mathcal{B}_\rho(x^*)$, and put

$$e_k := x - x_k.$$

From Proposition 11.2 follows that $\|e_k\|$ converges to some $\varepsilon \ge 0$. We are going to show that $\{e_k\}$ is a Cauchy sequence. Given $j \ge k$ we choose some integer l between k and j with

$$\|y - F(x_l)\| \le \|y - F(x_i)\| \qquad \text{for all } k \le i \le j. \tag{11.14}$$

We have

$$\|e_j - e_k\| \le \|e_j - e_l\| + \|e_l - e_k\| \tag{11.15}$$

and

$$\begin{aligned}\|e_j - e_l\|^2 &= 2\langle e_l - e_j, e_l \rangle + \|e_j\|^2 - \|e_l\|^2, \\ \|e_l - e_k\|^2 &= 2\langle e_l - e_k, e_l \rangle + \|e_k\|^2 - \|e_l\|^2.\end{aligned} \tag{11.16}$$

For $k \to \infty$, the last two terms on each of the right-hand sides of (11.16) converge to $\varepsilon^2 - \varepsilon^2 = 0$. We now apply (11.1) to show that $\langle e_l - e_j, e_l \rangle$ also tends to zero as

11.1. The Nonlinear Landweber Iteration

$k \to \infty$:

$$
\begin{aligned}
|\langle e_l - e_j, e_l \rangle| &= \left|\sum_{i=l}^{j-1} \langle F'(x_i)^*(y - F(x_i)), e_l \rangle\right| \leq \sum_{i=l}^{j-1} |\langle y - F(x_i), F'(x_i)(x - x_l) \rangle| \\
&\leq \sum_{i=l}^{j-1} \|y - F(x_i)\| \, \|F'(x_i)(x - x_i + x_i - x_l)\| \\
&\leq \sum_{i=l}^{j-1} \|y - F(x_i)\| \big(\|y - F(x_i) - F'(x_i)(x - x_i)\| \\
&\qquad + \|y - F(x_l)\| + \|F(x_l) - F(x_i) - F'(x_i)(x_l - x_i)\| \big) \\
&\leq (1 + \eta) \sum_{i=l}^{j-1} \|y - F(x_i)\| \, \|y - F(x_l)\| + 2\eta \sum_{i=l}^{j-1} \|y - F(x_i)\|^2 \\
&\leq (1 + 3\eta) \sum_{i=l}^{j-1} \|y - F(x_i)\|^2,
\end{aligned}
$$

where we have used (11.14) to obtain the last inequality. Similarly, one can show that

$$
|\langle e_l - e_k, e_l \rangle| \leq (1 + 3\eta) \sum_{i=k}^{l-1} \|y - F(x_i)\|^2.
$$

With these estimates it follows from (11.13) that the right-hand sides of (11.16) go to zero as $k \to \infty$, and we have shown, cf. (11.15), that $\{e_k\}$ and thus $\{x_k\}$ are Cauchy sequences. Finally, in view of the remark following Corollary 11.3, the limit of x_k as $k \to \infty$ must be a solution of $F(x) = y$. ∎

Note that the assumption that $F'(\cdot)$ be continuous has not been used in the proof of Theorem 11.4. F' may even be some weaker type of derivative, e.g., a Gateaux derivative.

We emphasize that, in general, the limit x of the nonlinear Landweber iteration is no x^*-minimum-norm solution as introduced in Section 10.1. Nevertheless, because of the special structure of the set of solutions to $F(x) = y$ due to assumption (11.6), an x^*-minimum-norm solution $x^\dagger$ can easily be obtained from the fact that $x^\dagger - x \in \mathcal{N}(F'(x))$.

Concerning perturbed data we employ the discrepancy principle as a stopping rule for the nonlinear Landweber iteration. As mentioned in the remark following Corollary 11.3 this results in a well defined stopping index $k(\delta, y^\delta)$ with

$$
\|y^\delta - F(x^\delta_{k(\delta, y^\delta)})\| \leq \tau\delta < \|y^\delta - F(x^\delta_k)\|, \qquad 0 \leq k < k(\delta, y^\delta) \tag{11.17}
$$

if the parameter τ is chosen subject to the constraint

$$
\tau > 2\frac{1 + \eta}{1 - 2\eta}. \tag{11.18}
$$

Here, η is the same as in (11.6). Our final result of this section shows that this stopping rule renders the Landweber iteration a regularization method.

Theorem 11.5. *Under the assumptions of Theorem 11.4, if the Landweber iteration applied to y^δ is stopped with $k(\delta, y^\delta)$ according to the discrepancy principle (11.17) with τ subject to (11.18), then $x^\delta_{k(\delta,y^\delta)}$ converges to a solution $x \in \mathcal{B}_\rho(x^*)$ of $F(x) = y$ as $\delta \to 0$.*

Proof. Let x be the limit of the Landweber iteration with precise data y, and δ_n, $n = 1, 2, \ldots$, be a sequence converging to zero as $n \to \infty$. Denote by $y_n := y^{\delta_n}$ a corresponding sequence of perturbed data, and by $k_n = k(\delta_n, y^{\delta_n})$ the stopping index determined from the discrepancy principle for the Landweber iteration applied to the pair (y_n, δ_n).

Assume first that k is a finite accumulation point of $\{k_n\}$. Without loss of generality, we can assume that $k_n = k$ for all $n \in \mathbb{N}$. Thus, from the definition of k_n it follows that

$$\|y_n - F(x_k^{\delta_n})\| \leq \tau \delta_n. \tag{11.19}$$

As k is fixed, x_k^δ depends continuously on y^δ, and hence, if we go to the limit $n \to \infty$ in (11.19), we obtain

$$x_k^{\delta_n} \to x_k, \quad F(x_k^{\delta_n}) \to F(x_k) = y \quad \text{as} \quad n \to \infty.$$

In other words, the kth iterate of the Landweber iteration with precise data is a solution of $F(x) = y$, and hence, the iteration terminates with $x = x_k$, and $x_{k_n}^{\delta_n} \to x$ for this subsequence $\delta_n \to 0$.

It remains to consider the case where $k_n \to \infty$ as $n \to \infty$. Without loss of generality, we may assume that k_n increases monotonically with n. Then, for $n > m$ Proposition 11.2 yields

$$\begin{aligned} \|x_{k_n}^{\delta_n} - x\| &\leq \|x_{k_n-1}^{\delta_n} - x\| \leq \ldots \leq \|x_{k_m}^{\delta_n} - x\| \\ &\leq \|x_{k_m}^{\delta_n} - x_{k_m}\| + \|x_{k_m} - x\|. \end{aligned} \tag{11.20}$$

Given $\varepsilon > 0$ it follows from Theorem 11.4 that we can fix some $m = m(\varepsilon)$ so large that the second term on the right-hand side of (11.20) is smaller than $\varepsilon/2$. Because of the stability of the nonlinear Landweber iteration we also have $\|x_{k_m}^{\delta_n} - x_{k_m}\| < \varepsilon/2$ as soon as $\delta_n < \delta_{n(\varepsilon)}$, showing that the left-hand side of (11.20) is smaller than ε for $n > n(\varepsilon)$. Thus, $x_{k_n}^{\delta_n} \to x$ as $n \to \infty$. ∎

In practice, it may often be hard or impossible to verify assumption (11.6). Even if (11.6) holds, a reasonably small η may be impossible to derive analytically. Still, the discrepancy principle (e.g., with τ around 2) may be a useful option for a stopping rule. For example, for inverse scattering problems as considered in Section 1.7 there is little hope that (11.6) is valid. Nevertheless, good numerical results have been obtained with the nonlinear Landweber iteration in [118], where some ad hoc parameters τ between 2 and 4 have been used for the discrepancy principle.

We hasten to add that although no requirement like (11.6) is required for the convergence analysis of nonlinear Tikhonov regularization there is still a need for a similar assumption when it comes to the numerical implementation. The determination of the actual minimum of the Tikhonov functional by standard minimization routines is likely to fail for general functions F. Chavent and Kunisch [38], for example, introduce the notion of a *weakly nonlinear problem*, and give a sufficient condition based on geometric considerations under which the numerical computation of the minimum of the Tikhonov functional for a weakly nonlinear problem is well-posed. It has been shown in [118] that this condition from [38] is stronger than the assumption (11.6) which is required for the above convergence analysis of the nonlinear Landweber iteration.

We finally mention that generalizations of the above convergence analysis have been carried out successfully in [244] for the method of steepest descent, and in [267] for a nonlinear generalization of Showalter's method (cf. Example 4.7).

11.2. Newton Type Methods

While Newton type methods are the usual choice for the solution of nonlinear problems, there are a number of difficulties arising in the ill-posed case. As a consequence, for the time being, there are only few convergence results available. The key idea of any Newton type method consists in repeatedly linearizing the operator equation $F(x) = y$ around some approximate solution x_k^δ, and then solving the linearized problem

$$F'(x_k^\delta)(x_{k+1}^\delta - x_k^\delta) = y^\delta - F(x_k^\delta) \qquad (11.21)$$

for x_{k+1}^δ. However, if F is continuous and compact, then $F'(x_k^\delta)$ is compact, and therefore, (11.21) is ill-posed, too, in general. Consequently, (11.21) will typically have no solution (even no least-squares solution), and to obtain reasonable approximate solutions some sort of regularization will again be necessary.

Tikhonov regularization, applied to (11.21), for example, would give the approximation

$$x_{k+1}^\delta = x_k^\delta + (F'(x_k^\delta)^* F'(x_k^\delta) + \alpha_k I)^{-1} F'(x_k^\delta)^* (y^\delta - F(x_k^\delta)). \qquad (11.22)$$

This method, known as the *Levenberg-Marquardt method*, can alternatively be derived by linearizing F around x_k^δ within the (nonlinear) Tikhonov functional (10.3) with $x^* = x_k^\delta$, i.e.,

$$\|y^\delta - F(x_k^\delta) - F'(x_k^\delta)(z - x_k^\delta)\|^2 + \alpha_k \|z - x_k^\delta\|^2, \qquad (11.23)$$

and then minimizing this quadratic functional for $z = x_{k+1}^\delta$. It is obvious that other regularized approximations of the solution of (11.21) are also possible.

It is an important open problem of how to choose the regularization parameters α_k in (11.22) appropriately. Since the "update direction" in (11.22) is a descent direction for the residual norm, one option is to choose the parameters α_k so as to impose

$$\|y^\delta - F(x_{k+1}^\delta)\| < \|y^\delta - F(x_k^\delta)\|$$

in each step. Note that this can be guaranteed by choosing α_k sufficiently large, since the update direction $x_{k+1}^\delta - x_k^\delta$ approaches $1/\alpha_k$ times the Landweber direction $F'(x_k^\delta)^*(y^\delta - F(x_k^\delta))$ for α_k large.

The original idea behind the Levenberg-Marquardt approach is to minimize $\|y^\delta - F(x)\|$ within a *trust region* $\|x - x_k^\delta\| \leq h_k$, in which case α_k is the corresponding Lagrange parameter. The notion trust region refers to the presumed validity of the linearization (11.21). Depending on the agreement of the actual residual with the "predicted residual", i.e.,

$$\|y^\delta - F(x_{k+1}^\delta)\| \qquad \text{vs.} \qquad \|y^\delta - F(x_k^\delta) - F'(x_k^\delta)(x_{k+1}^\delta - x_k^\delta)\|,$$

the parameter h_k is enlarged (good agreement) or reduced (bad agreement) by, e.g., a factor of two.

While a convergence theory of trust region methods is available for well-posed problems (cf., e.g., [53]), we do not know about any convergence result for ill-posed problems, even with exact data y. Obviously, in view of the analysis of the nonlinear Landweber iteration, we have to consider both k and α_k as regularization parameters, and we may be tempted to stop the iteration (i.e., choose k) according to some variant of the discrepancy principle.

The iteration (11.22) may also be seen as a nonlinear version of iterated Tikhonov regularization, cf. (5.37). Recall that a finite number k of "iterations" in iterated Tikhonov regularization for linear problems has a finite qualification $\mu_0 = k$, and hence any parameter choice method (including the discrepancy principle) can only yield limited accuracy. However, if the discrepancy principle is applied to choose k for some fixed parameter α in iterated Tikhonov regularization, then full order-optimal accuracy can be obtained, since it is easily checked that the qualification of this iterative method is $\mu_0 = \infty$. This yields another argument for choosing the regularization parameters α_k not too small, and then hoping for the regularizing properties of the iteration to obtain good accuracy.

Recently, a variant of (11.22) has been suggested by Bakushinskii [19], namely the iteration process $x_0 = x^*$ and, for $k \geq 0$,

$$x_{k+1}^\delta = x_k^\delta + (F'(x_k^\delta)^* F'(x_k^\delta) + \alpha_k I)^{-1} [F'(x_k^\delta)^*(y^\delta - F(x_k^\delta)) + \alpha_k(x^* - x_k^\delta)]. \quad (11.24)$$

The approximate solutions x_{k+1}^δ are now minimizers of the functional

$$\|y^\delta - F(x_k^\delta) - F'(x_k^\delta)(z - x_k^\delta)\|^2 + \alpha_k \|z - x^*\|^2. \quad (11.25)$$

Note that the only difference between (11.25) and (11.23) is in the penalty term. To motivate this we mention that the reuse of x^* in the penalty term – besides its use as an initial guess – may have a stabilizing effect on the minimizing elements: for instance, if α_k remains bounded away from zero during the iteration, then the iterates x_{k+1}^δ will never diverge to infinity in norm. While this is a purely heuristic argument for using (11.25) over (11.23), the rigorous analysis in [19] and [31] shows that certain regularizing properties of this modified iteration can in fact be verified.

11.2. Newton Type Methods

In [19], Bakushinskii proved local convergence of method (11.24), essentially under the smoothness condition

$$x^\dagger - x^* = (F'(x^\dagger)^* F'(x^\dagger))^\mu w, \quad \mu \geq 1 \qquad (11.26)$$

if F' is Lipschitz continuous and if

$$\alpha_k > 0, \quad 1 \leq \frac{\alpha_k}{\alpha_{k+1}} \leq r, \quad \lim_{k \to \infty} \alpha_k = 0$$

for some $r > 1$. Here, $x^\dagger$ is an x^*-minimum-norm solution of $F(x) = y$. For exact data $y \in \mathcal{R}(F)$ he also obtained the convergence rate

$$\|x_k - x^\dagger\| = O(\alpha_k) \quad \text{as} \quad k \to \infty.$$

It was shown in [31] that the range $\mu \geq 1$ in (11.26) can be extended to $1/2 \leq \mu < 1$, in which case the corresponding rate

$$\|x_k - x^\dagger\| = O(\alpha_k^\mu) \quad \text{as} \quad k \to \infty,$$

is attained. Unfortunately, so far, similar results can be proved for $0 \leq \mu < 1/2$ only under additional assumptions on F', namely

$$F'(\tilde{x}) = R(\tilde{x}, x) F'(x) + Q(\tilde{x}, x), \quad x, \tilde{x} \in B_\rho(x_0), \qquad (11.27)$$

with

$$\|I - R(\tilde{x}, x)\| \leq c_R, \quad \|Q(\tilde{x}, x)\| \leq c_Q \|F'(x^\dagger)(\tilde{x} - x)\|,$$

and ρ, c_R and c_Q sufficiently small. This condition is similar to the one mentioned in Section 11.1 to obtain convergence rates for the Landweber iteration. Since α_k can be chosen as $\alpha_k := 2^{-k}$, method (11.24) converges much faster than the Landweber iteration, but each iteration step is of course more expensive.

For noisy data y^δ the iteration has to be stopped, as always, after an appropriate number of $k(\delta)$ steps. Under the appropriate conditions one can show that

$$\|x_{k(\delta)}^\delta - x^\dagger\| = O(\delta^{\frac{2\mu}{2\mu+1}}), \quad 0 \leq \mu \leq 1$$

if $k(\delta)$ is chosen such that

$$\delta \sim \alpha_{k(\delta)}^{\mu + \frac{1}{2}}.$$

Under the more restrictive condition (11.27), one can show that the generalized discrepancy principle (11.17) (with $\tau > 1$ sufficiently large) yields the rates

$$\|x_{k(\delta, y^\delta)}^\delta - x^\dagger\| = \begin{cases} o(\delta^{\frac{2\mu}{2\mu+1}}), & \text{if } 0 \leq \mu < \frac{1}{2}, \\ O(\sqrt{\delta}), & \text{if } \mu = \frac{1}{2}. \end{cases}$$

Since the proofs are rather technical, they are omitted here. See [31] for details.

A. Appendix

A.1. Weighted Polynomial Minimization Problems

In the following we denote by Π_k the set of all polynomials of degree k, and in Π_k^0 we collect those polynomials p in Π_k with $p(0) = 1$. Any polynomial in Π_k^0 can be employed in iterative methods as residual polynomial, cf. Section 6.2. To design *efficient* iterative methods, however, additional properties are required. For well-posed problems, for example, where the spectrum of T^*T is contained in $[a, b]$ with $0 < a < b$ we only wish to consider residual polynomials p_k for which $\|p_k\|_{C[a,b]}$ is small. This naturally leads to the minimization problem

$$\|p_k\|_{C[a,b]} \to \min, \qquad p_k \in \Pi_k^0. \tag{A.1}$$

Rewriting the polynomial $p_k \in \Pi_k^0$ as $p_k = 1 - \lambda q_{k-1}$ with $q_{k-1} \in \Pi_{k-1}$ we observe, for $w(\lambda) = \lambda$,

$$\|p_k\|_{C[a,b]} = \|w\left(\frac{1}{\lambda} - q_{k-1}\right)\|_{C[a,b]},$$

and hence, the above minimization problem is equivalent to approximating $1/\lambda$ in Π_{k-1} with respect to a weighted infinity norm with weight function $w > 0$. There is a huge literature on problems of this type. Here we shall collect those results which are of major importance for our purposes. For a more detailed exposition we refer to the book of Akhiezer [2].

We shall start with the "Equal-Ripple Theorem". It states that there is a unique best-approximation $q_{k-1}^* \in \Pi_{k-1}$; q_{k-1}^* (defining $p_k^* = 1 - \lambda q_k^*$) is characterized by the existence of an alternation set $\mathcal{S}_k$ of $k+1$ points $\{\lambda_{j,k}\}_{j=0}^k \subset [a, b]$ in increasing order, such that

$$w(\lambda_{j,k})\left(\frac{1}{\lambda_{j,k}} - q_{k-1}^*(\lambda_{j,k})\right) = p_k^*(\lambda_{j,k}) = (-1)^j \|p_k^*\|_{C[a,b]}.$$

Notice that the value of p_k^* at the smallest element $\lambda_{0,k}$ of $\mathcal{S}_k$ must be positive, since $p_k^*(0) = 1$. The Equal-Ripple Theorem essentially goes back to Chebyshev; the minimizing polynomials p_k^* are shifted Chebyshev polynomials of the first kind. We denote the Chebyshev polynomials by

$$T_k(x) = \begin{cases} \cos(k \arccos(x)), & x \in [-1, 1], \\ \cosh(k \operatorname{Arcosh}(x)), & x > 1, \\ (-1)^k \cosh(k \operatorname{Arcosh}(-x)), & x < -1, \end{cases}$$

and find

$$p_k^*(\lambda) = \frac{T_k(x(\lambda))}{T_k(x(0))}; \qquad x(\lambda) = 1 - \frac{2}{b-a}(\lambda - a).$$

290   A. Appendix

Note that $T_k(x(\lambda))$ is equioscillating in [a,b], attaining its maximum modulus 1 on an alternation set S_k of $k+1$ points in $[a,b]$; thus we obtain

$$\|p_k^*\|_{C[a,b]} = T_k(x(0))^{-1} = 2\kappa^k(1+\kappa^{2k})^{-1} \le 2\kappa^k$$

with

$$\kappa = e^{-\mathrm{Arcosh}(x(0))} = \frac{\sqrt{b}-\sqrt{a}}{\sqrt{b}+\sqrt{a}} < 1.$$

Any oscillating polynomial $p \in \Pi_k^0$ can be used to estimate $\|p_k^*\|_{C[a,b]}$. This is the essence of the Theorem of de la Vallée-Poussin: if $\lambda_0 < \ldots < \lambda_k$ are $k+1$ points in $[a,b]$ with $p(\lambda_j)$ alternating in sign, then

$$\min_{0 \le j \le k} |p(\lambda_j)| \le \|p_k^*\|_{C[a,b]} \le \|p\|_{C[a,b]}.$$

In other words, if $p \in \Pi_k^0$ is almost equioscillating over $[a,b]$ with k zeros in $[a,b]$, then $\|p\|_{C[a,b]}$ is close to $\|p^*\|_{C[a,b]}$, and p is a reasonable candidate for a residual polynomial.

While $a > 0$ for well-posed problems and the numbers $\|p_k^*\|_{C[a,b]}$ decay geometrically with k, things become more complicated when the problem is ill-posed, i.e., when $a = 0$. In this case, $1/\lambda$ is no longer continuous over the considered interval, and the Equal-Ripple Theorem no longer applies. Note that the weighted polynomial wq_{k-1} – but not w/λ which is the constant 1 – has a root at the origin; consequently, $\|p_k\|_{C[0,b]} \ge 1$ for any $p \in \Pi_k^0$. The situation can be remedied by increasing the order of the root of w at $\lambda = 0$, e.g., by choosing

$$w(\lambda) = \lambda^{\nu+1}, \qquad \nu > 0.$$

We therefore consider the weighted minimization problem (the operator T can of course be rescaled such that $\|T\|^2 \le b = 1$):

$$\|\lambda^\nu p_k\|_{C[0,1]} = \|w(\frac{1}{\lambda} - q_{k-1})\|_{C[0,1]} \to \min. \tag{A.2}$$

In contrast to the former problem (A.1), the solution of this modified problem is explicitly known only when $\nu = 1/2$ or $\nu = 1$.

For $\nu = 1/2$ Nemirovskii and Polyak [209] observed that

$$p_k^*(\lambda) = \frac{(-1)^k}{2k+1} \frac{T_{2k+1}(\sqrt{\lambda})}{\sqrt{\lambda}}, \qquad 0 < \lambda \le 1, \tag{A.3}$$

is a polynomial in Π_k^0 and that $\sqrt{\lambda}\, p_k^*(\lambda)$ attains the values $\pm(2k+1)^{-1}$ with alternating signs at $k+1$ consecutive points (T_{2k+1} has $2k+2$ local extrema in $[-1,1]$, $k+1$ of which belong to $(0,1]$). Hence, according to the Equal-Ripple Theorem, p_k^* is the polynomial minimizing (A.3) when $\nu = 1/2$.

When $\nu = 1$, consider the smallest zero $t_{k+1} = -\cos(\pi/(2k+2))$ of T_{k+1}, and set

$$p_k^*(\lambda) = \frac{1}{\lambda} \frac{T_{k+1}\big((1-t_{k+1})\lambda + t_{k+1}\big)}{(1-t_{k+1})T'_{k+1}(t_{k+1})},$$

cf. Nevanlinna [222]. Obviously, λp_k^* is a shifted Chebyshev polynomial, with its argument shifted in such a way that the smallest zero comes to lie in the origin. Hence, λp_k^* has an alternation set $\mathcal{S}_k$ of $k+1$ points in $[0,1]$, and it follows that

$$\|\lambda p_k^*\|_{C[0,1]} = |(1-t_{k+1})T'_{k+1}(t_{k+1})|^{-1} = \frac{\pi}{4}k^{-2} + O(k^{-3}) \quad \text{as} \quad k \to \infty.$$

We observe that the above results for the two cases $\nu = 1/2$ and $\nu = 1$ reveal the asymptotic behaviour

$$\|\lambda^\nu p_k^*\|_{C[0,1]} \sim k^{-2\nu} \quad \text{as} \quad k \to \infty, \tag{A.4}$$

for these two values of ν. In fact, (A.4) remains true for all values $\nu > 0$ as shown in [111].

A.2. Orthogonal Polynomials

Let α be a nondecreasing distribution function in $[a,b]$ with $k_{\max} \leq \infty$ points of increase and $\int_a^b d\alpha(\lambda) < \infty$. Then (cf. Szegö [264] as a general reference in this section) there is a unique set of polynomials $\{\check{p}_k\}_{k=0}^{k_{\max}}$ with $\check{p}_k \in \Pi_k$, $\check{p}_k(\lambda) > 0$ for $\lambda > b$, and

$$\int_a^b \check{p}_k(\lambda)\check{p}_l(\lambda)\, d\alpha(\lambda) = \delta_{k,l}, \qquad 0 \leq k \leq k_{\max},\ 0 \leq l < k_{\max}.$$

Each $\check{p}_k$ has exact degree k and $\{\check{p}_0, \ldots, \check{p}_k\}$, $k < k_{\max}$, is an orthonormal basis (with respect to $d\alpha(\lambda)$) of Π_k. If $k_{\max} < \infty$ and $\lambda_1, \ldots, \lambda_{k_{\max}}$ are the finitely many points of increase of α, then we set $\check{p}_{k_{\max}}(\lambda) := \prod_{j=1}^{k_{\max}}(\lambda - \lambda_j)$, and any further polynomial of the form $p = q\check{p}_{k_{\max}}$, with q a polynomial, is orthogonal to all polynomials with respect to $d\alpha(\lambda)$. Throughout, we shall restrict the ˇ superscript to $\check{p}_{k_{\max}}$ (if $k_{\max} < \infty$), and to polynomials which are ortho*normal* with respect to the given $d\alpha(\lambda)$ and which are positive for large λ.

The polynomials $\check{p}_k$, $0 \leq k \leq k_{\max}$, have a number of remarkable properties, one of which concerns their computation: as an immediate consequence of the orthogonality relation, there exist $\{a_k, b_k\}_{k \geq 0}$ such that

$$\lambda \check{p}_k(\lambda) = a_{k+1}\check{p}_{k+1}(\lambda) + b_k\check{p}_k(\lambda) + a_k\check{p}_{k-1}(\lambda), \qquad 0 \leq k < k_{\max}, \tag{A.5}$$

where $\check{p}_0 = (\int_a^b d\alpha(\lambda))^{-\frac{1}{2}}$, and where we have set $\check{p}_{-1} = 0$ for notational convenience. Thus, when the numbers a_k and b_k are known, $\check{p}_k$ can be computed with simple two-term recursions.

In this book we are mainly concerned with the behaviour of $\check{p}_k$ in $[a,b]$. For $1 \leq k < k_{\max}$, $\check{p}_k$ has k distinct zeros in the open interval (a,b) and the zeros of $\check{p}_k$ interlace with those of $\check{p}_{k+1}$. In other words, $\check{p}_k$ is oscillating in $(0,1)$. We elaborate further

A. Appendix

on that by studying the Gegenbauer polynomials. The Gegenbauer or ultraspherical polynomials $P_k^{(\gamma)}$, $\gamma > -1/2$, are orthogonal with respect to $d\alpha(x) = w(x)dx$, where

$$w(x) = (1-x^2)^{\gamma-\frac{1}{2}}, \qquad -1 < x < 1. \tag{A.6}$$

The corresponding orthonormal polynomials $\check{P}_k^{(\gamma)}$ satisfy

$$\check{P}_k^{(\gamma)} = 2^{\gamma-\frac{1}{2}}\left(\frac{(k+\gamma)\Gamma(k+1)}{\Gamma(k+2\gamma)}\right)^{\frac{1}{2}}\frac{\Gamma(\gamma)}{\sqrt{\pi}} P_k^{(\gamma)},$$

and the corresponding recursion coefficients of (A.5) are

$$a_k = \frac{1}{2}\left(\frac{k(k+2\gamma-1)}{(k+\gamma)(k+\gamma-1)}\right)^{\frac{1}{2}}, \qquad b_k = 0. \tag{A.7}$$

If $\varepsilon > 0$ is a small, fixed number, the oscillation of $\check{P}_k^{(\gamma)}$ is manifested by the so-called *Hilb-type asymptotic formula* ([264, Theorem 8.21.13]):

$$\check{P}_k^{(\gamma)}(x) = A_w(x)\left[\cos(k\theta + \gamma(\theta - \frac{\pi}{2})) + O\left(\frac{1}{k\sqrt{1-x^2}}\right)\right], \tag{A.8}$$

where $x = \cos\theta$, $|x| \leq 1 - \varepsilon k^{-2}$, and

$$A_w(x) = \sqrt{\tfrac{2}{\pi}}(1-x^2)^{-\frac{\gamma}{2}} = \sqrt{\tfrac{2}{\pi}}\left(\frac{1}{w(x)\sqrt{1-x^2}}\right)^{\frac{1}{2}}.$$

The remainder term on the right-hand side of (A.8) is always $O(1)$ for x in the given range, and when x is restricted to a compact subinterval of $(-1,1)$ then the error term goes uniformly to zero like $O(k^{-1})$. This shows, that $\check{P}_k^{(\gamma)}$ is rapidly oscillating – like $\cos k\theta$ – in any compact subinterval of $[-1,1]$, and the amplitude (or envelope) of the oscillation is asymptotically given by A_w. This amplitude is intimately connected to the weight function w, but we emphasize the role of the extra factor $\sqrt{1-x^2}$ that appears in the definition of A_w. This factor originates from the so-called arcsine distribution (i.e., (A.6) with $\gamma = 0$) which corresponds to the equilibrium distribution of a positive charge on $[-1,1]$, i.e., the distribution of a charge for which the energy with respect to the logarithmic potential is minimized (cf., e.g., [260, Appendix] for a survey of complex potential theory and [166, Chapter 12] for the computation of this particular equilibrium distribution). The Chebyshev polynomials of the first kind – which are equioscillating – are orthogonal with respect to the arcsine distribution; in fact, for the normalized Chebyshev polynomials we obtain the constant amplitude function $A_w(x) = 2/\pi$ in (A.8). For other values of γ the weight function has to be adjusted accordingly to obtain the corresponding envelope function.

We finally point out that for the Chebyshev polynomials of the first kind the Hilb-type formula (A.8) is true for all $x \in [-1,1]$ without remainder term. Another example with no remainder term is $\gamma = 1$: $P_k^{(1)}$ is the kth Chebyshev polynomial of the second kind, and the right-hand side of (A.8) becomes

$$\sqrt{\tfrac{2}{\pi}}(1-x^2)^{-\frac{1}{2}}\cos((k+1)\theta - \tfrac{\pi}{2}) = \sqrt{\tfrac{2}{\pi}}\,\frac{\sin(k+1)\theta}{\sin\theta} = \check{P}_k^{(1)}.$$

For several other continuous weight functions, results analogous to (A.8) can be obtained. In particular, Hilb-type formulas exist for all other Jacobi weights $w(x) = (1-x)^\alpha(1+x)^\beta$, with $\alpha,\beta > -1$. This takes us back to the topic of the preceding paragraph, where we have been interested in weighted polynomials $\lambda^\nu p_k(\lambda)$, $\lambda > 0$, which are equioscillating over $[0,1]$. In view of (A.8), natural candidates for $\{p_k\}_{k\in\mathbb{N}}$ would be orthonormal polynomials corresponding to $A_w(\lambda) = \lambda^{-\nu}$ over $\lambda \in [0,1]$. We are therefore led to consider the shifted Jacobi weight with $\alpha = 2\nu - 1/2$ and $\beta = -1/2$, i.e.,

$$w_\nu(\lambda) := \frac{\lambda^{2\nu}}{\sqrt{\lambda(1-\lambda)}}, \qquad 0 < \lambda < 1. \tag{A.9}$$

The corresponding orthonormal polynomials $\{\check{p}_k^{(\nu)}\}$ are shifted Jacobi polynomials, but they can also be expressed in terms of Gegenbauer polynomials, an approach which will turn out more convenient for our purposes: using that $P_{2k}^{(2\nu)}$ is even, $P_{2k}^{(2\nu)}(\sqrt{1-\lambda})$ is a polynomial of degree k in λ, and it is easy to check that

$$\check{p}_k^{(\nu)}(\lambda) = (-1)^k \check{P}_{2k}^{(2\nu)}(\sqrt{1-\lambda}), \qquad 0 \le \lambda \le 1.$$

Now, from (A.8) we readily obtain

$$\check{p}_k^{(\nu)}(\lambda) = (-1)^k\sqrt{\tfrac{2}{\pi}}\,\lambda^{-\nu}[\cos(k\theta + \nu(\theta - \pi)) + O(\tfrac{1}{k\sqrt{\lambda}})], \tag{A.10}$$

valid for $\varepsilon k^{-2} \le \lambda \le 1$ with $\lambda = \sin^2(\theta/2)$, $\theta \in (0,\pi]$.

Recall that $\check{p}_k^{(\nu)}$ has no root at the origin, hence there is a unique polynomial $p_k^{(\nu)} \in \Pi_k^0$ such that

$$\check{p}_k^{(\nu)}(\lambda)^2 = \pi_k^{(\nu)} p_k^{(\nu)}(\lambda)^2,$$

i.e.,

$$\pi_k^{(\nu)} = \check{p}_k^{(\nu)}(0)^2 = \check{P}_{2k}^{(2\nu)}(1)^2 = \frac{\Gamma(2k+4\nu)(k+\nu)}{\Gamma(2k+1)}\,\frac{4^{-2\nu+1}}{\Gamma^2(2\nu+\tfrac{1}{2})}.$$

From Stirling's formula we obtain

$$\pi_k^{(\nu)} = \frac{2}{\Gamma^2(2\nu+\tfrac{1}{2})}\,k^{4\nu} + O(k^{4\nu-1}) \quad\text{as}\quad k\to\infty, \tag{A.11}$$

and thus we conclude from (A.10):

$$|p_k^{(\nu)}(\lambda)| = O((k^2\lambda)^{-\nu}), \qquad \varepsilon k^{-2} \le \lambda \le 1. \tag{A.12}$$

294   A. Appendix

On the other hand, it is known that $\check{P}_{2k}^{(2\nu)}(x)$ attains its maximum absolute value over $[0, 1]$ at $x = 1$. Hence, $|\check{p}_k^{(\nu)}|$ is bounded on $[0, 1]$ by its value at $\lambda = 0$, i.e., by 1, and we conclude that there exists $c_\nu > 0$ such that

$$|\check{p}_k^{(\nu)}(\lambda)| \leq c_\nu (1 + k^2 \lambda)^{-\nu}, \qquad \lambda \in [0, 1], \ k \in \mathbb{N}_0.$$

Unfortunately, however, sharp constants c_ν are difficult to obtain.

In the case when $\nu = 1/2$ we find $\check{p}_k^{(\frac{1}{2})}$ from $\check{P}_{2k}^{(\gamma)}$ with $\gamma = 1$. As shown above, the Hilb-type formula (A.10) is valid for all $\lambda \in [0, 1]$ with no remainder term when $\gamma = 1$. Thus, $\sqrt{\lambda}\,\check{p}_k^{(\frac{1}{2})}(\lambda)$ is equioscillating over $[0, 1]$ and $\check{p}_k^{(\frac{1}{2})}$ coincides with the polynomial p_k^* that we have obtained for $\nu = 1/2$ in the preceding paragraph. In other words, the above approach for finding approximate minimizers of (A.2) yields the exact minimizer when $\nu = 1/2$.

We now turn to the derivative of $\check{p}_k^{(\nu)}$. Fortunately, the study of this derivative requires no additional tools, since the derivative of a Gegenbauer polynomial is itself a Gegenbauer polynomial, namely we have [264, (4.7.14)]:

$$\frac{d}{dx} P_k^{(\gamma)}(x) = 2\gamma P_{k-1}^{(\gamma+1)}(x).$$

Going through all renormalizations, this yields

$$\begin{aligned}\frac{d}{d\lambda}\check{p}_k^{(\nu)}(\lambda) &= (-1)^{k+1}\sqrt{k(k+2\nu)}\,\frac{\check{P}_{2k-1}^{(2\nu+1)}(\sqrt{1-\lambda})}{\sqrt{1-\lambda}} \quad &\text{(A.13)}\\ &= (-1)^{k+1} k \sqrt{\frac{2}{\pi}}\,\frac{\lambda^{-\nu}}{\sqrt{\lambda(1-\lambda)}}\,[\sin(k\theta + \nu(\theta - \pi)) + O(\frac{1}{k\sqrt{\lambda}})],\end{aligned}$$

where the latter identity holds for $\varepsilon k^{-2} \leq \lambda < 1$ with the same meaning of θ as in (A.10). We note that for $\lambda \in [0, \varepsilon k^{-2}]$ we have

$$|\frac{d}{d\lambda}\check{p}_k^{(\nu)}(\lambda)| \leq \sqrt{k(k+2\nu)}\,\frac{\check{P}_{2k-1}^{(2\nu+1)}(1)}{\sqrt{1-\lambda}} = O(k^{2\nu+2}) \quad \text{as} \quad k \to \infty.$$

Since $\check{p}_k^{(\nu)}$ is a convex function in $[0, \lambda_{1,k}]$ (where $\lambda_{1,k}$ denotes the smallest root of $\check{p}_k^{(\nu)}$) we can now conclude from (A.11) the important result

$$|\check{p}_k^{(\nu)}(\lambda)| \geq \frac{1}{2}|\check{p}_k^{(\nu)}(0)|, \qquad 0 \leq \lambda \leq \varepsilon_0 k^{-2}, \qquad \text{(A.14)}$$

with $\varepsilon_0 > 0$ sufficiently small. In particular, this – together with (A.10), see below – implies that the smallest root $\lambda_{1,k}$ of $\check{p}_k^{(\nu)}$ satisfies

$$\lambda_{1,k} \sim k^{-2} \quad \text{as} \quad k \to \infty. \qquad \text{(A.15)}$$

A.3. Christoffel Functions

The Christoffel functions $\Lambda_n(\lambda; d\alpha)$, $n \in \mathbb{N}_0$, associated with the nondecreasing distribution function $\alpha(\lambda)$ over $[a, b]$ are defined as

$$\Lambda_n(\lambda; d\alpha) := \left(\sum_{k=0}^{n-1} \breve{p}_k^2(\lambda)\right)^{-1}, \tag{A.16}$$

where $\{\breve{p}_k\}$ are the corresponding orthonormal polynomials, cf. Freud [87], or Nevai [221]. An equivalent expression, known as formula of Christoffel-Darboux, is

$$\Lambda_n(\lambda; d\alpha) = \frac{1}{a_n} (\breve{p}_n'(\lambda)\breve{p}_{n-1}(\lambda) - \breve{p}_{n-1}'(\lambda)\breve{p}_n(\lambda))^{-1}, \tag{A.17}$$

where a_n has the same meaning as in the recurrence relation (A.5).

Christoffel functions play an important role in numerical analysis, namely in Gaußian quadrature rules. Denote by $\{\lambda_{j,n}\}_{j=1}^n$ the roots of $\breve{p}_n$: the Gaußian quadrature rule associated with $d\alpha(\lambda)$ is given by

$$\int_a^b f(\lambda) \, d\alpha(\lambda) \approx \sum_{j=1}^n \Lambda_n(\lambda_{j,n}; d\alpha) f(\lambda_{j,n}),$$

with equality for all $f \in \Pi_{2n-1}$. When λ_0 belongs to (a, b) but is different from the zeros of $\breve{p}_n$, $\Lambda_n(\lambda_0; d\alpha)$ can still be interpreted as the corresponding weight for a Gauß-type quadrature rule for $d\alpha(\lambda)$ with one prescribed node at $\lambda = \lambda_0$; this quadrature rule is exact for polynomials of degree $2n - 2$ (or $2n - 3$, respectively, if λ_0 is a root of $\breve{p}_{n-1}$), see [87, Section I.3] for details. This latter property can be used to derive the so-called *Markov-Stieltjes inequalities*, a corollary of which yields the following bound [87, Section I.5]:

$$\Lambda_n(\lambda; d\alpha) \leq \int_{\lambda_{j-1,n}}^{\lambda_{j+2,n}} d\alpha(\lambda), \quad \text{when } \lambda_{j,n} \leq \lambda < \lambda_{j+1,n}. \tag{A.18}$$

Below, we will use (A.18) to determine bounds for the asymptotic behaviour of $\Lambda_n(\lambda; d\alpha)$.

Another remarkable property of the Christoffel functions arises from an $\mathcal{L}^2$-analogue of the minimization problem (A.2): fix $\lambda^+ \in \mathbb{R}$ and consider the solution p_n^+ of the problem

$$\int_a^b p^2(\lambda) \, d\alpha(\lambda) \to \min, \quad p \in \Pi_n, \ p(\lambda^+) = 1. \tag{A.19}$$

As follows immediately from an expansion of p_n^+ in terms of the orthonormal polynomials $\breve{p}_k$, $0 \leq k \leq n$, the minimum of (A.19) is the $(n+1)$st Christoffel function evaluated at λ^+, i.e.,

$$\int_a^b p_n^+(\lambda)^2 \, d\alpha(\lambda) = \Lambda_{n+1}(\lambda^+; d\alpha). \tag{A.20}$$

A. Appendix

In particular, if $d\alpha(\lambda) = \lambda^{2\nu}d\lambda$, $\lambda \in [0,1]$, then we have

$$\min_{p_n \in \Pi_n^0} \|\lambda^\nu p_n\|_{\mathcal{L}^2[0,1]} = (\Lambda_{n+1}(0; \lambda^{2\nu}))^{\frac{1}{2}}.$$

Using (A.16) it is possible to evaluate the minimum (note that $\breve{p}_k$ and its derivative are shifted Jacobi polynomials):

$$\min_{p_n \in \Pi_n^0} \|\lambda^\nu p_n\|_{\mathcal{L}^2[0,1]} \sim n^{-2\nu-1}. \quad (A.21)$$

Comparing this with (A.4) we see that the mimimum with respect to the $\mathcal{L}^2$-norm decays faster than the minimum with respect to the $\mathcal{L}^\infty$-norm by an extra factor of the order of n^{-1} as $n \to \infty$.

We finally turn to the behaviour of $\Lambda_n(\lambda; w_\nu)$ in $[0,1]$. The reciprocal of the Christoffel function is the sum of n oscillating functions with almost the same envelope $A_{w_\nu}^2(\lambda) = 2\lambda^{-2\nu}/\pi$, but with increasing frequency; we may therefore expect that the oscillations asymptotically equilibrate and that

$$n\Lambda_n(\lambda; w_\nu) \approx (\tfrac{1}{2}A_{w_\nu}^2(\lambda))^{-1}. \quad (A.22)$$

In fact, using (A.17) and the Hilb-type formulas (A.13) and (A.10), it is not difficult to show that (A.22) becomes an identity in the limit $n \to \infty$, uniformly in compact subintervals of $(0,1)$. For our purposes, however, estimates are required which hold uniformly in the *entire* interval $[0,1]$. The estimate we are looking for is due to Nevai [220]:

$$\Lambda_n(\lambda; w_\nu) \sim \frac{1}{n}(n^{-2} + \lambda)^{2\nu}, \qquad 0 \leq \lambda \leq 1. \quad (A.23)$$

While the lower bound of (A.23) for $\Lambda_n(\lambda; w_\nu)$ can be obtained by inserting the estimate (A.10) for the polynomials $\breve{p}_k^{(\nu)}$ in (A.16) (cf. the proof of Theorem 7.71.2 in [264]), the upper bound requires more sophisticated tools, namely application of the Markov-Stieltjes inequality (A.18). We sketch the proof for the reader's convenience since Nevai's paper is written in Russian.

Denote the roots of $\breve{p}_n^{(\nu)}$ by $\lambda_{j,n} = \sin^2(\theta_{j,n}/2)$. We choose $c > 0$ sufficiently large such that the remainder term in (A.10) is at most $1/2$ in absolute value, provided $cn^{-2} \leq \lambda \leq 1$. This implies that each root $\lambda_{j,n}$ in the interval $[cn^{-2}, 1]$ satisfies

$$\left|\theta_{j,n} - \frac{j + \nu - \frac{1}{2}}{n + \nu}\pi\right| \leq \frac{\frac{1}{6}}{n+\nu}\pi.$$

In particular, it follows that any two consecutive roots $\lambda_{j,n}$ and $\lambda_{j+1,n}$ in the interval $[cn^{-2}, 1]$ satisfy

$$\frac{\frac{2}{3}}{n+\nu}\pi \leq |\theta_{j,n} - \theta_{j+1,n}| \leq \frac{\frac{4}{3}}{n+\nu}\pi.$$

We increase c further such that at least the two smallest zeros and the two largest zeros of $\breve{p}_n^{(\nu)}$ do not belong to $[cn^{-2}, 1-cn^{-2}]$. The Markov-Stieltjes inequality (A.18) and the Mean Value Theorem then yield

$$\Lambda_n(\lambda; w_\nu) \leq \int_{\theta_{j-1,n}}^{\theta_{j+2,n}} \sin^{4\nu}\frac{\theta}{2}d\theta \sim \frac{1}{n}\lambda^{2\nu}, \qquad cn^{-2} \leq \lambda \leq 1 - cn^{-2}.$$

A.3. Christoffel Functions

Now, consider $\lambda \in [0, cn^{-2}]$. By (A.14), there exists $0 < \varepsilon < 1$ such that for $k \leq \varepsilon n$

$$|\breve{p}_k^{(\nu)}(0)| \geq |\breve{p}_k^{(\nu)}(\lambda)| \geq \frac{1}{2}|\breve{p}_k^{(\nu)}(0)|, \qquad 0 \leq \lambda \leq cn^{-2}.$$

From the optimality property (A.20) of the Christoffel functions, together with their definition (A.16), we thus conclude

$$\Lambda_n(\lambda; w_\nu) \leq \Lambda_k(\lambda; w_\nu) \sim \Lambda_k(0; w_\nu).$$

By (A.11) we have $\Lambda_k(0; w_\nu) \sim k^{-4\nu-1}$. Hence, choosing $k \sim n$ with $k \leq \varepsilon n$ we obtain

$$\Lambda_n(\lambda; w_\nu) \leq O(n^{-4\nu-1}) \quad \text{as} \quad n \to \infty.$$

To complete the proof of (A.23) it remains to consider $\Lambda_n(\lambda; w_\nu)$ for $\lambda \in [1-cn^{-2}, 1]$; the argument is essentially the same as for λ close to 0.

Bibliography

[1] R. ACAR AND C. VOGEL, *Analysis of bounded variation penalty methods for ill-posed problems*, Inverse Problems 10 (1994), 1217–1229.

[2] N. I. ACHIESER, *Vorlesungen über Approximationstheorie*, Akademie-Verlag, Berlin, 1967.

[3] K. ADAMIAK, *On Fredholm integral equations of the first kind occuring in synthesis of electromagnetic fields*, Internat. J. Numer. Methods Engrg. 17 (1981), 1187–1200.

[4] H.-M. ADORF, *Hubble space telescope image restoration in its fourth year*, Inverse Problems 11 (1995), 639–653.

[5] G. ALESSANDRINI, *Stable determination of conductivity by boundary measurements*, Appl. Anal. 27 (1988), 153–172.

[6] O. M. ALIFANOV AND S. V. RUMJANCEV, *On the stability of iterative methods for the solution of linear ill-posed problems*, Soviet Math. Dokl. 20 (1979), 1133–1136.

[7] A. ALLERS AND F. SANTOSA, *Stability and resolution analysis of a linearized problem in electrical impedance tomography*, Inverse Problems 7 (1991), 515–533.

[8] B. K. ALPERT, *Wavelets and other bases for fast numerical linear algebra*, in: C. K. Chui, ed., Wavelets: A Tutorial in Theory and Applications, Academic Press, Boston, New York, London, 1992, 181–216.

[9] U. AMATO AND W. HUGHES, *Maximum entropy regularization of Fredholm integral equations of the first kind*, Inverse Problems 7 (1991), 793–808.

[10] R. S. ANDERSSEN, *The linear functional strategy for improperly posed problems*, in [36], 11–30.

[11] R. S. ANDERSSEN AND P. BLOOMFIELD, *Numerical differentiation procedures for non-exact data*, Numer. Math. 22 (1974), 157–182.

[12] R. S. ANDERSSEN AND F. R. DE HOOG, *Finite difference methods for the numerical differentiation of non-exact data*, Computing 33 (1984), 259–267.

[13] G. ANGER, *Inverse Problems in Differential Equations*, Akademieverlag, Berlin, 1990.

[14] G. ANGER, R. GORENFLO, H. JOCHMANN, H. MORITZ, AND W. WEBERS, eds., *Inverse Problems: Principles and Applications in Geophysics, Technology, and Medicine*, Akademieverlag, Berlin, 1993.

[15] Y. E. ANIKOV, *Multidimensional Inverse and Ill-Posed Problems for Differential Equations*, VSP, Utrecht, 1995.

[16] H. BABOVSKY, *An inverse model problem in kinetic theory*, Inverse Problems 11 (1995), 555–570.

[17] G. BACKUS AND F. GILBERT, *Numerical applications of a formalism for geophysical inverse problems*, Geophys. J. R. Astron. Soc. 13 (1967), 247–276.

[18] A. B. BAKUSHINSKII, *Remarks on choosing a regularization parameter using the quasi-optimality and ratio criterion*, USSR Comp. Math. Math. Phys. 24,4 (1984), 181–182.

[19] ———, *The problem of the convergence of the iteratively regularized Gauss-Newton method*, Comput. Math. Math. Phys. 32 (1992), 1353–1359.

[20] A. BAKUSHINSKY AND A. GONCHARSKY, *Ill-Posed Problems: Theory and Applications*, Kluwer, Dordrecht, 1995.

[21] H. BANKS AND K. KUNISCH, *Parameter Estimation Techniques for Distributed Systems*, Birkhäuser, Boston, 1989.

[22] J. BAUMEISTER, *Stable Solution of Inverse Problems*, Vieweg, Braunschweig, 1987.

[23] J. BECK, B. BLACKWELL, AND C. S. CLAIR, *Inverse Heat Conductions*, Wiley, Sussex, 1985.

[24] M. BERTERO, P. BRIANZI, AND E. R. PIKE, *Super-resolution in confocal scanning microscopy*, Inverse Problems 3 (1987), 195–212.

[25] G. BEYLKIN, R. COIFMAN, AND V. ROKHLIN, *Fast wavelet transforms and numerical algorithms I*, Comm. Pure Appl. Math. 44 (1991), 141–183.

[26] H. BIALY, *Iterative Behandlung linearer Funktionalgleichungen*, Arch. Rational Mech. Anal. 1 (1959/60), 166–176.

[27] A. BINDER, H. W. ENGL, C. W. GROETSCH, A. NEUBAUER, AND O. SCHERZER, *Weakly closed nonlinear operators and parameter identification in parabolic equations by Tikhonov regularization*, Appl. Anal. 55 (1994), 215–234.

[28] A. BINDER, H. W. ENGL, AND S. VESSELLA, *Some inverse problems for a nonlinear parabolic equation connected with continuous casting of steel: stability estimates and regularization*, Numer. Funct. Anal. Optim. 11 (1990), 643–672.

[29] A. BJÖRCK, *Least squares methods*, in: P. G. Ciarlet and J. L. Lions, eds., Handbook of Numerical Analysis, Vol. I: Finite Difference Methods. Solutions of Equations in $\mathbb{R}^n$, Elsevier North-Holland, 1990.

[30] B. BLASCHKE, H. W. ENGL, W. GREVER, AND M. V. KLIBANOV, *An application of Tikhonov regularization to phase retrieval*, Nonlinear World 3 (1996), 771–786.

[31] B. BLASCHKE, A. NEUBAUER, AND O. SCHERZER, *On convergence rates for the iteratively regularized Gauss-Newton method*, IMA Journal of Numer. Anal. 17 (1997), 421–436.

[32] F. BLOOM, *Ill-Posed Problems for Integrodifferential Equations in Mechanics and Electromagnetic Theory*, SIAM, Philadelphia, 1981.

[33] J. BORWEIN AND C. LEWIS, *Convergence of best entropy estimates*, SIAM J. Optim. 1 (1991), 191–205.

[34] H. BRAKHAGE, *On ill-posed problems and the method of conjugate gradients*, in [72], 165–175.

[35] L. BREGMAN, *The relaxation method of finding the common point of convex sets and its application to the solution of problems in convex programming*, USSR Comp. Math. Math. Phys. 7,3 (1967), 200–217.

[36] J. CANNON AND U. HORNUNG, eds., *Inverse Problems*, Birkhäuser, Basel, 1986.

[37] Y. CENSOR, A. DE PIERRO, AND A. IUSEM, *Optimization of Burg's entropy over linear constraints*, Appl. Numer. Math. 7 (1991), 151–165.

[38] G. CHAVENT, *Strategies for the regularization of nonlinear least squares problems*, in [83], 217–232.

[39] G. CHAVENT AND K. KUNISCH, *Convergence of Tikhonov regularization for constrained ill-posed inverse problems*, Inverse Problems 10 (1994), 63–73.

[40] F. COLONIUS AND K. KUNISCH, *Stability for parameter estimation in two point boundary value problems*, J. Reine Angew. Math. 370 (1986), 1–29.

[41] ———, *Output least squares stability in elliptic systems*, Appl. Math. Optim. 19 (1989), 33–63.

[42] D. COLTON, *Analytic Theory of Partial Differential Equations*, Pitman, Boston, 1980.

[43] D. COLTON, R. EWING, AND W. RUNDELL, eds., *Inverse Problems in Partial Differential Equations*, SIAM, Philadelphia, 1990.

[44] D. COLTON AND R. KRESS, *Integral Equation Methods in Scattering Theory*, Wiley, New York, 1983.

[45] ——, *Inverse Acoustic and Electromagnetic Scattering Theory*, Springer, Berlin, 1992.

[46] D. COLTON AND P. MONK, *A novel method for solving the inverse scattering problem for time-harmonic acousting waves in the resonance region I*, SIAM J. Appl. Math. 45 (1985), 1039–1053.

[47] ——, *A novel method for solving the inverse scattering problem for time-harmonic acousting waves in the resonance region II*, SIAM J. Appl. Math. 46 (1986), 506–523.

[48] ——, *The inverse scattering problem for time-harmonic acoustic waves in an inhomogenous medium: numerical experiments*, IMA J. Appl. Math. 42 (1989), 77–95.

[49] C. E. CREFFIELD, E. G. KLEPFISH, E. R. PIKE, AND S. SANKAR, *Spectral weight functions for the half-filled Hubbord model: a singular value decomposition approach*, Phys. Rev. Lett. 75 (1995), 517–520.

[50] R. DAUTRAY AND J.-L. LIONS, *Mathematical Analysis and Numerical Methods for Science and Technology, Vol. 3: Spectral Theory and Applications*, Springer, Berlin, Heidelberg, New York, 1990.

[51] M. DEFRISE AND C. DE MOL, *A note on stopping rules for iterative regularization methods and filtered SVD*, in: P. C. Sabatier, ed., Inverse Problems: An Interdisciplinary Study, Academic Press, London, Orlando, San Diego, New York, 1987, 261–268.

[52] ——, *Super-resolution in confocal scanning microscopy: generalized inversion formulae*, Inverse Problems 8 (1992), 175–185.

[53] J. E. DENNIS AND R. B. SCHNABEL, *Numerical Methods for Unconstrained Optimization and Nonlinear Equations*, Prentice-Hall, Englewood Cliffs, 1983.

[54] R. A. DEVORE AND G. G. LORENTZ, *Constructive Approximation*, Springer-Verlag, Berlin, Heidelberg, New York, 1993.

[55] D. DOBSON AND F. SANTOSA, *An image enhancement technique for electrical impedance tomography*, Inverse Problems 10 (1994), 317–334.

[56] D. L. DONOHO, *Nonlinear solution of linear inverse problems by wavelet-vagulette decomposition*, Appl. Comput. Harmon. Anal. 2 (1995), 101–126.

[57] K. DRESSLER, *Inverse problems in linear transport theory*, European J. Mech. B Fluids 8 (1989), 351–372.

[58] N. DUNFORD AND J. T. SCHWARTZ, *Linear Operators, Part 1: General Theory*, Interscience Publishers, New York, 1967.

[59] P. M. EDIC, G. J. SAULNIER, J. C. NEWELL, AND D. ISAACSON, *A real-time electrical impedance tomography*, IEEE Trans. Biomed. Engrg. 42 (1995), 849–859.

[60] P. P. B. EGGERMONT, *Maximum entropy regularization for Fredholm integral equations of the first kind*, SIAM J. Math. Anal. 24 (1993), 1557–1576.

[61] B. EICKE, *Iteration methods for convexly constrained ill-posed problems in Hilbert space*, Numer. Funct. Anal. Optim. 13 (1992), 413–429.

[62] B. EICKE, A. K. LOUIS, AND R. PLATO, *The instability of some gradient methods for ill-posed problems*, Numer. Math. 58 (1990), 129–134.

[63] L. ELDÉN, *Algorithms for the regularization of ill-conditioned least squares problems*, BIT 17 (1977), 134–145.

[64] ———, *A weighted pseudoinverse, generalized singular values, and constrained least squares problems*, BIT 22 (1982), 487–502.

[65] ———, *An efficient algorithm for the regularization of ill-conditioned least squares problems with triangular Toeplitz matrix*, SIAM J. Sci. Statist. Comput. 5 (1984), 229–236.

[66] ———, *A note on the computation of the Generalized Cross-Validation function for ill-conditioned least squares problems*, BIT 24 (1984), 467–472.

[67] T. ELFVING, *An algorithm for maximum entropy image reconstruction from noisy data*, Math. Comput. Modelling 12 (1989), 729–745.

[68] H. W. ENGL, *Necessary and sufficient conditions for convergence of regularization methods for solving linear operator equations of the first kind*, Numer. Funct. Anal. Optim. 3 (1981), 201–222.

[69] ———, *On least-squares collocation for solving linear integral equations of the first kind with noisy right-hand side*, Boll. Geodesia Sc. Aff. 41 (1982), 291–313.

[70] H. W. ENGL AND H. GFRERER, *A posteriori parameter choice for general regularization methods for solving linear ill-posed problems*, Appl. Numer. Math. 4 (1988), 395–417.

[71] H. W. ENGL AND W. GREVER, *Using the L-curve for determining optimal regularization parameters*, Numer. Math. 69 (1994), 25–31.

[72] H. W. ENGL AND C. W. GROETSCH, eds., *Inverse and Ill-Posed Problems*, Academic Press, Orlando, 1987.

[73] H. W. ENGL AND V. ISAKOV, *On the identifiability of steel reinforcement bars in concrete from magnetostatic measurements*, European J. Appl. Math. 3 (1992), 255–262.

[74] H. W. ENGL, K. KUNISCH, AND A. NEUBAUER, *Convergence rates for Tikhonov regularization of nonlinear ill-posed problems*, Inverse Problems 5 (1989), 523–540.

[75] H. W. ENGL AND G. LANDL, *Convergence rates for maximum entropy regularization*, SIAM J. Numer. Anal. 30 (1993), 1509–1536.

[76] ———, *Maximum entropy regularization of nonlinear ill-posed problems*, in: V. Lakshmikantham, ed., Proceedings of the First World Congress of Nonlinear Analysts, Vol. I, de Gruyter, Berlin, 1996, 513–525.

[77] H. W. ENGL AND T. LANGTHALER, *Control of the solidification front by secondary cooling in continuous casting of steel*, in: H. W. Engl, H. Wacker, and W. Zulehner, eds., Case Studies in Industrial Mathematics, Teubner/Kluwer, Stuttgart/Dordrecht, 1988.

[78] H. W. ENGL, A. K. LOUIS, AND W. RUNDELL, eds., *Inverse Problems in Geophysics*, SIAM, Philadelphia, 1996.

[79] ———, eds., *Inverse Problems in Medical Imaging and Nondestructive Testing*, Springer, Wien, New York, 1996.

[80] H. W. ENGL AND P. MANSELLI, *Stability estimates and regularization for an inverse heat conduction problem in semi-infinite and finite time intervals*, Numer. Funct. Anal. Optim. 10 (1989), 517–540.

[81] H. W. ENGL AND J. MCLAUGHLIN, eds., *Inverse Problems and Optimal Design in Industry*, Teubner, Stuttgart, 1994.

[82] H. W. ENGL AND A. NEUBAUER, *Convergence rates for Tikhonov regularization of implicitly defined nonlinear inverse problems with an application to inverse scattering*, in: S. Kubo, ed., Inverse Problems, Techn. Publ., Atlanta, 1993, 90–98.

[83] H. W. ENGL AND W. RUNDELL, eds., *Inverse Problems in Diffusion Processes*, SIAM, Philadelphia, 1995.

[84] H. W. ENGL, W. RUNDELL, AND O. SCHERZER, *A regularization scheme for an inverse problem in age-structured populations*, J. Math. Anal. Appl. 182 (1994), 658–679.

[85] H. W. ENGL, O. SCHERZER, AND M. YAMAMOTO, *Uniqueness and stable determination of forcing terms in linear partial differential equations with overspecified boundary data*, Inverse Problems 10 (1994), 1253–1276.

[86] A. FARIDANI, *Abtastbedingungen und Auflösung in der Beugungstomographie*, PhD thesis, Universität Münster, 1988.

[87] G. FREUD, *Orthogonal Polynomials*, Pergamon Press, Budapest, 1971.

[88] V. FRIDMAN, *Methods of successive approximations for Fredholm integral equations of the first kind*, Uspekhi Mat. Nauk 11,1 (1956), 233–234, in Russian.

[89] A. FRIEDMAN AND V. ISAKOV, *On the uniqueness in the inverse conductivity problem with one measurement*, Indiana Univ. Math. J. 38 (1989), 563–579.

[90] Z. GAO AND T. MURA, *On the inversion of residual stresses from surface displacements*, ASME Ser. E J. Appl. Mech. 56 (1989), 508–513.

[91] M. GEHATIA AND D. WIFF, *Solution of Fujita's equation for equilibrium sedimentation by applying Tikhonov's regularizing functions*, J. Polymer Sci. A-2 8 (1970), 2039–2050.

[92] H. GFRERER, *An a-posteriori parameter choice for ordinary and iterated Tikhonov regularization of ill-posed problems leading to optimal convergence rates*, Math. Comp. 49 (1987), 507–522 and S5–S12.

[93] S. F. GILYAZOV, *Iterative solution methods for inconsistent operator equations*, Moscow Univ. Comput. Math. Cybernet. 3 (1977), 78–84.

[94] D. A. GIRARD, *A fast 'Monte-Carlo Cross-Validation' procedure for large least squares problems with noisy data*, Numer. Math. 56 (1989), 1–23.

[95] G. GLADWELL, *Inverse Problems in Vibrations*, Nijhoff, Dordrecht, 1986.

[96] G. H. GOLUB AND C. F. VAN LOAN, *Matrix Computations*, The Johns Hopkins University Press, Baltimore, London, 1989.

[97] R. GORENFLO AND S. VESSELLA, *Abel Integral Equations: Analysis and Applications*, Lectures Notes in Math. 1461, Springer, Berlin, 1991.

[98] F. GORI AND G. GUATTARI, *Signal restoration for linear systems with weighted inputs. Singular value analysis for two cases of low-pass filtering*, Inverse Problems 1 (1985), 67–85.

[99] W. GREVER, A. BINDER, H. W. ENGL, AND K. MÖRWALD, *Optimal cooling strategies in continuous casting of steel with variable casting speed*, Inverse Problems Engrg. 2 (1997), 289–300.

[100] R. D. GRIGORIEFF AND R. PLATO, *On a minimax equality for seminorms*, Linear Algebra Appl. 221 (1995), 227–243.

[101] C. W. GROETSCH, *Generalized Inverses of Linear Operators: Representation and Approximation*, Dekker, New York, 1977.

[102] ——, *Comments on Morozov's discrepancy principle*, in: G. Hämmerlin and K. H. Hoffmann, eds., Improperly Posed Problems and Their Numerical Treatment, Birkhäuser, Basel, 1983, 97–104.

[103] ——, *The Theory of Tikhonov Regularization for Fredholm Equations of the First Kind*, Pitman, Boston, 1984.

[104] ——, *Uniform convergence of regularization methods for Fredholm equations of the first kind*, J. Austral. Math. Soc. Ser. A 39 (1985), 282–286.

[105] ——, *Remarks on some iterative methods for an integral equation in Fourier optics*, in: P. Nelson et al., eds., Transport Theory, Invariant Imbedding, and Integral Equations, Marcel Dekker, Inc., New York, Basel, 1989, 313–324.

[106] ——, *Differentiation of approximately specified functions*, Amer. Math. Monthly 48 (1991), 847–850.

[107] ——, *Inverse Problems in the Mathematical Sciences*, Vieweg, Braunschweig, 1993.

[108] C. W. GROETSCH AND A. NEUBAUER, *Convergence of a general projection method for an operator equation of the first kind*, Houston J. Math. 14 (1988), 201–208.

[109] ——, *Regularization of ill-posed problems: optimal parameter choice in finite dimensions*, J. Approx. Theory 58 (1989), 184–200.

[110] C. W. GROETSCH AND C. R. VOGEL, *Asymptotic theory of filtering for linear operator equations with discrete noisy data*, Math. Comp. 49 (1987), 499–506.

[111] M. HANKE, *Accelerated Landweber iterations for the solution of ill-posed equations*, Numer. Math. 60 (1991), 341–373.

[112] ——, *Regularization with differential operators: an iterative approach*, Numer. Funct. Anal. Optim. 13 (1992), 523–540.

[113] ——, *Conjugate Gradient Type Methods for Ill-Posed Problems*, Longman Scientific & Technical, Harlow, Essex, 1995.

[114] ——, *The minimal error conjugate gradient method is a regularization method*, Proc. Amer. Math. Soc. 123 (1995), 3487–3497.

[115] ——, *Limitations of the L-curve method in ill-posed problems*, BIT 36 (1996), 287–301.

[116] M. HANKE AND H. W. ENGL, *An optimal stopping rule for the ν-method for solving ill-posed problems using Christoffel functions*, J. Approx. Theory 79 (1994), 89–108.

[117] M. HANKE AND P. C. HANSEN, *Regularization methods for large-scale problems*, Surveys Math. Indust. 3 (1993), 253–315.

[118] M. HANKE, F. HETTLICH, AND O. SCHERZER, *The Landweber iteration for an inverse scattering problem*, in: K. W. Wang et al., eds., Proceedings of the 1995 Design Engineering Technical Conferences Vol. 3 Part C, ASME, New York, 1995, 909–915.

[119] M. HANKE AND J. G. NAGY, *Restoration of atmospherically blurred images by symmetric indefinite conjugate gradient techniques*, Inverse Problems 12 (1996), 157–173.

[120] M. HANKE, A. NEUBAUER, AND O. SCHERZER, *A convergence analysis of the Landweber iteration for nonlinear ill-posed problems*, Numer. Math. 72 (1995), 21–37.

[121] M. HANKE AND T. RAUS, *A general heuristic for choosing the regularization parameter in ill-posed problems*, SIAM J. Sci. Comput. 17 (1996), 956–972.

[122] P. C. HANSEN, *Analysis of discrete ill-posed problems by means of the L-curve*, SIAM Rev. 34 (1992), 561–580.

[123] ———, *Numerical tools for analysis and solution of Fredholm integral equations of the first kind*, Inverse Problems 8 (1992), 849–872.

[124] P. C. HANSEN AND D. P. O'LEARY, *The use of the L-curve in the regularization of discrete ill-posed problems*, SIAM J. Sci. Comput. 14 (1993), 1487–1503.

[125] E. HEINZ, *Beiträge zur Störungstheorie der Spektralzerlegung*, Math. Ann. 123 (1951), 425–438.

[126] J. HEJTMANEK, *The problem of reconstructing objects from projections as an inverse problem in scattering theory of the linear transport operator*, in: G. Herman and F. Natterer, eds., Mathematical Aspects of Computerized Tomography, Springer Lecture Notes in Medical Informatics, Berlin, 1981, 28–35.

[127] G. HELMBERG, *Introduction to Spectral Theory in Hilbert Spaces*, North Holland, Amsterdam, 1969.

[128] M. R. HESTENES AND E. STIEFEL, *Methods of conjugate gradients for solving linear systems*, J. Research Nat. Bur. Standards 49 (1952), 409–436.

[129] E. HEWITT AND K. STROMBERG, *Real and Abstract Analysis*, Springer, New York, 1965.

[130] A. HOBSON, *Concepts in Statistical Mechanics*, Gordon and Breach Science Publishers, New York, 1971.

[131] B. HOFMANN, *Regularization for Applied Inverse and Ill-Posed Problems*, Teubner, Leipzig, 1986.

[132] J. HONERKAMP, *Ill-posed problems in rheology*, Rheol. Acta 28 (1989), 363–371.

[133] M. IKEHATA, *Inversion formulas for the linearized problem for an inverse boundary value problem in elastic prospection*, SIAM J. Appl. Math. 50 (1990), 1635–1644.

[134] V. ISAKOV, *On uniqueness of recovery of discontinuous conductivity coefficients*, Comm. Pure Appl. Math. 47 (1988), 864–876.

[135] ———, *Inverse Source Problems*, Math. Surveys Monographs 34, Amer. Math. Soc., Providence, 1990.

[136] ———, *Inverse Problems in Partial Differential Equations*, Springer, Heidelberg, Berlin, New York, 1998.

[137] V. ISAKOV AND J. POWELL, *On the inverse conductivity problem with one measurement*, Inverse Problems 6 (1990), 311–318.

[138] Y. ISO AND M. YAMAMOTO, eds., *Advances in Analysis for Ill-Posed Problems*, VSP, Utrecht, 1996.

[139] A. IUSEM, *Convergence analysis for a multiplicatively relaxed EM algorithm*, Math. Methods Appl. Sci. 14 (1991), 573–593.

[140] E. T. JAYNES, *Information theory and statistical mechanics I*, Phys. Rev. 106 (1957), 620–630.

[141] ———, *Information theory and statistical mechanics II*, Phys. Rev. 108 (1957), 171–190.

[142] ———, *Prior probabilities*, IEEE Trans. Syst. Sci. Cybernetics SSC-4 (1968), 227–241.

[143] I. D. JOHNSON AND B. S. HUDSON, *Environmental modulation of M13 caot protein tryptophan fluorescence dynamics*, Biochemistry 28 (1989), 6392–6400.

[144] R. W. JOHNSTON AND J. W. SHORE, *Which is the better entropy expression for speech processing: -S log S or log S?*, IEEE Trans. Acoust., Speech Signal Proc. 32 (1984), 129–136.

[145] L. K. JONES, *Approximation-theoretic derivation of logarithmic entropy principles for inverse problems and unique extension of the maximum entropy method to incorporate prior knowledge*, SIAM J. Appl. Math. 49 (1989), 650–661.

[146] T. KAILATH AND J. CHUN, *Generalized displacement structure for block-Toeplitz, Toeplitz-block, and Toeplitz-derived matrices*, SIAM J. Matrix Anal. Appl. 15 (1994), 114–128.

[147] W. J. KAMMERER AND M. Z. NASHED, *Iterative methods for best approximate solutions of linear integral equations of the first and second kinds*, J. Math. Anal. Appl. 40 (1972), 547–573.

[148] ———, *On the convergence of the conjugate gradient method for singular linear operator equations*, SIAM J. Numer. Anal. 9 (1972), 165–181.

[149] A. KATZ, *Principles of Statistical Mechanics*, Freeman and Company, San Francisco, 1967.

[150] J. KELLER, *Inverse problems*, Amer. Math. Monthly 83 (1976), 107–118.

[151] J. T. KING, *A minimal error conjugate gradient method for ill-posed problems*, J. Optim. Theory Appl. 60 (1989), 297–304.

[152] J. T. KING AND A. NEUBAUER, *A variant of finite-dimensional Tikhonov regularization with a-posteriori parameter choice*, Computing 40 (1988), 91–109.

[153] A. KIRSCH, R. KRESS, P. MONK, AND A. ZINN, *Two methods for solving the inverse acoustic scattering problem*, Inverse Problems 4 (1988), 749–770.

[154] A. KIRSCH, B. SCHOMBURG, AND G. BERENDT, *The Backus-Gilbert method*, Inverse Problems 4 (1988), 771–783.

[155] M. KLAUS AND R. SMITH, *A Hilbert space approach to maximum entropy reconstruction*, Math. Methods Appl. Sci. 10 (1988), 397–406.

[156] R. J. KNOPS, ed., *Symposium on Non-Well-Posed Problems and Logarithmic Convexity*, Lecture Notes in Math. 316, Springer, New York, 1973.

[157] J. KÖHLER AND U. TAUTENHAHN, *Error bounds for regularized solutions of nonlinear ill-posed probelms*, J. Inverse Ill-Posed Probl. 3 (1995), 47–74.

[158] R. V. KOHN AND M. VOGELIUS, *Determining conductivity by boundary measurements II: Interior results*, Comm. Pure Appl. Math. 38 (1985), 643–667.

[159] ———, *Relaxation of a variational method for impedance computed tomography*, Comm. Pure Appl. Math. 40 (1987), 745–777.

[160] A. KONDOR, *Method of convergent weights – an iterative method of solving Fredholm's integral equations of the first kind*, Nuclear Instrum. Methods 216 (1983), 177–181.

[161] E. KOSAREV, *Applications of integral equations of the first kind in experiment physics*, Comput. Phys. Comm. 20 (1980), 69–75.

[162] M. A. KRASNOSELSKII, P. P. ZABREIKO, E. I. PUSTYLNIK, AND P. E. SBOLEVSKII, *Integral Operators in Spaces of Summable Functions*, Noordhoff International Publishing, Leyden, 1976.

[163] C. KRAVARIS AND J. H. SEINFELD, *Identification of parameters in distributed parameter systems by regularization*, SIAM J. Control Optim. 23 (1985), 217–241.

[164] S. G. KREIN AND J. I. PETUNIN, *Scales of Banach spaces*, Russian Math. Surveys 21 (1966), 85–160.

[165] R. KRESS, *Linear Integral Equations*, Springer, Berlin, 1989.

[166] V. I. KRYLOV, *Approximate Calculation of Integrals*, Macmillan, New York, 1962.

[167] S. KUBO, *Inverse problems related to the mechanics and fracture of solids and structures*, Japan Soc. Mech. Eng. Int. J., Ser. I 31 (1988), 157–166.

[168] K. KUNISCH, *Inherent identifiability of parameters in elliptic differential equations*, J. Math. Anal. Appl. 132 (1988), 453–472.

[169] K. KUNISCH AND G. GEYMAYER, *Convergence rates for regularized nonlinear ill-posed problems*, in: Modelling and Inverse Problems of Control for Distributed Parameter Systems (Laxenburg, 1989), Lecture Notes in Control and Inform. Sci. 154, Springer, Berlin, 1991, 81–92.

[170] R. L. LAGENDIJK AND J. BIEMOND, *Iterative Identification and Restoration of Images*, Kluwer, Boston, Dordrecht, London, 1991.

[171] H. J. LANDAU AND H. WIDOM, *Eigenvalue distribution of time and frequency limiting*, J. Math. Anal. Appl. 77 (1980), 469–481.

[172] G. LANDL, *The Maximum Entropy Method: Convergence Analysis and an Application to a Problem from Physical Chemistry*, PhD thesis, Johannes Kepler Universität Linz, January 1992, appeared in Verlag der Wissenschaftlichen Gesellschaften Österreichs, Wien, 1992.

[173] G. LANDL AND R. S. ANDERSSEN, *Non-negative differentially constrained entropy-like regularization*, Inverse Problems 12 (1996), 35–53.

[174] G. LANDL, H. W. ENGL, J. CHEN, AND K. ZEMAN, *Optimal strategies for the cooling of steel strips in hot strip mills*, Inverse Problems Engrg. 2 (1995), 103–118.

[175] G. LANDL, T. LANGTHALER, H. W. ENGL, AND H. F. KAUFFMANN, *Distribution of event times in time-resolved fluorescence: the exponential series approach – algorithm, regularization, analysis*, J. Comput. Phys. 95 (1991), 1–28.

[176] L. LANDWEBER, *An iteration formula for Fredholm integral equations of the first kind*, Amer. J. Math. 73 (1951), 615–624.

[177] M. LAVRENTIEV, *Some Improperly Posed Problems of Mathematical Physics*, Springer, New York, 1967.

[178] A. S. LEONOV, *On the choice of regularization parameters by means of the quasi-optimality and ratio criteria*, Soviet Math. Dokl. 19 (1978), 537–540.

[179] ———, *On the accuracy of Tikhonov regularizing algorithms and quasioptimal selection of a regularization parameter*, Soviet Math. Dokl. 44 (1991), 711–716.

[180] J. L. LIONS AND E. MAGENES, *Non-Homogeneous Boundary Value Problems and Applications I*, Springer, Berlin, Heidelberg, New York, 1972.

[181] J. LOCKER AND P. PRENTER, *Regularization with differential operators I: General theory*, J. Math. Anal. Appl. 74 (1980), 504–529.

[182] A. K. LOUIS, *Convergence of the conjugate gradient method for compact operators*, in [72], 177–183.

[183] ———, *Inverse und schlecht gestellte Probleme*, Teubner, Stuttgart, 1989.

[184] ———, *Approximate inverse for linear and some nonlinear problems*, Inverse Problems 12 (1996), 175–190.

[185] A. K. LOUIS AND P. MAASS, *A mollifier method for linear operator equations of the first kind*, Inverse Problems 6 (1990), 427–440.

[186] G. R. LUECKE AND K. R. HICKEY, *Convergence of approximate solutions of an operator equation*, Houston J. Math. 11 (1985), 345–353.

[187] M. A. LUKAS, *Asymptotic optimality of generalized cross-validation for choosing the regularization parameter*, Numer. Math. 66 (1993), 41–66.

[188] P. MAASS, *Wavelet-projection methods for inverse problems*, in: E. Schock, ed., Beiträge zur Angewandten Analysis und Informatik, Helmut Brakhage zu Ehren, Shaker, Aachen, 1994, 213–224.

[189] M. McIVER, *An inverse problem in electromagnetic crack detection*, IMA J. Appl. Math. 47 (1991), 127–145.

[190] D. W. MEAD, *Determination of molecular weight distributions of linear flexible polymers from linear viscoelastic material functions*, J. Rheol. 38 (1994), 1797–1827.

[191] A. A. MELKMAN AND C. A. MICCHELLI, *Optimal estimation of linear operators in Hilbert spaces from inaccurate data*, SIAM J. Numer. Anal. 16 (1979), 87–105.

[192] H. MORITZ, *Advanced Physical Geodesy*, Wichmann, Karlsruhe, 1980.

[193] V. A. MOROZOV, *On the solution of functional equations by the method of regularization*, Soviet Math. Dokl. 7 (1966), 414–417.

[194] ———, *Methods for Solving Incorrectly Posed Problems*, Springer, New York, Berlin, Heidelberg, 1984.

[195] ———, *Regularization Methods for Ill-Posed Problems*, CRC Press, Boca Raton, Florida, 1993.

[196] U. MOSCO, *Convergence of convex sets and of solutions of variational inequalities*, Adv. Math. 3 (1969), 510–585.

[197] H. MÜLTHEI AND B. SCHORR, *On properties of the iterative maximum likelihood reconstruction method*, Math. Methods Appl. Sci. 11 (1989), 331–342.

[198] D. A. MURIO, *The Mollification Method and the Numerical Solution of Ill-Posed Problems*, Wiley, New York, 1993.

[199] D. A. MURIO, Y. LIU, AND H. ZHENG, *Numerical experiments in multidimensional IHCP on bounded domains*, in [83], 151–180.

[200] A. NACHMAN, *Global uniqueness for the two-dimensional inverse boundary value problem*, Ann. of Math. 142 (1995), 71–96.

[201] J. G. NAGY, *Fast inverse QR factorization for Toeplitz matrices*, SIAM J. Sci. Comput. 14 (1993), 1174–1193.

[202] M. Z. NASHED, ed., *Generalized Inverses and Applications*, Academic Press, New York, 1976.

[203] ———, *Inner, outer, and generalized inverses in Banach and Hilbert spaces*, Numer. Funct. Anal. Optim. 9 (1987), 261–325.

[204] M. Z. NASHED AND G. WAHBA, *Convergence rates of approximate least squares solutions of linear integral and operator equations of the first kind*, Math. Comp. 28 (1974), 69–80.

[205] F. NATTERER, *Regularisierung schlecht gestellter Probleme durch Projektionsverfahren*, Numer. Math. 28 (1977), 329–341.

[206] ———, *Error bounds for Tikhonov regularization in Hilbert scales*, Appl. Anal. 18 (1984), 29–37.

[207] ———, *The Mathematics of Computerized Tomography*, Teubner, Stuttgart, 1986.

[208] A. NEMIROVSKII, *The regularizing properties of the adjoint gradient method in ill-posed problems*, USSR Comp. Math. Math. Phys. 26,2 (1986), 7–16.

[209] A. S. NEMIROVSKII AND B. T. POLYAK, *Iterative methods for solving linear ill-posed problems under precise information I*, Engrg. Cybernetics 22,3 (1984), 1–11.

[210] ———, *Iterative methods for solving linear ill-posed problems under precise information II*, Engrg. Cybernetics 22,4 (1984), 50–56.

[211] A. NEUBAUER, *Tikhonov-Regularization of Ill-Posed Linear Operator Equations on Closed Convex Sets*, PhD thesis, Johannes Kepler Universität Linz, November 1985, appeared in Verlag der Wissenschaftlichen Gesellschaften Österreichs, Wien, 1986.

[212] ———, *Finite-dimensional approximation of constrained Tikhonov-regularized solutions of ill-posed linear operator equations*, Math. Comp. 48 (1987), 565–583.

[213] ———, *An a-posteriori parameter choice for Tikhonov-regularization in Hilbert scales leading to optimal convergence rates*, SIAM J. Numer. Anal. 25 (1988), 1313–1326.

[214] ———, *An a-posteriori parameter choice for Tikhonov regularization in the presence of modelling error*, Appl. Numer. Math. 4 (1988), 507–519.

[215] ———, *Tikhonov regularization of ill-posed linear operator equations on closed convex sets*, J. Approx. Theory 53 (1988), 304–320.

[216] ———, *When do Sobolev spaces form a Hilbert scale?*, Proc. Amer. Math. Soc. 103 (1988), 557–562.

[217] ———, *Tikhonov regularization for nonlinear ill-posed problems: optimal convergence and finite-dimensional approximation*, Inverse Problems 5 (1989), 541–557.

[218] ———, *On converse and saturation results for regularization methods*, in: E. Schock, ed., Beiträge zur Angewandten Analysis und Informatik, Helmut Brakhage zu Ehren, Shaker, Aachen, 1994, 262–270.

[219] A. NEUBAUER AND O. SCHERZER, *Finite-dimensional approximation of Tikhonov regularized solutions of nonlinear ill-posed problems*, Numer. Funct. Anal. Optim. 11 (1990), 85–99.

[220] P. G. NEVAI, *Orthogonal polynomials on the real line associated with the weight $|x|^\alpha \exp(-|x|^\beta)$, I*, Acta Math. Acad. Sci. Hungar. 24 (1973), 335–342, in Russian.

[221] ———, *Géza Freud, orthogonal polynomials and Christoffel functions. a case study*, J. Approx. Theory 48 (1986), 3–167.

[222] O. NEVANLINNA, *Convergence of Iterations for Linear Equations*, Birkhäuser, Basel, Boston, Berlin, 1993.

[223] L. PÄIVÄRINTA AND E. SOMERSALO, eds., *Inverse Problems in Mathematical Physics*, Springer, Berlin, Heidelberg, 1993.

[224] W. M. PATTERSON, *Iterative Methods for the Solution of a Linear Operator Equation in Hilbert Space – A Survey*, Springer, Berlin, Heidelberg, New York, 1974.

[225] L. E. PAYNE, *Improperly Posed Problems in Partial Differential Equations*, SIAM, Philadelphia, 1975.

[226] D. L. PHILLIPS, *A technique for the numerical solution of certain integral equations of the first kind*, J. Assoc. Comput. Mach. 9 (1962), 84–97.

[227] R. PLATO, *Optimal algorithms for linear ill-posed problems yield regularization methods*, Numer. Funct. Anal. Optim. 11 (1990), 111–118.

[228] ———, *Iterative and parametric methods for linear ill-posed problems*, Habilitationsschrift, TU Berlin, 1995.

[229] R. PLATO AND G. VAINIKKO, *On the regularization of projection methods for solving ill-posed problems*, Numer. Math. 57 (1990), 63–79.

[230] J. PÖSCHL AND E. TRUBOWITZ, *Inverse Spectral Theory*, Academic Press, Boston, 1987.

[231] J. RADON, *Über die Bestimmung von Funktionen durch ihre Integralwerte längs gewisser Mannigfaltigkeiten* (reprinted), in: P. M. Gruber, E. Hlawka, W. Nöbauer, and L. Schmetterer, eds., J. Radon, Gesammelte Abhandlungen – Collected Works, Band 2, Verlag d. Österr. Akad. d. Wiss., Birkhäuser, Basel, 1987.

[232] A. RAMM, *Scattering by Obstacles*, Reidel, Dordrecht, 1986.

[233] T. RAUS, *Residue principle for ill-posed problems*, Acta et comment. Univers. Tartuensis 672 (1984), 16–26, in Russian.

[234] ——, *Residue principle for ill-posed problems with nonselfadjoint operators*, Acta et comment. Univers. Tartuensis 715 (1985), 12–20, in Russian.

[235] ——, *About regularization parameter choice in case of approximately given error bounds of data*, Acta et comment. Univers. Tartuensis 932 (1992), 77–89.

[236] T. REGIŃSKA, *A regularization parameter in discrete ill-posed problems*, SIAM J. Sci. Comput. 17 (1996), 740–749.

[237] V. ROKHLIN, *Rapid solution of integral equations of classical potential theory*, J. Comput. Phys. 60 (1985), 187–207.

[238] V. G. ROMANOV AND S. I. KABANIKHIN, *Inverse Problems for Maxwell's Equations*, VSP, Utrecht, 1994.

[239] L. I. RUDIN AND S. OSHER, *Total variation based restoration with free local constraints*, Proc. of IEEE Int. Conf. on Image Processing, Austin, Tx., 1994, 31-35.

[240] P. C. SABATIER, *A few geometrical features of inverse and ill-posed problems*, in [72], 1–18.

[241] ——, ed., *Some Topics on Inverse Problems*, World Scientific, Singapore, 1988.

[242] ——, ed., *Inverse Methods in Action*, Springer, Berlin, Heidelberg, New York, 1990.

[243] O. SCHERZER, *The use of Tikhonov regularization in the identification of electrical conductivities from overdetermined boundary data*, Results Math. 22 (1992), 598–618.

[244] ——, *A convergence analysis of a method of steepest descent and a two-step algorithm for nonlinear ill-posed problems*, Numer. Funct. Anal. Optim. 17 (1996), 197–214.

[245] O. SCHERZER, H. W. ENGL, AND R. S. ANDERSSEN, *Parameter identification from boundary measurements in a parabolic equation arising from geophysics*, Nonlinear Anal. 20 (1993), 127–156.

[246] O. SCHERZER, H. W. ENGL, AND K. KUNISCH, *Optimal a-posteriori parameter choice for Tikhonov regularization for solving nonlinear ill-posed problems*, SIAM J. Numer. Anal. 30 (1993), 1796–1838.

[247] E. SCHOCK, *Approximate solution of ill-posed equations: arbitrarily slow convergence vs. superconvergence*, in: G. Hämmerlin and K. H. Hoffmann, eds., Constructive Methods for the Practical Treatment of Integral Equations, Birkhäuser, Basel, 1985, 234–243.

[248] ——, *Ritz-Regularization versus least-square-regularization. Solution methods for integral equations of the first kind*, Z. Anal. Anwendungen 4 (1985), 277–284.

[249] ——, *Semi-iterative methods for the approximate solution of ill-posed problems*, Numer. Math. 50 (1987), 263–271.

[250] ——, *Pointwise rational approximation and iterative methods for ill-posed problems*, Numer. Math. 54 (1988), 91–103.

[251] T. SCHRÖTER AND U. TAUTENHAHN, *Tikhonov regularization in Hilbert scales*, Numer. Funct. Anal. Optim. 15 (1994), 155–168.

[252] T. I. SEIDMAN, *Nonconvergence results for the application of least-squares estimation to ill-posed problems*, J. Optim. Theory Appl. 30 (1980), 535–547.

[253] T. I. SEIDMAN AND C. R. VOGEL, *Well-posedness and convergence of some regularization methods for nonlinear ill-posed problems*, Inverse Problems 5 (1989), 227–238.

[254] L. A. SHEPP AND Y. VARDI, *Maximum likelihood reconstruction for emission tomography*, IEEE Trans. Medical Imaging MI-1 (1982), 113–122.

[255] D. SLEPIAN, *Some comments on Fourier analysis, uncertainty and modelling*, SIAM Rev. 25 (1985), 379–393.

[256] C. R. SMITH AND W.T GRANDY, eds., *Maximum-Entropy and Bayesian Methods in Inverse Problems*, Fundamental Theories of Physics, Reidel, Dordrecht, 1985.

[257] E. SOMERSALO, M. CHENEY, D. ISAACSON, AND E. ISAACSON, *Layer stripping: a direct numerical method for impedance imaging*, Inverse Problems 7 (1991), 899–926.

[258] E. SOMERSALO, D. ISAACSON, AND M. CHENEY, *Existence and uniqueness for electrode models for electric current computed tomography*, SIAM J. Appl. Math. 52 (1992), 1023–1040.

[259] ——, *A linearized inverse boundary value problem for Maxwell's equations*, J. Comput. Appl. Math. 42 (1992), 123–136.

[260] H. STAHL AND V. TOTIK, *General Orthogonal Polynomials*, Cambridge University Press, Cambridge, 1992.

[261] J. STOER AND R. BULIRSCH, *Introduction to Numerical Analysis*, Springer, New York, 1983.

[262] J. SYLVESTER, *An anisotropic inverse boundary value problem*, Comm. Pure Appl. Math. 43 (1990), 201–232.

[263] J. SYLVESTER AND G. UHLMANN, *Inverse boundary value problems at the boundary - Continuous dependence*, Comm. Pure Appl. Math. 41 (1988), 197–219.

[264] G. SZEGÖ, *Orthogonal Polynomials*, Amer. Math. Soc. Colloq. Publ. 23, Amer. Math. Soc., Providence, Rhode Island, 1975.

[265] G. TALENTI, ed., *Inverse Problems*, Lecture Notes in Math. 1225, Springer, Berlin, 1986.

[266] D. B. TATA, M. FORESTI, J. CORDERO, P. TOMASHEFSKY, M. A. ALFANO, AND R. R. ALFANO, *Fluorescence polarization spectroscopy and time-resolved fluorescence kinetics of native cancerous and normal rat kidney tissues*, Biophys J. 50 (1986), 463–469.

[267] U. TAUTENHAHN, *On the asymptotical regularization method for nonlinear ill-posed problems*, in: D. D. Ang et al., eds., Inverse Problems and Applications to Geophysics, Industry, Medicine and Technology, HoChiMinh City Math. Soc., HoChiMinh City, 1995, 158–169.

[268] ———, *Tikhonov regularization for identification problems in differential equations*, in: J. Gottlieb and P. DuChateau, eds., Parameter Identification and Inverse Problems in Hydrology, Geology and Ecology, Kluwer, Dordrecht, 1997, 261–270.

[269] A. N. TIKHONOV, *Regularization of incorrectly posed problems*, Soviet Math. Dokl. 4 (1963), 1624–1627.

[270] ———, *Solution of incorrectly formulated problems and the regularization method*, Soviet Math. Dokl. 4 (1963), 1035–1038.

[271] A. N. TIKHONOV AND V. ARSENIN, *Solutions of Ill-Posed Problems*, Wiley, New York, 1977.

[272] A. N. TIKHONOV AND V. B. GLASKO, *Use of the regularization method in non-linear problems*, USSR Comp. Math. Math. Phys. 5,3 (1965), 93–107.

[273] A. N. TIKHONOV, A. GONCHARSKY, V. STEPANOV, AND A. YAGOLA, *Numerical Methods for the Solution of Ill-Posed Problems*, Kluwer, Dordrecht, 1995.

[274] G. M. VAINIKKO, *The discrepancy principle for a class of regularization methods*, USSR Comp. Math. Math. Phys. 22,3 (1982), 1–19.

[275] ———, *The critical level of discrepancy in regularization methods*, USSR Comp. Math. Math. Phys. 23,6 (1983), 1–9.

[276] ———, *On the optimality of methods for ill-posed problems*, Z. Anal. Anwendungen 6 (1987), 351–362.

[277] ——, *On the discretization and regularization of ill-posed problems with noncompact operators*, Numer. Funct. Anal. Optim. 13 (1992), 381–396.

[278] G. M. VAINIKKO AND A. Y. VERETENNIKOV, *Iteration Procedures in Ill-Posed Problems*, Nauka, Moscow, 1986, in Russian.

[279] A. VAN HOCK, K. VOS, AND A. J. W. G. VISSER, *Ultrasensitive time-resolved polarized fluorescence spectroscopy as a tool in biology and medicine*, IEEE J. Quantum Electronics QE-23 (1987), 1812–1820.

[280] C. F. VAN LOAN, *Computational Frameworks for the Fast Fourier Transform*, SIAM, Philadelphia, 1992.

[281] R. S. VARGA, *Matrix Iterative Analysis*, Prentice-Hall, Englewood Cliffs, New Jersey, 1962.

[282] V. V. VASIN AND A. L. AGEEV, *Ill-Posed Problems with A-priori Information*, VSP, Utrecht, 1995.

[283] V. V. VOEVODIN, *The method of regularization*, USSR Comp. Math. Math. Phys. 9,3 (1969), 228–232.

[284] C. R. VOGEL, *Optimal choice of a truncation level for the truncated SVD solution of linear first kind integral equations when data are noisy*, SIAM J. Numer. Anal. 23 (1986), 109–117.

[285] ——, *Non-convergence of the L-curve regularization parameter selection method*, Inverse Problems 12 (1996), 535–547.

[286] G. WAHBA, *Spline Models for Observational Data*, SIAM, Philadelphia, 1990.

[287] A. WEXLER, B. FRY, AND R. NEUMANN, *Impedance-computed tomography: algorithm and system*, Appl. Optics 24 (1985), 3985–3992.

[288] G. M. WING, *A Primer on Integral Equations of the First Kind: The Problem of Deconvolution and Unfolding*, SIAM, Philadelphia, 1992.

[289] M. YAMAGUTI, K. HAYAKAWA, Y. ISO, M. MORI, T. NISHIDA, K. TOMOEDA, AND M. YAMAMOTO, eds., *Inverse Problems in Engineering Sciences*, Springer, Tokyo, Berlin, Heidelberg, 1991.

[290] E. ZEIDLER, *Nonlinear Functional Analysis and its Applications I, Fixed Point Theorems*, Springer, New York, 1986.

Index

a-priori information, 7, 20, 22, 135, 140, 241
Abel integral equation, 9, 40
approximate inverse, 113, 115
asymptotic regularization, 77, 83, 96, 285
attainability, 32, 84, 89, 241, 247

Backus-Gilbert method, 115
bandlimited, 14, 15, 17, 154
best-approximate solution, 32, 49, 140
bidiagonalization, 229
Born approximation, 26

Chebyshev method, 166
Chebyshev polynomials, 195, 289, 292
Christoffel function, 168, 173, 193, 295
compact operators, 36, 242
complementation condition, 197
confocal scanning microscopy, 17
conjugate gradient method, 16, 177, 208
    convergence, 184
    convergence rate, 189, 192
    discrepancy principle, 189
    numerical realization, 239
    stability, 181
    ultimate termination index, 179
continuous casting of steel, 21
convex constraint, 140, 160, 249
cross-validation, 105, 237
    Monte Carlo, 108, 239

deblurring, 13
deconvolution, 10, 13, 237
degree of ill-posedness, 40, 215, 253
differentiation, 4, 9, 24, 40
Dirichlet-to-Neumann map, 25
discrepancy principle, 84, 121, 152, 157, 165, 167, 186, 218, 234, 249, 283, 287

dual least-squares method, 67

entropy, 138
expectation minimization, 138

functional calculus, 42

generalized inverse, 33
    boundedness, 34
    unboundedness, 38, 199
    weighted, 199, 224
Gerchberg-Papoulis algorithm, 15, 154
Gramian matrix, 222

Hammerstein integral equation, 260
heat equation
    backwards, 21, 40
    sideways, 19
high frequency errors, 7, 14, 20, 42
Hilbert scale, 211, 213, 215, 253, 258

ill-posed, 31, 241, 242
    mildly, 40
    severely, 40
image reconstruction, 12, 137, 140
integral equation of the first kind, 9, 11, 17, 22, 30, 41, 68
interpolation inequality, 47, 211
inverse potential problems, 11
inverse scattering, 25, 284
iterated Tikhonov regularization, 123, 169, 286
iteration polynomial, 161

Jacobi polynomials, 167, 192, 293, 296

Kullback-Leibler number, 138

L-curve, 109, 236, 240
Landweber iteration, 77, 155, 207
    constrained, 160
    convergence, 155, 282
    convergence rate, 159

INDEX

discrepancy principle, 157, 283
  monotonicity, 157, 281
  nonlinear, 277
  numerical realization, 238
Laplace inversion, 10
least-squares collocation, 67, 222
least-squares projection, 63
least-squares solution, 32, 35
Levenberg-Marquardt method, 285
linear functional strategy, 75, 112
logarithmic convexity, 22
logarithmic stability, 20

Markov's inequality, 163
maximum entropy method, 15, 136, 262
  convergence, 139, 273
  convergence rate, 139, 274
  stability, 139, 272
maximum entropy solution, 139, 273
maximum likelihood, 138
Melkman-Micchelli formula, 75
modelling error, 50
mollifier method, 113
moment discretization, 67, 115, 222
moment problem, 115
Moore-Penrose inverse, 33

Nemytskii operators, 263
Newton type methods, 285
noise level, 49
nondestructive testing, 8, 12, 25
nonnegativity, 10, 14, 139, 142, 160, 263
normal equation, 35
ν-method, 168
  numerical realization, 238
  order optimal rule, 175
  qualification, 168
Nyquist rate, 17

operation count, 222
orthogonal polynomials, 162, 172, 180, 291
oversmoothing, 210

parameter choice rule, 50
  a-posteriori, 51, 83, 91, 98, 251
  a-priori, 51, 74, 216, 250, 253, 256
  heuristic, 52, 100
parameter identification, 23, 257, 259, 279
Picard criterion, 39, 58
point spread function, 13
preconditioning, 196
prior distribution, 135, 262
projection method, 63, 127
  convergence, 64
  convergence rate, 66
  non-compact operators, 67
  nonconvergence, 64

qualification, 63, 76, 80, 85, 217
quasioptimality criterion, 125, 236

Radon transform, 8, 40
reconstruction kernel, 114
regularization, 50, 67, 71
  by iteration, 154
  by projection, 63
  in Hilbert scales, 215, 253, 256
  with differential operators, 197
regularization method, 50
  arbitrarily slow convergence, 57, 163
  convergence, 52, 67, 72
  convergence rate, 55, 87, 93, 216, 218
  converse result, 80, 164
  nonconvergence, 52, 72
  optimal, 60, 95
  optimal order, 60, 75, 85, 94, 98
  qualification, 63
  selfadjoint operators, 73
  stability, 73
  uniform convergence, 79
regularization operator, 50, 67, 71
regularized projection method, 127
  convergence rate, 131
  non-compact operators, 131

numerical realization, 223
remote sensing, 9
residual polynomial, 162, 180, 289
rheology, 11
Riemann-Lebesgue Lemma, 13

saturation, 76, 80, 120, 168, 220
semiconvergence, 157
semiiterative methods, 161
    arbitrarily slow convergence, 163
    convergence, 162
    discrepancy principle, 165
    optimal rate, 165
Shannon Sampling Theorem, 17
Showalter's method, 77, 83, 96, 285
signal processing, 12, 154
singular system, 36
singular value decomposition, 37, 224
singular value expansion, 37
    truncated, 14, 69, 78, 83, 133
smoothness condition, 58
source condition, 58, 139, 145, 205, 215, 253, 258, 260, 261, 274, 277
spectral family, 45
spectral theory, 42
steepest descent, 285
stopping rule, 154, 157

Tikhonov functional, 117, 126, 197, 215, 243, 253, 285
Tikhonov regularization, 117, 207
    and projection, 127, 223
    constrained, 143
    convergence, 118, 131, 145, 244
    convergence rate, 120, 123, 131, 146, 149, 245, 247, 252, 253, 256
    discrepancy principle, 87, 121, 150, 249
    numerical realization, 221, 228
    of nonlinear problems, 243, 250
    order optimal rule, 122, 133, 233, 251

qualification, 83
saturation, 120
variational characterization, 117
tomography, 7, 25, 40
total variation regularization, 140

update polynomial, 173

wavelets, 15, 238
weakly closed, 241
well-posed, 31, 242

x^*-minimum-norm solution, 241, 253, 283, 287

Printed in Great Britain
by Amazon.co.uk, Ltd.,
Marston Gate.